Springer Series in Statistics

Springer Series in Statistics

Andersen/Borgan/Gill/Keiding: Statistical Models Based on Counting Processes.
Atkinson/Riani: Robust Diagnostic Regression Analysis.
Atkinson/Riani/Cerioli: Exploring Multivariate Data with the Forward Search.
Berger: Statistical Decision Theory and Bayesian Analysis, 2nd edition.
Borg/Groenen: Modern Multidimensional Scaling: Theory and Applications.
Brockwell/Davis: Time Series: Theory and Methods, 2nd edition.
Chan/Tong: Chaos: A Statistical Perspective.
Chen/Shao/Ibrahim: Monte Carlo Methods in Bayesian Computation.
Coles: An Introduction to Statistical Modeling of Extreme Values.
David/Edwards: Annotated Readings in the History of Statistics.
Devroye/Lugosi: Combinatorial Methods in Density Estimation.
Efromovich: Nonparametric Curve Estimation: Methods, Theory, and Applications.
Eggermont/LaRiccia: Maximum Penalized Likelihood Estimation, Volume I: Density Estimation.
Fahrmeir/Tutz: Multivariate Statistical Modelling Based on Generalized Linear Models, 2nd edition.
Fan/Yao: Nonlinear Time Series: Nonparametric and Parametric Methods.
Farebrother: Fitting Linear Relationships: A History of the Calculus of Observations 1750–1900.
Federer: Statistical Design and Analysis for Intercropping Experiments, Volume I: Two Crops.
Federer: Statistical Design and Analysis for Intercropping Experiments, Volume II: Three or More Crops.
Ghosh/Ramamoorthi: Bayesian Nonparametrics.
Glaz/Naus/Wallenstein: Scan Statistics.
Good: Permutation Tests: A Practical Guide to Resampling Methods for Testing Hypotheses, 2nd edition.
Gouriéroux: ARCH Models and Financial Applications.
Gu: Smoothings Spline ANOVA Models.
Györfi/Kohler/Krzyżak/Walk: A Distribution-Free Theory of Nonparametric Regression.
Haberman: Advanced Statistics, Volume I: Description of Populations.
Hall: The Bootstrap and Edgeworth Expansion.
Härdle: Smoothing Techniques: With Implementation in S.
Harrell: Regression Modeling Strategies: With Applications to Linear Models, Logistic Regression, and Survival Analysis.
Hart: Nonparametric Smoothing and Lack-of-Fit Tests.
Hastie/Tibshirani/Friedman: The Elements of Statistical Learning: Data Mining, Inference, and Prediction.
Hedayat/Sloane/Stufken: Orthogonal Arrays: Theory and Applications.
Heyde: Quasi-Likelihood and its Application: A General Approach to Optimal Parameter Estimation.
Huet/Bouvier/Poursat/Jolivet: Statistical Tools for Nonlinear Regression: A Practical Guide with S-PLUS and R Examples, 2nd edition.
Ibrahim/Chen/Sinha: Bayesian Survival Analysis.

(continued after index)

Sam C. Saunders

Reliability, Life Testing and the Prediction of Service Lives

For Engineers and Scientists

Springer

Sam C. Saunders
Kirkland, WA 98033
USA
smru@comcast.net

ISBN-13: 978-1-4419-2188-8
e-ISBN-10: 0-387-48538-4
e-ISBN-13: 978-0-387-48538-6

Printed on acid-free paper.

9 8 7 6 5 4 3 2 1

springer.com

Preface

The prerequisite for reading this text is a calculus-based course in Probability and Mathematical Statistics, along with the usual curricular mathematical requirements for every science major. For graduate students from disciplines other than mathematical sciences much advantage, viz., both insight and mathematical maturity, is gained by having had experience quantifying the assurance for safety of structures, operability of systems or health of persons. It is presumed that each student will have some familiarity with Mathematica or Maple or better yet also have available some survival-analysis software such as S-Plus or R, to handle the computations with the data sets.

This material has been selected under the conviction that the most practical aid any investigator can have is a good theory. The course is intended for persons who will, during their professional life, be concerned with the 'theoretical' aspects of applied science. This implies consulting with industrial mathematicians/statisticians, lead engineers in various fields, physcists, chemists, material scientists and other technical specialists who are collaborating to solve some difficult technological/scientific problem. Accordingly, there are sections devoted to the deportment of applied mathematicians during consulting. This corresponds to the 'bedside manner' of physicians and is a important aspect of professionalism.

While Henri Poincaré lectured successively in: capillarity, elasticity, thermodynamics, optics, electricity, telegraphy, cosmogeny, not to name all; very few of us can be such *universalists*. But he was an expert in each of these fields because he could understand the mathematical problems at the foundations of each. That is what we hope, in small measure, to foster here: To present the basic methods for application of probability and statistics to the ubiquitous task of calculating the reliability, or its equivalent, for some of the engineered systems in modern civilization.

Remembering the sense of satisfaction I obtained as a student when I discovered an oversight in a textbook, I have not sought, exhaustively, to deprive the readers of this text from experiencing that same private exhilaration.

> The beginner... should not be discouraged if... he finds he does not have the prerequisites for reading the prerequisites.
>
> Paul Halmos

> Science is not a collection of facts anymore than a heap of stones is a house,
>
> Henri Poincaré

Acknowledgements

An acknowledgement is owed to influential teachers and exemplars; the former category includes Professors Ralph Badgely, Ivan Niven, and Z.W. Birnbaum and in the latter are Carl Allendoerfer and, Edwin Hewitt. The Mathematical Analysis group, headed by Burton Colvin, at the erstwhile Boeing Scientific Research Laboratories contained notable colleagues Frank Proschan, George Marsaglia, Albert Marshall, Gordon Crawford and James Esary. Z.W.(Bill) Birnbaum said this group rivaled the Analysis Group led by Lev Sierspinski at Lwöw, Poland, when he studied there with Stephan Banach. Students have helped in organizing material, in correcting my errors and suggesting clarifications. In particular Prof. Juhn-Hsiong Wong, Prof. Jung Soo Woo and Dr. Jonathan Martin are owed a debt of gratitude.

Of course we are all influenced by our genealogy: $\cdots$ Ferdinand Lindemann begat David Hilbert who begat Hugo Steinhaus who begat Z. W. Birnbaum who begat my mathematical siblings Ron Pyke and Albert Marshall, who have both remained life-long colleagues and friends.

> Every man who rises above the common level has received two educations: the first from his teachers; the second, more personal and important, from himself.
>
> Edward Gibbon

> If I have seen farther than others it was because I was standing on the shoulders of giants.
>
> Sir Isaac Newton

Vörtrekkers

1. *Statistical Theory of Reliability and Life Testing;* Richard Barlow and Frank Proschan, Holt, Rinehart & Winston, 1981, reprinted SIAM 1996.
2. *Probabilistic Models of Cumulative Damage;* J.L. Bogdanoff and F. Kozin, John Wiley & Sons Inc., 1985.

Glossary

- as *means* "almost surely or with probability one"
- arv *means* " associated random variable or vector"
- asas *means* "after some algebraic simplification"
- cdf *means* "cumulative distribution function"
- pdf *means* "probability density function"
- sdf *means* "survival distribution function"
- edf *means* the same as ecdf or "empirical (cumulative) distribution function"
- esf *means* the same as "empirical survival distribution function"
- iff *means* "if and only if"
- iid *means* "independent and identically distributed"
- K-M *means* "Kaplan - Meier" e.g. as an affix to edf
- mle *means* "maximum likelihood estimator"
- nasc *means* " necessary and sufficient condition"
- NB *means* Nota Bene, Latin for "It should be well noted that"
- rhs (or lhs) *means* "right-hand side" (left-hand side)
- rwt *means* "random waiting time"
- rv or rv's *means* "random variable or random vector and its plural"
- sp *means* "stochastic process"
- tidpat *means* "Thus it doth plainly appear that" (Lagrange's phrase) but it is often paraphrased as "This is difficult, paradoxical and tedious."
- wrt *means* " with respect to"
- wlog *means* " without loss of generality"
- wp *means* " with probability"
- $:=$ *means* "is defined to be equal to"
- $\doteq$ *means* "is closely approximated by"
- $\asymp$ *means* "is asymptotically equal to"
- $\ll$ *means* "is much less than "
- $\preceq$, *means* "is stochastically less than"
- $\downarrow$ ($\uparrow$) *means* " non-increasing" (non-decreasing); so $F \in\uparrow$ means F is non-decreasing.
- $\perp$ *means* "mutually, stochastically independent"
- $\sim$ *means* " has the distribution or is distributed by"
- $\square$ *means* the same as *quod erat demonstrandum* and marks the end of a proof.
- $\imath = \sqrt{-1}$ is the unit of imaginary numbers
- $\Re$ denotes the real line, viz., $\{x : -\infty < x < \infty\}$
- $\mathrm{I}(x \pi y)$ is the indicator of the relation $x \pi y$ taking the value 1 if true, 0 otherwise

Admonitions

· · · Mathematical ideas originate in empirics, although the genealogy is sometimes long and obscure. But, once they are so conceived, the subject begins to live a peculiar life of its own and is better compared to a creative one, governed by almost entirely aesthetical motivations, than to anything else, in particular, to an empirical science. There is, however, a further point which, I believe, needs stressing. As a mathematical discipline travels far from its empirical source, or still more, if it is a second and third generation only indirectly inspired by ideas coming from 'reality', it is beset with very grave dangers.It becomes more and more purely aestheticising, more and more purely *l'art pour l'art.* This need not be bad, if the field is surrounded by correlated subjects, which still have closer empirical connections, or if the discipline is under the influence of men with an exceptionally well-developed taste. But there is a grave danger that the subject will develop along the line of least resistance, that the stream, so far from its source, will separate into a multitude of insignificant branches and that the discipline will become a disorganized mass of details and complexities. In other words, at a great distance from its empirical source, or after much 'abstract' inbreeding a mathematical subject is in danger of degeneration.

John von Neumann

An explanation is satisfactory only if we are able to reconstruct it logically from our previous knowledge and apply that understanding to circumstances different from those in which it was originally offered. That is why science teachers, to the chagrin of many students in the humanities, put heavy emphasis on problem solving. In order to demonstrate that (s)he has understood a scientific principle, a student is expected to be able to apply this understanding to situations different from the ones in which it was first learned. Similarly, a mathematics student is deficient who knows a theorem, in general, but cannot apply it in an unfamiliar context. Neither memorizing nor reproducing what one has seen or heard in a lecture ensures understanding.

Roger O. Newton

Let no one who is ignorant of Geometry (mathematics) enter here (proceed farther).

Written at the entrance to Plato's Academy

Mathematicians are like Frenchmen: whatever you say to them they translate into their own jargon and thenceforth it becomes something entirely different.

Johann W. von Goethe

We have therefore the equation of condition

$$F(x) = \int dq \, Q(q) \cos(qx).$$

If we substitute for Q any function of q and conduct the integration from $q = 0$ to $q = \infty$, we should find a function of x; it is required to solve the inverse problem, that is to say, to ascertain what function of q, after being substituted for Q, gives as a result the specified function $F(x)$, a remarkable problem whose solution demands attentive examination.

Joseph Fourier

Contents

CHAPTER 1

Requisites

1.1. Why Reliability Is Important

All artifacts of mankind will eventually fail in service or be discarded because of wear or obsolescence. This was as true for the roads and aquaducts of Rome as it is today for the infrastructure of America. All constructs bearing the hallmark of civilization, from the tomahawk to the cruise missle of the same name, suffer from material weakness or imperfection. This applies *a fortiori* to our electronic systems, computers, and video communication as well as the military's smart-bombs and the Concorde's avionics. Palliative efforts include preventative maintenance for machines and medicines for humankind. But always "an ounce of prevention is better than a pound of cure" and not only because it is earlier and cheaper. The origins of ubiquitious failure are manifold; the designer has neglected or been unaware of the severity of some of the factors of the invironment in which the system/structure must operate; the owner-operator of the system has needfully operated it outside its design envelope; the manufacturer failed to eliminate minor defects from the system either during construction or inspection; the supplier has substituted inferior material, causing inherent weaknesses in a component. Such imperfections can cause the early and unexpected failure of the system, or just the incapability of the system to perform its function during its warranteed life. Such practice may lead to hazards both to the operators and to the public weal.

The primary source of failure is the gradual impairment of structural components caused through repetition of their designed duty cycle. These include friction or abrasive wear, metallic fatigue, stress corrosion or chemical degradation. Failure is caused by mistakes (accidents) aggravated by operation or insufficient maintenance, more often than the confluence of unhappy circumstance, labeled "acts of God."

Most often, system or structural failure is the result of many coincident factors. The failure of a dynamically loaded structure may, for example, be the result of a small defect in a critical component undetected because of the insufficient quality control during production. This leads to crack initiation, and a growing fatigue crack accelerated by a corrosive environment; thus, ultimately an extreme random load exceeds the residual strength of the component. Who was at fault? Were the designers, the users, or the nature or all three?

Unfortunately, all failures in service have undesirable economic consequences and not always to those responsible. For example, the destruction by bomb of a Boeing 747 over Lockerbie, Scotland, resulted in the bankruptcy of Pan-American

Airlines; the fading of a new environmentally safe paint from UV-radiation caused the bankruptcy of the Studebaker Avanti automobile corporation. Fortunately, most system failures necessitate only repair or component replacement with its entailed disruption of service. Sometimes such failures may cause an interruption in manufacturing among secondary users. Sometimes dramatic failures force a system redesign or even concept abandonment.

As a rule of thumb, the expenditure of funds for the maintenance of systems, structures, machinery, or equipment amounts to about half the initial investment cost before obsolescence forces replacement. For well-designed, long-life items maintenance expenses may be much more than the initial cost. It is estimated that about 6–8% of the gross national product (GNP) is spent annually for maintenace. This may be small in comparison with the loss of production attributable to unwarranted in-service failures.

Today many purchasers of equipment are aware of the cost of subsequent maintenance. Consequently, they consider not only purchase price but the total life-cycle costs, including maintenance and repair. This is now routinely done for the evaluation of system proposals by industry for the military and certain governmental divisions, such as the federal highways. For political reasons these costs are often underestimated. In the 18th century the costs for the construction and operation of the frigate *Constitution*, "Old Ironsides," were underestimated by about the same percentage as they were in the 20th century for the Stealth Bomber; and with the same perceived political reasons for such mendacity.

For many large complex systems/structures, such as high-rise buildings, nuclear power plants, off-shore oil structures, aircraft, and life-support medical equipment, the consequences of unreliability involve public welfare and safety. Of course, the failure of comparatively minor products can, besides just being an annoyance, have serious consequences for the public, for example, the failure of an electric razor or a battery in a fire-alarm system.

Issues concerning safety have, during the last few decades, come increasingly to public attention and hence they become more important to the design engineer. One of the most controversial is nuclear power (since Chernobyl and Three-Mile Island), one of the most dramatic is in-flight safety, while the much higher frequency of death from automobile accidents is, relatively, of small concern. Governmental requirements for safety analysis of systems is increasing, as it should. Legislation aportioning responsibility for product liability is proceeding, as it should, to help avoid the excessively large compensation claims awarded by the American tort system, whenever insurance for the manufacturer is available or the industry is large enough to have "deep pockets."

All of this will increase the need for industry to perform systematic studies for the identification and reduction of causes of failures (with hope for their virtual elimination). These studies must be performed by persons who (i) can identify and quantify the modes of failure, (ii) know how to obtain and analyze the statistics of failure occurrences, and (iii) can construct mathematical models of failure that depend upon, for example, the parameters of material strength or design quality, fatigue or wear resistance, and the stochastic nature of the anticipated duty cycle.

Only then can procedures for optimal design be implemented in parallel with a study of the economic consequences for each failure mode so as to reduce, in an optimal way, the probability of their occurrence.

The purpose of this book is to help supply information about the mathematical and statistical aspects of calculating the reliability in order to make a valid service-life prediction for the use of cognizant persons and to help provide them with the capability to utilize or develop analytical techniques specific to their usage and the needs of their employers.

> The beginner ... should not be discouraged if ... he finds that he does not have the prerequisites for reading the prerequisites.
>
> Paul Halmos

1.2. Valuable Concepts

1.2.1. Concepts from Probability

Consider the outcome of a well-defined experiment, the result of which can not be exactly anticipated, except it will terminate in some measured quantity, denoted by X, within a known set of possible outcomes, labeled $\mathcal{X}$ and called the *sample space*. This measured quantity X is called a *random* or *stochastic variable* (which we abbreviate by rv). We will use X, Y, Z, with or without affixes, to denote rv's. We presume the sample space of each rv is a subset of the real line $\Re$, or of the product of real lines, $\Re^n$ for some n. Associated with each experiment is the relative frequency of the different outcomes, which would occur if the experiment could be replicated indefinitely. This probabilistic behavior of the repeated determinations of X is summarized in a mathematical function, called the *cumulative distribution function*, abbreviated cdf and usually denoted by F_X (without affix when no ambiguity results), which maps $\Re$, or $\Re^n$, respectively, onto the unit interval. This function may often be classified into one of the two cases, viz., *discrete* or *continuous*, according to whether as F_X is a step function or is absolutely continuous with density $F'_X := f_X$. In the latter case, that "X is continuous," by which we mean the the cdf is absolutely continuous, it is given by

$$\Pr[X \le x] := F_X(x) = \int_{-\infty}^{x} f_X(t)\,dt \qquad \text{for all } x \in \Re. \tag{1.1}$$

The *support* of a continuous rv X is the set $\mathcal{X} = \{x \in \Re \mid f_X(x) > 0\}$. When X is discrete, the support is some countable set, say $\mathcal{X} = \{x_1, x_2, \ldots\}$, where

$$\Pr[X = x] = \begin{cases} p_i > 0 & \text{if } x = x_i \text{ for some } i = 1, 2, \ldots, \\ 0 & \text{otherwise.} \end{cases} \tag{1.2}$$

Here p_i is the height of the ith saltus of the cdf F_X and

$$\Pr[X \le x] := F_X(x) = \sum_{x_i \le x} p_i \qquad \text{for all } x \in \Re. \tag{1.3}$$

If Y be a constant rv, i.e., the event $[Y = c]$ occurs with probability one for some $c \in \Re$, then we write the cdf F_Y in terms of the indicator function of an event, viz., $\mathrm{I}(x \in A) = 0$ iff $x \in A$ and $\mathrm{I}(x \in A) =)$ iff $x \notin A$. When the interval is $[c, \infty)$, we write

$$F_Y(y) = \mathrm{I}(y \geq c) = \begin{cases} 1 \text{ for } y \geq c, \\ 0 \text{ for } y < c. \end{cases} \quad \text{for all } y \in \Re. \tag{1.4}$$

In physics, $\varepsilon(y) := \mathrm{I}(y \geq 0)$ is called the *Heaviside* function. The "derivative" of $\varepsilon(y)$ is the *Dirac delta-function*, which is often used in heuristic arguments, especially when the correctness of the result can be verified empirically.

Thus, the distribution of any discrete rv, as defined in eqn (1.3), can be written

$$F_X(x) = \sum_{i=1}^{\infty} p_i \mathrm{I}(x \geq x_i) \qquad \text{for all } x \in \Re.$$

We say that X is a *mixed* rv iff for some $\gamma \in (0, 1)$ we have, for every $x \in \Re$, and some denumerable subset $\{x_1, x_2, \ldots\}$,

$$F_X(x) = \gamma \sum_{i=1}^{\infty} p_i \mathrm{I}(x \geq x_i) + (1 - \gamma) \int_{-\infty}^{x} f_X(t)\, dt. \tag{1.5}$$

Familiar Discrete Densities and Distributions

A *Bernoulli* rv, X, is a binary rv that takes the values 0 or 1. Thus, its pdf, i.e., probability distribution function, is defined by

$$\mathrm{P}[X = x] = f(x; p) = p^x q^{1-x} \qquad \text{for} \quad x = 0, 1; 0 < p < 1, q = 1 - p.$$

A *De Moivre* rv, say X, usually called "Binomial," has pdf defined by

$$\mathrm{P}[X = x] = f(x; n, p) = \tbinom{n}{x} p^x q^{n-x} \qquad \text{for} \quad x = 0, 1, \ldots, n; 0 < p < 1, \; q = 1 - p.$$

A *Poisson* rv, X, has pdf defined by

$$\mathrm{P}[X = x] = f(x; \lambda) = \frac{e^{-\lambda}\lambda^x}{x!} \qquad \text{for} \quad x = 0, 1, \cdots; 0 < \lambda.$$

This distribution had its origin as the "law of small numbers" since it can be derived from the De Moivre distribution as the limit when $n \to \infty$, $p \to 0$ with $np = \lambda$, a constant. Consequently, we have

$$\binom{n}{x} p^x q^{n-x} \doteq \frac{e^{-np}(np)^x}{x!} \qquad \text{when } n \text{ is large and } p \text{ is small.}$$

A *Pascal* rv, say X, has pdf defined, again letting $q = 1 - p$, by

$$\mathrm{P}[X = x] = f(x; r, p) = \binom{r + x - 1}{x} p^r q^x \quad \text{for} \quad x = 0, 1, \cdots, n; 0 < p < 1,$$

which is the probability that the rth success in a sequence of independent Bernoulli trials occurs at the $(r+x)$th trial. It is often called a 'Negative Binomial' since $f(x; r, p) = \binom{-r}{x} p^r (-q)^x$.

A *multivariate De Moivre* rv, say $(X_1, \ldots, X_n)$, has pdf

$$P[X_1 = x_1, \cdots, X_n = x_n] = n! \prod_{i=1}^{n} \frac{p_i^{x_i}}{x_i!} \quad \text{for} \quad x_i \geq 0;\ 0 < p_i < 1,$$
$$\text{where } \sum\nolimits_{i=1}^{m} x_i = m,\ \sum\nolimits_{i=1}^{m} p_i = 1.$$

Another multivariate rv, $(X_1, \ldots, X_n)$, which is continuous, has the *Dirichlet* distribution on the simplex $\mathcal{S} = \{(x_1, x_2, \cdots, x_k) : 0 \leq x_i \leq 1, \sum_1^k x_i \leq 1\}$ when it has density

$$\frac{(n+k)!}{\prod_{j=1}^{k+1} (n_j)!} \left(\prod_{j=1}^{k} x_j^{n_j} \right) \left[1 - \sum\nolimits_1^k x_j \right]^{n_{k+1}},$$

where $n_j \geq 0$ and $n = \sum_1^{k+1} n_j$.

The Mathematical Expectation

Let $X \sim F_X$ be a real rv on $\Re$. By the *expectation* of the quantity $\varphi(X)$ we mean the integral

$$\mathrm{E}\varphi(X) = \int_{-\infty}^{\infty} \varphi(x)\, dF_X(x). \tag{1.6}$$

Here we mean the integral is the Stieltjes integral. If this concept is unfamiliar, please read the section in Chapter 15.

By the *moments* of X we mean the quantities

$$\mathrm{E}X^k \quad \text{for } k = 1, 2, \ldots \tag{1.7}$$

for all integral values of k for which the integrals, $\mathrm{E}\{|X|^k\}$ for $k = 1, 2, \ldots$, are finite. Of special interest are the two parameters called the *mean,* often denoted by μ, and the *variance,* often denoted by σ^2. They are defined, repectively, by

$$\mu := \mathrm{E}[X], \quad \sigma^2 := \mathrm{Var}[X] = \mathrm{E}(X - \mu)^2 = \mathrm{E}[X^2] - \mu^2. \tag{1.8}$$

The first moment, say μ, corresponds to the center of gravity or *centroid*, of the probability "mass" represented by the density. It is a measure of central tendency; the variance corresponds to its moment of inertia about that centroid and so is a measure of dispersion.

Multivariate Random Variables

Let X, Y be a pair of rv's with *joint distribution* $F_{X,Y}$ and *marginal distributions* F_X and F_Y, respectively defined, for all $x, y \in \Re$, by

$$F_{X,Y}(x, y) = \Pr[X \leq x, Y \leq y],$$

with

$$F_X(x) = \Pr[X \leq x] = F_{X,Y}(x, \infty), \quad F_Y(y) = \Pr[Y \leq y] = F_{X,Y}(\infty, y).$$

Moreover the *conditional distribution* of X given that $[Y = y]$ is defined by

$$F_{X|Y}(x|y) = \lim_{h \to 0} \frac{\Pr[X \leq x, y - h < Y \leq y + h]}{F_Y(y + h) - F_Y(y - h)},$$

when it exists. If the conditional distribution $F_{X|Y}$ exists, then we can obtain the cdf of the sum $S = X + Y$ and the product $V = XY$, respectively, as

$$F_S(s) = \int_{-\infty}^{\infty} F_{X|Y}(s - y|y) dF_Y(y) \quad \text{for all } s \in \Re \tag{1.9}$$

and for all $v \in \Re$

$$F_V(v) = \int_0^{\infty} F_{X|Y}(v/y|y) dF_Y(y) + \int_{-\infty}^{0} [1 - F_{X|Y}(v/y|y)] dF_Y(y). \tag{1.10}$$

The *Fourier transform* of the pair (X, Y) is defined, letting $\imath = \sqrt{-1}$, by

$$c_{X,Y}(t, s) := Ee^{\imath tX + \imath sY} \quad \text{for all } (t, s) \in \Re^2.$$

Iff X and Y are *independent*, i.e., $X \perp Y$, does it follow that

$$F_{X,Y}(x, y) = F_X(x) \cdot F_Y(y) \quad \text{for all } (x, y) \in \Re^2,$$

and in this case $F_{X|Y} = F_X$, $F_{Y|X} = F_Y$, and we see $c_{X,Y} = c_X \cdot c_Y$

1.2.2. Concepts from Statistics

We now recall some results from the theory of statistics that will be useful in what follows: write $\tilde{\theta}_n = \tilde{\theta}(x_1, \ldots, x_n)$ for any estimator of $\theta \in \Theta$, based on a sample of size n.

Definition 1. The estimator $\tilde{\theta}_n$ is *unbiased* for θ iff $\mathrm{E}\tilde{\theta}_n = \theta$ for all $n \in \aleph, \theta \in \Theta$.

Definition 2. The estimator $\tilde{\theta}_n$ is *consistent* for θ iff $\tilde{\theta}_n \xrightarrow{P} \theta$ as $n \to \infty$.

Definition 3. The estimator $\tilde{\theta}_n$ is *strongly consistent* for θ iff $\tilde{\theta}_n \xrightarrow{a.s.} \theta$ as $n \to \infty$.

A *sufficient statistic* for a parameter is one that carries all the information in the sample about that parameter.

Definition 4. A statistic T whose value $T(\boldsymbol{x})$ can be computed from data $\boldsymbol{x}$ without knowledge of θ, such that the observed value of T is sufficient to determine the likelihood $\mathcal{L}(\theta|\boldsymbol{x})$, up to a constant of proportionality, is a *sufficient statistic* for θ.

We have the result

Theorem 1. *The statistic T is sufficient for θ iff the density can be factored appropriately*

$$f_{\boldsymbol{X}}(\boldsymbol{x};\theta) = g(\boldsymbol{x}) \cdot h(T(\boldsymbol{x});\theta),$$

where $g(\cdot)$ is not a function of θ.

Another formulation serves as an alternate definition.

Theorem 2. *The conditional distribution of outcomes $\boldsymbol{X}$, given a sufficient statistic T, does not depend upon θ.*

PROOF. Let

$$f_T(t;\theta) = \int_{\{\boldsymbol{x}:T(\boldsymbol{x})=t\}} g(\boldsymbol{x})\,d\boldsymbol{x} \cdot h(t;\theta) = \psi(t) \cdot h(t;\theta).$$

Thus, we see

$$f_{X|T}(\boldsymbol{x}|t) = \frac{f_X(\boldsymbol{x};\theta)}{f_T(t;\theta)} = \frac{g(\boldsymbol{x})}{\psi(t)}$$

is independent of the parameter θ. □

Exercise Set 1.A

1. Show that for any rv $X \sim F$, discrete, continuous, or mixed:

$$\mathrm{E}[X] = \int_0^{\infty} [1 - F(t)]\,dt - \int_{-\infty}^{0} F(t)\,dt.$$

2. Much is made in applied mathematics about the nature of the Dirac delta function e.g., even though $\delta(t) = 0$ for $t \neq 0$, it is argued by physical reasoning, involving dimensionality, that $\delta(at+b) \neq \delta(t+\frac{b}{a})$ with $a \neq 0$, since by a change of variable of integration we see, when legitimate,

$$\int \phi(t)\delta(at+b)dt = \int \phi\left(\frac{x-b}{a}\right)\frac{\delta(x)}{|a|}dx = \frac{\phi(-b/a)}{|a|}.$$

So, it is argued $\delta(at+b) = \delta(t+\frac{b}{a})/|a|$. Show that this "anomaly" disappears by using the Heaviside distribution $\varepsilon(t) = \mathrm{I}(t \geq 0)$ in the Stieltjes integral $\int \phi(t)d\,\varepsilon(at+b)$.

3. Using the Schwarz inequality, viz., $\mathrm{E}^2|XY| \le \mathrm{E}|X|^2 \cdot \mathrm{E}|Y|^2$, show that $\ln \mathrm{E}|X|^r$ is a convex function of r. Recall that for all x, y in an open interval, we have convexity of ϕ, viz., $\phi(\frac{x+y}{2}) \ge \frac{1}{2}\phi(x) + \frac{1}{2}\phi(y)$, plus continuity, is equivalent with

$$\phi[tx + (1-t)y] \ge t\phi(x) + (1-t)\phi(y) \quad \text{for all } t \in (0,1).$$

4. Show that $\mathrm{E}^{\frac{1}{r}}|X|^r$ is a nondecreasing function of r.

5. The characteristic function of the rv X is given by $c(t) = \exp(-t^2/2)$. Find the density function of this random variable.

6. If X is a random variable with characteristic function c, then find, in terms of c, the characteristic function of the linear transformation $aX + b$ for any constants a and b.

7. Show that the cumulants of X and $Y = X + a$, for any constant a, are the same. Cumulants are sometimes called the *semi-invariants*.

8. If the bivariate density of (X,Y) is $f(x, y) = e^{-y}$ for $0 < x < y < \infty$, find the density of $X + Y$.

9. If X and Y are dependent, but $F_{X|Y}$ exists, find formulas for the distributions of $U = \max(X, Y)$, $W = \min(X, Y)$, and $Z = X/Y$.

10. Evaluate E[X] and Var[X] when $X \sim F$ in the two cases where:
(a)

$$F(x) = \Phi[\xi(x/\beta)/\alpha] \quad \text{for } x > 0; \alpha, \beta > 0,$$

with Φ defined in eqn (2.25) and ξ defined in eqn (2.35) in Chapter 2.
(b)

$$F'(x) = \frac{\beta}{x^2\Phi(\alpha)}\varphi\left(\frac{\beta}{x} - \alpha\right) \quad \text{for } x > 0 : \alpha, \beta > 0.$$

Remember that $\varphi(x)$ is a transcendental function.

11. * Let the pair of rv's (X, Y) have the density f defined on its support by

$$f(x, y) = [1 + xy(x^2 - y^2)]/4 \quad \text{for all } |x| \le 1 \text{ and } |y| \le 1, \text{ and zero elsewhere.}$$

Show that
(a) $f_X \cdot f_Y \ne f$,
(b) $f_{X+Y}(z) = (2 - |z|)/4$ for $0 \le |z| \le 2$, and zero elsewhere.
(c) Find c_X, c_Y, c_{X+Y}.
(d) Can it be true that $c_X(t) \cdot c_Y(t) = c_{X+Y}(t)$ for dependent X and Y?

12. The "Law of the Unconscious Statistician" refers to the oversight that occurs when $X \sim F$, with F absolutely continuous, its expectation defined by $\mathrm{E}X := \int_{-\infty}^{\infty} xF'(x)\,dx$, when, without proof, it is assumed, for any (measurable) transformation $g(X)$, its expectation is $\mathrm{E}g(X) = \int_{-\infty}^{\infty} g(x)F'(x)\,dx$. Using the definition $\mathrm{E}X := \int_{-\infty}^{\infty} x\,dF(x)$ show that if $Y = g(X)$ for any monotone g, that $\mathrm{E}Y = \int_{-\infty}^{\infty} g(x)\,dF(x)$.

Persons who do not understand mathematics are not truly human; they are, at best, a tolerable subspecies that has learned to wear shoes, bathe and not make messes in the house.

Lazarus Long: A character of Robert Heinlein

I tell them that if they will occupy themselves with the study of Mathematics they will find in it the best remedy aginst the lusts of the flesh.

Thomas Mann: in the *Magic Mountain*

I had a feeling once about Mathematics—that I saw it all. Depth beyond depth was revealed to me—the Byss and the Abyss. I saw—as one might see the transit of Venus or even the Lord Mayor's show—a quantity passing through infinity and changing its sign from plus to minus. I saw exactly how it happened and why the tergiversation was inevitable—but it was just after dinner and I let it all go.

Winston S. Churchill

The supposed advantage of having all those humanities courses taught on campus is more than counterbalanced by the general dopiness of the people who study them.

Richard P. Feynman, on why he left Cornell to go to Cal. Tech.

CHAPTER 2

Elements of Reliability

2.1. Properties of Life Distributions

Reliability studies are concerned with an assessment of the rate of wear, deterioration, or accumulating damage to a structure or system and the entailed distribution of useful service life, i.e., until it can no longer, safely or profitably, perform its operational mission. Since damage during service occurs in a known manner but at unpredictable times, the waiting time until failure occurs is also a random variable which must be nonnegative. Let $T \geq 0$ be an rv denoting *life length;* then we write $T \sim F$ when F is its cdf. NB that F has support on $[0, \infty)$, i.e., $F(t) = 0$ for $t < 0$. The corresponding *survival distribution,* denoted by *sdf*, is defined by

$$\bar{F}(t) := 1 - F(t) = \Pr[T > t] \quad \text{for } t \geq 0. \tag{2.1}$$

It is also called, in many applications, the *reliability* function.

Remark. *We know the following statements:*

If T is discrete, *then $\bar{F}$ is a decreasing step function.*
If T is continuous, *then $\bar{F}$ is continuously decreasing and $F' := f$.*
If T is mixed, *then $\bar{F}$ is decreasing with at least one saltus.*

NB the convention that an rv being 'continuous' means its cdf is absolutely continuous.

Any function, say $\bar{G}$, is a reliability function for some life-length variate iff it has the following three properties:

(i) $\bar{G}(0) = \delta$ for some $0 < \delta \leq 1$, $\quad (ii)$ $t_1 \leq t_2$ implies $\bar{G}(t_1) \geq \bar{G}(t_2)$,
(iii) $\lim_{t\to\infty} \bar{G}(t) = 0$.

Consider the probability of failure for a life-length $T \sim F$ during the time interval $(t, t + x]$:

$$\Pr[t < T \leq t + x] = F(t + x) - F(t), \tag{2.2}$$

and the conditional probability of failure, given survival to time $t > 0$:

$$F[t + x|t] := \Pr[t < T \leq t + x|T > t] = \frac{F(t + x) - F(t)}{1 - F(t)}. \tag{2.3}$$

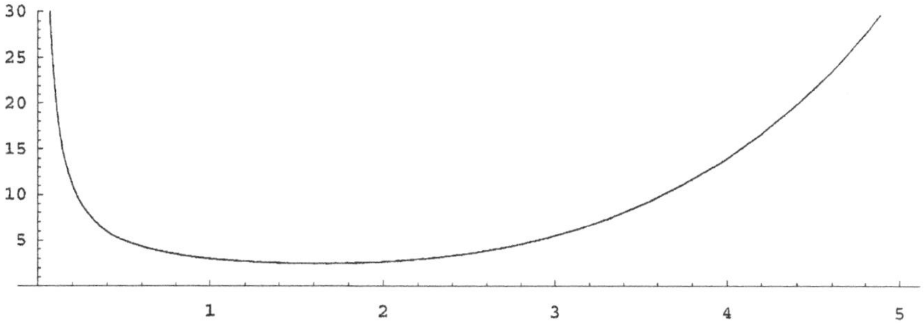

Figure 2.1. The bathtub-shaped hazard rate.

The *hazard rate*, label it h when it exists, is defined as

$$h(t) := \lim_{x \downarrow 0} \frac{F[t+x|t]}{x} = \frac{F'(t)}{1-F(t)}; \tag{2.4}$$

this function is also called the *failure rate*, and in actuarial usage it is called *the force of mortality*.

The mortality tables of both humans and animals exhibit a characteristic behavior; the force of mortality initially decreases and then remains virtually constant for a time and finally increases. See Figure 2.1. Failure data for machines often exhibit the same behavior. This bathtub shape is explained by the operation of three independent failure modes, namely, (i) manufacturing or assembly error causing early failure, (ii) failure due to accidents while in service, and (iii) cumulative damage (wearout or fatigue) failures, which are manifested during late service life.

Hazard rates are sometimes easier to determine from physical considerations than are densities. In fact, the Gompertz–Makeham law of human mortality [38, 1825; 67, 1860], was among the first statistical models ever applied, predating statistical inference. For a given cdf F with density f we have the corresponding hazard rate, say h, given by the relation of eqn (2.4) and its integral, the *hazard function* (call it H), given by:

$$H(x) = \int_0^x h(t)\,dt = -\ln[1-F(x)] \quad \text{for } x > 0. \tag{2.5}$$

Note that we can define $H := -\ln[1-F]$, which always exists.

Thus three properties determine a hazard function for a life variate, viz., (i) $H(0) = 0$, (ii) H is nondecreasing, and (iii) $H(\infty) = \infty$. Moreover, we can write the sdf as

$$\bar{F}(t) = e^{-H(t)} = \exp\left[-\int_0^t h(x)\,dx\right] \quad \text{for all } t > 0. \tag{2.6}$$

Denote the conditional reliability of a unit of age t during a time x by

$$\bar{F}(x|t) := \frac{\bar{F}(t+x)}{\bar{F}(t)} \qquad \text{if } \bar{F}(t) > 0.$$

Thus, we make the important

Definition. A cdf F is IHR, which stands for *increasing hazard rate*, iff F satisfies

$$\bar{F}(x|t) \quad \text{is decreasing in } 0 < t < \infty \text{ for each } x \geq 0. \tag{2.7}$$

NB that a distribution can be IHR without H' existing!

It follows, when a density $f(t)$ exists for an IHR cdf, that

$$h(t) = \lim_{x \downarrow 0} \frac{\left[1 - \bar{F}(x|t)\right]}{x} \quad \text{is increasing in } t \geq 0.$$

Conversely, when the hazard rate $h(t)$ is increasing, then $\bar{F}(x|t)$ is decreasing in $t \geq 0$ for each $x \geq 0$. Thus, we see that when a density exists, the IHR definition is equivalent with the hazard rate being an increasing function. We also define a DHR distribution as one for which the words "increasing" and "decreasing" are interchanged in the preceeding definition.

There is another concept that has intuitive meaning and can be used in modeling life distributions in reliability. For a given $0 < T \sim F$ which is right-continuous, we consider the *mean residual-life function*, say m, defined by

$$m(t) := \mathrm{E}[X - t | X > t], \quad \text{when } \bar{F}(t) > 0, \quad \text{and} \quad = 0 \quad \text{otherwise}.$$

NB that we can write for any t such that $\bar{F}(t) > 0$,

$$m(t) = \int_0^\infty \bar{F}(x|t)dx = \int_t^\infty \frac{\bar{F}(x)}{\bar{F}(t)}dx.$$

If $F' = f$ exists, then we also have

$$m(t) + t = \frac{1}{\bar{F}(t)} \int_t^\infty u f(u) du, \tag{2.8}$$

from which the following relationship can be deduced:

$$m'(t) + 1 = m(t) \cdot h(t). \tag{2.9}$$

Let $\mu = \mathrm{E}X$. So if $F(0) = 0$, then $m(0) = \mu$; and if $F(0) > 0$, then $m(0) = \mu / \bar{F}(0)$.

Let us presume that $F(0) = 0$; then we have the representation by setting $F^{-1}(1) = \sup\{t > 0 | F(t) < 1\}$,

$$\bar{F}(t) = \begin{cases} \frac{m(0)}{m(t)} \exp\left\{- \int_0^t \frac{1}{m(u)} du\right\} & \text{for } 0 \leq t < F^{-1}(1), \\ 0 & \text{for } t \geq F^{-1}(1). \end{cases} \tag{2.10}$$

Thus, we have the following classifications:

Definition. A distribution F has a *decreasing mean residual life* (DMRL) iff its mean residual-life function is a decreasing function.

Definition. A life length rv $T \sim F$ is *new better than used in expectation* (NBUE) iff $m(0) \geq m(t)$ for all $t > 0$.

Exercise Set 2.A

1. The hazard rate for human mortality suggested by Gompertz was $h(t) = \alpha$ and the "Gompertz–Makeham law" was $h(t) = \alpha + \lambda e^{\gamma t}$ with $\alpha, \lambda, \gamma > 0$, for $t > 0$. Why do you think the GM law was more successful in applications?

2. The "logistic distribution," in standard form, is given by

$$f(x) = \frac{e^x}{(1+e^x)^2} \quad \text{for} \quad -\infty < x < \infty,$$

   Add scale and location parameters and discuss its behavior as a model for life length.

3. A component as produced, has a life length, say T, which is an rv with hazard rate h. Each component is subjected to a *burn-in* of length τ and passed iff it did not fail during burn-in. Express the hazard rate of the passed component, say T_τ, in terms of h.

4. Let the cost for each burn-in test be \$$c$, when the component passes, and \$$C$, when the component fails. The gain is \$$D$ per unit of increase of expected service-life, which is obtained by testing. If the hazard rate h is bathtub-shaped, derive a formula for an optimum burn-in period and the distribution of life obtained.

5. Show that the expectation of the life-length rv T is its accumulated reliability, i.e.,

$$\mu = E(T) = \int_0^\infty \bar{F}(t)\,dt, \quad \text{if} \quad \lim_{t\to\infty} \bar{F}(t)/h(t) = 0.$$

   Hint: Evaluate, via L'Hopital's rule, conditions for $\lim_{y\to\infty} y\bar{F}(y) = 0$. Is this a mathematical or practical consideration?

6. If $\nu_r(T) = [E(T)^r]^{1/r}$, for some life-length variate T, exists for all $r > 0$, what would be the values of $\lim_{r\to 0} \nu_r(T)$ and $\lim_{r\to\infty} \nu_r(T)$?

7. Define $W(t) = \int_t^\infty \bar{F}(x)\,dx$ for $t > 0$, note that $\mu = W(0)$, and show that

$$E[T^2] = 2\int_0^\infty W(t)\,dt, \quad \text{provided that} \quad \lim_{t\to\infty} t^2\bar{F}(t) = \lim_{t\to\infty} tW(t) = 0.$$

   NB that $W(t)/\mu$ is itself a survival distribution.

8. Check the conditions of Exercise 3 for the case when

$$h(t) = \frac{1}{1+at} \quad \text{for } t > 0; \text{ for some } a > 0.$$

2.2. Useful Parametric Life Distributions

2.2.1. The Epstein (Exponential) Distribution

One of the simplest, yet most useful distributions, was studied in detail by Benjamin Epstein in the 1940s. It is often called the *exponential,* or *negative exponential,* distribution. However, during the 19th century "exponential" was the name given to what we now call the Gaussian (Normal) distribution. In statistics we have "the exponential family" and to avoid confusion we utilize his proper name. We write $T \sim \text{Ep}(\lambda)$ whenever the hazard rate of the rv T is constant, namely,

$$h(t) = \lambda \quad \text{for all } t > 0;\ \lambda > 0.$$

The density and survival distribution are then given, respectively, by

$$f(t) = \lambda e^{-t\lambda}, \qquad \bar{F}(t) = e^{-t\lambda} \quad \text{for } t > 0.$$

An alternate parameterization is in terms of the mean $\mu = 1/\lambda$. In this case the notation often used is $T \sim \text{Exp}(\mu)$; so it follows that $E(T) = \mu, \quad \text{Var}(T) = \mu^2$.

The "two-parameter exponential" distribution results when the Epstein origin is changed. It is denoted here as the shifted-Epstein, $T \sim \text{Shep}(\lambda, \nu)$, with the density and sdf given by

$$f(t) = \lambda e^{-\lambda(t-\nu)^+}, \qquad \bar{F}(t) = e^{-\lambda(t-\nu)^+} \quad \text{for } t \in \Re; \lambda > 0, \nu \in \Re. \tag{2.11}$$

There is another distribution related to the exponential, called the two-sided exponential or the *Laplace distribution*. This is denoted by $T \sim \text{Lap}(\lambda, \mu)$ whenever

$$f(t) = \frac{\lambda}{2} e^{-\lambda|t-\mu|} \quad \text{for } t \in \Re; \lambda > 0, \mu \in \Re. \tag{2.12}$$

Exercise Set 2.B

1. Let $T_1, \ldots, T_n$ be iid $\text{Ep}(\eta)$ and set $X_n = \max_{i=1}^n T_i$. Find the distribution of X_n and show its mean and variance satisfy

$$\text{E}X_n = \frac{1}{\eta}\sum_{j=1}^{n}\frac{1}{j}, \qquad \text{Var}X_n = \frac{1}{\eta^2}\sum_{j=1}^{n}\frac{1}{j^2}. \qquad \text{What happens as as } n \to \infty.?$$

2. If $Y_i \sim \text{Ep}(i\eta)$ for $i = 1, \ldots, n$ are independent, what are the mean and variance of $\sum_{i=1}^n Y_i$?

3. Suppose a device consists of m components each of which has an Epstein life with hazard rate proportional to load. If each surviving component shares the total imposed load equally, what is the distribution of life of this device if it fails whenever the critical kth element fails for some fixed $1 \le k \le m$.

2.2.2. The Gamma Distribution

We say T has a Gamma distribution and write $T \sim \mathcal{G}\text{am}(\alpha, \beta)$ whenever its cdf transform and density are given, respectively, by

$$F^{\dagger}(t) = (1 - t\beta)^{-\alpha}, \quad F'(t) = \frac{t^{\alpha-1}e^{-t/\beta}}{\beta^{\alpha}\Gamma(\alpha)} \quad \text{for } t > 0;\ \alpha, \beta > 0. \tag{2.13}$$

Here $\Gamma(\alpha)$ denotes the *gamma function*, an Eulerian integral of the second kind, which serves as the normalizing factor in the density. There are two special cases: Because of its application in early studies in telephone traffic by Erlang, a Bell Telephone employee, the distribution $\mathcal{G}\text{am}(k, \beta)$ when $k \in \aleph = \{1, 2, \ldots\}$ is called the *Erlang-k* distribution. Also because of its early utilization in classical statistics the distribution $\mathcal{G}\text{am}(n/2, 2)$ is called the *Chi-square distribution*, and this is denoted as

$$\chi_n^2 \sim \mathcal{G}\text{am}(n/2, 2).$$

For $n \geq 30$ the chi-square distribution may be well approximated by the Normal distribution, to be identified subsequently.

We have the reliability of $X \sim \mathcal{G}\text{am}(\alpha, \beta)$ given by

$$\bar{G}(t) = \frac{1}{\beta^{\alpha}\Gamma(\alpha)} \int_t^{\infty} u^{\alpha-1}e^{-u/\beta}\, du \tag{2.14}$$

$$= e^{-t/\beta} \sum_{j=0}^{\alpha-1} \frac{(t/\beta)^j}{j!} \quad \text{when } \alpha \in \aleph = \{1, 2, \cdots\}. \tag{2.15}$$

The corresponding hazard rate h is best studied in terms of its reciprocal:

$$\frac{1}{h(t)} = \int_t^{\infty} (x/t)^{\alpha-1}e^{-(x-t)/\beta}dx = \int_0^{\infty} (1 + \frac{u}{t})^{\alpha-1}e^{-u/\beta}du \tag{2.16}$$

$$= \frac{\beta}{\alpha} \sum_{j=1}^{\alpha} (\alpha^{j\downarrow})(\beta/t)^{j-1} \quad \text{whenever } \alpha \in \aleph. \tag{2.17}$$

Note that again we have made use of the factorial power. See Figure 2.2.

2.2.3. The Pareto Distribution

A distribution originally derived by an Italian economist, Vilfredo Pareto (1848–1923), to describe the distribution of income within a population has proved to be of considerable interest in other areas of application, including reliability. We write $T \sim \mathcal{P}\text{ar}(\alpha, \beta)$ whenever the survival distribution is given by:

$$\bar{F}(t) = e^{-\alpha \ln(1+\frac{t}{\beta})} = \left(1 + \frac{t}{\beta}\right)^{-\alpha} \quad \text{for all } t > 0; \alpha, \beta > 0, \tag{2.18}$$

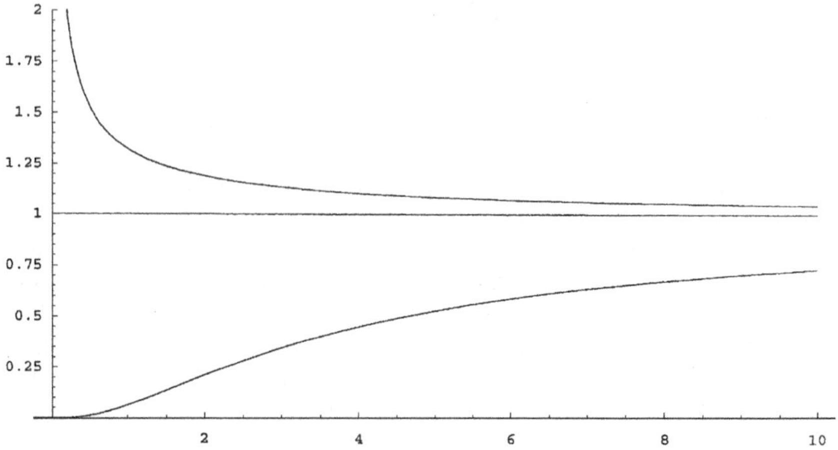

Figure 2.2. Gamma-Hazard-Rate with $\beta = 1$, for $\alpha = 1/2, \alpha = 1, \alpha = 4$.

and the corresponding hazard rate is given by

$$h(t) = \frac{\alpha}{\beta + t} \quad \text{for } t > 0. \tag{2.19}$$

2.2.4. The Gaussian or Normal Distribution

The rv X has a Gaussian or Normal distribution with location and scale parameters μ and σ, respectively, when the density is given by

$$f_X(t) = \frac{1}{\sqrt{2\pi}\sigma} e^{-\frac{(x-\mu)^2}{2\sigma^2}} \quad \text{for } x \in \Re. \tag{2.20}$$

We write $X \sim \mathcal{N}(\mu, \sigma^2)$ for any $\sigma > 0$ and $\mu \in \Re$. In *standardized form* one writes $Z \sim \Phi$ iff $Z \sim \mathcal{N}(0, 1)$. Here we define the standard distribution and density, respectively, by

$$\Phi(x) := \frac{1}{\sqrt{2\pi}} \int_{-\infty}^{x} e^{-t^2/2}\, dt, \qquad \Phi'(x) := \varphi(x) \ \text{ for } x \in \Re. \tag{2.21}$$

Thus, corresponding to the density given in eqn (2.20), $X \sim F_X$ would imply that $E(X) = \mu$ and $E(X - \mu)^2 = \sigma^2$ with the cdf and sdf given, respectively, for any $t \in \Re$ by

$$F_X(t) = \Phi\left(\frac{t - \mu}{\sigma}\right), \quad \text{with} \quad \bar{F}_X(t) = \Phi\left(\frac{\mu - t}{\sigma}\right). \tag{2.22}$$

A classical asymptotic expansion called *Mill's ratio*, see p. 932 of Abramowitz and Stegun [2] , which we here denote by $m(t)$ is

$$m(t) := \frac{t\bar{\Phi}(t)}{\varphi(t)} \asymp 1 - \frac{1}{t^2} + \frac{1 \cdot 3}{t^4} - \cdots + \frac{(-1)^n}{t^{2n}} \frac{(2n!)}{2(n!)} + \rho_n(t) \quad \text{for } t \gg 0, \tag{2.23}$$

where the error is always less than the absolute remainder; the remainder is

$$\rho_n(t) = (-1)^n \prod_{j=1}^{n} (2j-1) \times t \int_t^\infty \varphi(x) x^{-2n-2}\, dx. \tag{2.24}$$

The reciprocal of the Gaussian hazard rate, say h, is given by

$$\frac{1}{h(x)} = \frac{\bar{\Phi}(x)}{\varphi(x)} = e^{x^2/2} \int_x^\infty e^{-u^2/2}\, du \asymp \frac{1}{x} - \frac{1}{x^3} + \frac{1 \cdot 3}{x^5} - \cdots \quad \text{for } x \gg 0. \tag{2.25}$$

Computation of $h(x)$ for $x < 0$ is aided by using the identity

$$m(x) = \frac{x^-}{\varphi(x)} + m(|x|), \quad \text{where } x^- = \min(x, 0). \tag{2.26}$$

2.2.5. Transformations to Normality

The Gaussian, or normal, distribution with its unique and useful statistical properties is the distribution assumed most frequently in virtually every field. However, the support of any normal density is $\Re$ and so the probability of a negative value is always positive. Since many physical variables are nonnegative, if one adopts a Gaussian model of, say, life-length which implies there is a nonnegligible probability of being negative, a nonsensical result may occur. Moreover, it is not surprising that there are many practical problems that are "solved" by merely introducing a simple transformation of the data to normality (the logarithm is a popular choice). After finding the answer one transforms it back to the original sample space. Such procedures must be used with caution (avoided?), when they cannot be shown to give approximately correct answers.

The Truncated Normal Distribution

If $X \sim \mathcal{N}(\mu, \sigma^2)$ then the conditional rv $T = [X | X > 0]$ has the $\mathcal{TN}(\mu, \sigma^2)$ distribution with cdf given by

$$F(t) = \frac{\Phi\left(\frac{t-\mu}{\sigma}\right) - \Phi\left(\frac{-\mu}{\sigma}\right)}{\Phi(\frac{\mu}{\sigma})} \quad \text{for } t > 0. \tag{2.27}$$

Another related distribution, that has proven to be of practical utility in tool-life studies, see [51], is the *Alpha* distribution with density given by

$$f(x) = \frac{\beta}{x^2 \Phi(\alpha)} \varphi\left(\frac{\beta}{x} - \alpha\right) \quad \text{for } 0 < x < \infty. \tag{2.28}$$

The Log-Normal Distribution

This distribution is sometimes called the *Law of Galton*, because of its utilization in the pioneering efforts of Francis Galton (1822–1911) to construct an empirical and conceptual methodology in statistics that was based on probability. An rv $T > 0$ has the *log-normal* distribution whenever its logarithm has a normal distribution, i.e.,

$$T \sim \mathcal{LN}(\mu, \sigma^2) \quad \text{iff} \quad (\ln T) \sim \mathcal{N}(\mu, \sigma^2).$$

Thus, one finds the cdf is given by

$$F_T(t) = \Phi\left(\frac{\ln t - \mu}{\sigma}\right) \quad \text{for } t > 0, \tag{2.29}$$

and we find the mean and variance, respectively, to be

$$\mathrm{E}(T) = e^{\mu + \frac{\sigma^2}{2}} \quad \text{and} \quad \mathrm{Var}(T) = e^{2\mu + \sigma^2}(e^{\sigma^2} - 1). \tag{2.30}$$

The assumption that a variate has the log-normal distribution can be easily checked visually by merely taking the logarithm of the observations in a sample and seeing if the cumulative plot of the ordered observations departs significantly from a straight line when plotted on normal-probability paper.

> Every data-set they have is considered Normal unless its scatter covers several orders of magnitude and then they assume it is Log-normal.
>
> Bob McCarty - on engineering practice

Exercise Set 2.C

1. Show if $0 < \alpha \le 1$ then the gamma distribution is DHR, but for $\alpha \ge 1$ is IHR.
2. Find the density and first two moments of the Weibull distribution.
3. The rv X has the hazard rate $h(x) = (a - x)^{-1}$ for $0 < x < a$; what is the distribution of X?
4. Show that if $T \sim \mathrm{Par}(\alpha, \beta)$, that

$$\mathrm{E}(T) = \frac{\beta}{\alpha - 1} \quad \text{if } \alpha > 1: \qquad \mathrm{Var}(T) = \frac{\alpha\beta^2}{(\alpha-1)^2(\alpha-2)} \quad \text{if } \alpha > 2.$$

5. Show that if we have a gamma mixture of Epstein sdfs, namely

$$\bar{F}(t) = \int_0^\infty e^{-\lambda t} dG_\Lambda(\lambda),$$

where $\Lambda \sim \text{Gam}(\alpha, \beta)$ of eqn (2.13) that the resulting sdf is the Pareto distribution of eqn (2.18).

6. Derive the asymptotic extension of the hazard rate of a Gaussian distribution to negative values of its argument using Mill's ratio.

7. Prove that a Gaussian rv is IHR and graph its hazard rate using *Mathematica.*

8. Assume that the intelligence of human beings is normally distributed across the present population of 6.25 billion persons on earth, as measured by an IQ test score, $X \sim \mathcal{N}(\mu, \sigma^2)$, where $\mu = 100$ and $\sigma = 15$. What is your estimate of the number of persons now living who have IQ's over 190 (such as Galileo, Newton, and Einstein are thought to have had)?

9. Find the hazard rate of a Galton distribution and sketch its behavior. What can one infer about the advisability of log-normal models for life-length in reliability studies?

10. Describe exactly how one should code the abscissa and ordinate (x, y) in terms of a set of iid observations $t_1, \ldots, t_n$ which are assumed to follow the relation given in eqn (2.34) so as to check the distributional assumption.

11. If T has a Galton distribution, show that
    (a) Any power or multiple of T does also, and find the change in the parameters under such transformations.
    (b) Verify eqn (2.30) and let υ be the median, i.e., $F_T(\upsilon) = \frac{1}{2}$ confirm that $\text{E}\ln T = \ln \upsilon$ and $\upsilon = e^{\mu}$ and prove that $\upsilon = \text{E}T(1 + \nu^2)^{-1/2}$ where ν is the coefficient of variation of T is $\sqrt{\text{Var}(T)}/\text{E}(T)$.

The Xi-Normal Family

Let ξ be any *known* monotone increasing function mapping $\Re_+$ onto $\Re$ such that for any r.v. $T > 0$ within the given class there exist two parameters for which the standardizing transformation is

$$\frac{1}{\alpha}\xi\,(T/\beta) \sim \mathcal{N}(0, 1). \tag{2.31}$$

We say that T is Xi-normal, with parameters $\alpha, \beta > 0$, and we write

$$T \sim \xi\mathcal{N}(\alpha, \beta) \quad \text{iff} \quad T = \beta\xi^{-1}(\alpha Z) \;\text{ where }\; Z \sim \Phi.$$

In this case, since ξ is known, the distributional properties of T are easily determined. Moreover the adequacy of distribution to approximate the dataset can be checked as follows; use a computer to plot the ordinate and abcissa, in standard values, and see if the data are sufficiently close to a straight line. But samples must be large; see Figure 2.3.

NB knowing the transformation ξ from physical or engineering considerations is not at all the same as using a computer to examine a large number of analytic expressions and select the one, which after fitting two parameters, that best approximates the data. The latter is nothing more than data summarization, and its power for extrapolation, thereby, is limited.

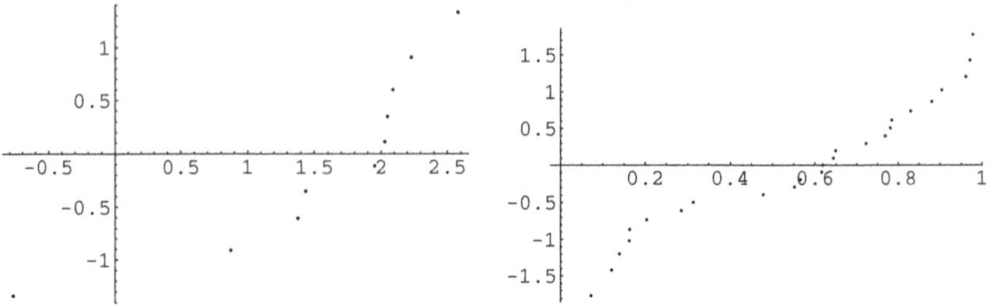

Figure 2.3. Typical plots of Xi-normal data for $n = 10, 25$.

Are there any functions ξ, satifying eqn (2.31) that are of statistical interest? The answer is "yes," since $\xi(x) = \ln x$ is such a function. But are there any other cases of interest? Consider what follows.

2.2.6. The Fatigue-Life Distribution

Let the ξ-function be defined by

$$\xi(x) := \sqrt{x} - \frac{1}{\sqrt{x}} = 2\sinh(\ln\sqrt{x}) \quad \text{for} \quad x > 0. \tag{2.32}$$

In this case ξ is montone and log-like since

$$\xi(x) = -\xi(\frac{1}{x}) \quad \text{for} \quad x > 0 \quad \text{with} \quad \xi(0) = -\infty \quad \xi(\infty) = \infty \quad \text{and} \quad \xi(1) = 0. \tag{2.33}$$

In this case we write

$$T \sim \mathcal{FL}(\alpha, \beta) \quad \text{iff} \quad F_T(t) = \Phi[\frac{1}{\alpha}\xi(t/\beta)] \quad \text{for} \quad t > 0.$$

The derivation of this complicated-looking distribution, in terms of duty cycles and load fluctuations, is given in an upcoming section.

2.2.7. The Inverse-Gaussian Distribution

This distribution was first discovered in 1915, as the first passage time for Brownian motion, by the Austrian physicist Erwin Schrödinger (1887–1961) who is famed for his wave equation in quantum theory. The same density was independently rediscovered later by Wald [117, 1944] and then later by Tweedie [114, 1945] and [115, 1957].

In the parametric form adopted by Tweedie, the density is given as

$$f(x; \mu, \lambda) = \sqrt{\frac{\lambda}{2\pi x^3}}\exp[-\frac{\lambda(x-\mu)^2}{2\mu^2 x}] \quad \text{for} \quad x > 0; \mu, \lambda > 0. \tag{2.34}$$

This parametric form is the one also used by Chhikara and Folks in [26, 1989]. If X has the density given in eqn (2.34), we shall write, following their notation (but which violates our convention of parametric order) $X \sim \mathcal{IG}(\mu, \lambda)$. It can be shown using mgfs; see Problem 3 in Exercise Set 5.A, that if X has the density given in eqn (2.34), then

$$\mathrm{E}(X) = \mu, \quad \text{and} \quad \mathrm{Var}(X) = \frac{\mu^3}{\lambda}. \tag{2.35}$$

If we reparameterize this distribution in the manner of Wald, by setting $\alpha = \sqrt{\frac{\mu}{\lambda}}$ and $\beta = \mu$, where we write, for *Wald's distribution*,

$$Y \sim G \simeq \mathcal{W}(\alpha, \beta),$$

then the cdf is given by, setting $\psi(x) := \sqrt{x} + \frac{1}{\sqrt{x}} = 2\cosh[\ln\sqrt{x}]$,

$$G(y) = \Phi[\tfrac{1}{\alpha}\xi(\tfrac{y}{\beta})] + e^{2/\alpha^2}\bar{\Phi}[\tfrac{1}{\alpha}\psi(\tfrac{y}{\beta})]. \tag{2.36}$$

Here ξ was defined in eqn (2.32) for the fatigue-life distribution. Thus, the corresponding density is given by the simple equation

$$g(y) = \frac{\sqrt{\beta}}{\alpha} y^{-3/2}\varphi[\frac{1}{\alpha}\xi(y/\beta)] \quad \text{for } y > 0, \tag{2.37}$$

from which it follows from eqn (2.35) that

$$\mathrm{E}(Y) = \beta \quad \text{and} \quad \mathrm{Var}(Y) = (\alpha\beta)^2. \tag{2.38}$$

Thus, α is the *coefficient of variation* and β is the *scale parameter and mean.*

2.2.8. The Extreme-Value Distribution of Minima

There is a class of distributions, derived by E. Gumbel as asymptotic distributions of extreme values, that have found application in many environmental phenomena. The first type is an asymptotic distribution of minima called *Gumbel's minimum law.* It is applicable even when the minimum may be negative, such as temperature, or compressive stress.

The hazard rate is given by the exponential function

$$h(t) = \frac{\alpha}{\beta}e^{t/\beta} \quad \text{for } t \in \Re; \alpha, \beta > 0, \tag{2.39}$$

and so the sdf can be written as

$$\bar{F}(t) = \exp\{-\alpha e^{(t/\beta)}\} = \exp\{-\exp[\frac{t - \ln\alpha}{\beta}]\} \quad \text{for } t \in \Re. \tag{2.40}$$

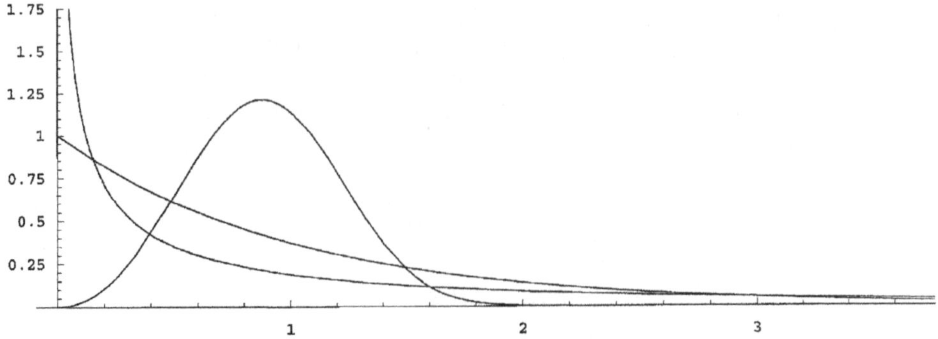

Figure 2.4. Densities for the Weibull distribution with $\beta = 1$ for $\alpha < 1, \alpha = 1, \alpha > 1$.

The Weibull Distribution

The second type is an asymptotic distribution, but of nonnegative mimima, which has become known as a *Weibull distribution* because of the attention that was directed to it by the wide applications made by the Swedish scientist, Wallodi Weibull (1887–1979), and because of its usefulness in calculating when the first failures in a fleet of systems in service will appear (see Figure 2.4). We write

$$T \sim \mathcal{W}\text{ei}(\alpha, \beta)$$

whenever the hazard is given by the power law

$$H(t) = (t/\beta)^{\alpha} \quad \text{for } t > 0; \alpha, \beta > 0. \tag{2.41}$$

Here α is a shape parameter, and β is a scale parameter often termed the *characteristic life*. This distribution has become one of the most useful in reliability studies. Today it rivals the Gaussian distribution in frequency of application.

2.2.9. Some Other Distributions

Many service-life applications required hazard rates that exhibited nonconstant behavior. One such distribution, utilized by Lord Rayleigh in his studies, and sometimes called the *Rayleigh* distribution, has a linear hazard rate given, for constants $\lambda_1, \lambda_2 > 0$, by

$$h(t) = \lambda_0 + \lambda_1 t \quad \text{for } t > 0,$$

but his name is now applied to all hazard rates that are polynomials of the form

$$h(t) = \sum_{i=0}^{p} \lambda_i t^i \quad \text{for } t > 0. \tag{2.42}$$

The Hjorth distribution see [44] was introduced as a model especially for machine-tool life. The survival distribution is given by:

$$\bar{F}(t) = \frac{e^{-\alpha t^2/2}}{(1+\beta t)^{\gamma/\beta}} \quad \text{for } t > 0; \alpha, \beta, \gamma > 0, \tag{2.43}$$

with the hazard rate given by

$$h(t) = \frac{\gamma}{1+t\beta} + \alpha t \quad \text{for } t > 0. \tag{2.44}$$

Note that this hazard rate is the sum of a convex decreasing and a linearly increasing term so it does model a "bath-tub shaped" hazard rate that is encountered in so many instances.

"A generalized inverse-Gaussian density" was studied by Jørgensen in [48] where

$$f(x) = \frac{(\chi/\psi)^{\lambda/2} x^{\lambda-1}}{2K_\lambda(\sqrt{\chi\psi})} \exp\{-\frac{1}{2}\left(\frac{\chi}{x} + \psi x\right)\} \quad \text{for } x > 0, \tag{2.45}$$

where $\lambda \in \Re$, $\chi, \psi > 0$ (the asymptotic cases where either χ or ψ equal zero are here excluded) and K_λ is the modified Bessel function of the third kind with index λ.

Exercise Set 2.D

12. Consider the alpha distribution:
    (a) Show the reliability for this distribution is given by

$$\frac{\Phi(\frac{\beta}{t} - \alpha) - \Phi(-\alpha)}{\Phi(\alpha)} \quad \text{for } t > 0.$$

    (b) If T has a truncated normal distribution, then prove $\frac{1}{T}$ has an alpha distribution.
    (c) Sketch the density of the alpha distribution and find its mode.
    (d) Find the hazard rate for the alpha distribution.

13. Show that the moments of the Rayleigh distribution have no closed-form algebraic expressions for $p > 1$, but if $p = 1$ then the mean is

$$\mu = \sqrt{\frac{2\pi}{\lambda_1}} e^{(\lambda_0/\lambda_1)^2} \bar{\Phi}(\lambda_0/\sqrt{\lambda_1})$$

14. In [48, p. 101] it said of the density given in eqn (2.45) that "Unfortunately (this) distribution appears not to be suitable for use in reliability and life testing situations ...." What do you think led to this supposition?

15. Show that, in eqn (2.45), if we set $\beta = \sqrt{\chi/\psi}$ and $\alpha = (\chi\psi)^{-1/4}$ we obtain the form

$$f(x; \alpha, \beta, \lambda) := \frac{x^{\lambda-1}}{J_\alpha(\lambda)\beta^\lambda} \exp\{-\frac{1}{2\alpha^2}\xi^2(x/\beta)\} \quad \text{for } x > 0,$$

where ξ was defined in eqn (2.32). The normalizing constant is difficult to compute since

$$J_\alpha(\lambda) = \int_0^\infty f(y : \alpha, 1, \lambda)\, dy.$$

(a) What does this paramaterization achieve?
(b) Which of the parameters can be readily estimated using mle.?
(c) Identify what practical conditions would lead one to consider such a distribution?
(d) Show that $J_\alpha(\lambda) = J_\alpha(-\lambda)$ and so wlog we can assume $\lambda > 0$.
(e) Show that integration by parts gives $J_\alpha(\lambda + 1) = 2\alpha^2 \lambda J_\alpha(\lambda) + J_\alpha(\lambda - 1)$.
(f) Tidpat

$$J_\alpha(\lambda) = \frac{J_\alpha(\lambda) + J_\alpha(-\lambda)}{2} = 4 \int_0^\infty \cosh(2\lambda x) \exp\{-\frac{2}{\alpha^2} \sinh^2(x)\}\, dx.$$

(g) Show that for $\lambda + \frac{1}{2} = n$, a positive integer, we have

$$J_\alpha(\lambda) = \sqrt{2\pi} \sum_{i=0}^{n-1} \frac{(n-1+j)!}{j!(n-1-j)!} \left(\alpha^2/2\right)^j.$$

16. Prove that if $T \sim \xi\mathcal{N}(\alpha, \beta)$, where ξ satisfies $\xi(x) = -\xi(1/x)$ for all $x > 0$, that
(a) $\frac{1}{T} \sim \xi\mathcal{N}(\alpha, \frac{1}{\beta})$ is a NASC.
(b) $aT \sim \xi\mathcal{N}(\alpha, a\beta)$ for any $a > 0$.
(c) $\mathrm{E}(\ln T) = \ln \beta$, $\mathrm{Var}(\ln T) = \mathrm{E}[\ln \xi^{-1}(\alpha Z)]^2$ when $Z \sim \mathcal{N}(0, 1)$.

17. If $T \sim \mathcal{FL}(\alpha, \beta)$, show that
(a) $\mathrm{E}T = \beta(1 + \frac{\alpha^2}{2})$.
(b) $\mathrm{Var}(T) = (\alpha\beta)^2(1 + \frac{5\alpha^2}{4})$.
(c) $\frac{2}{\alpha} \sinh\left(\frac{\ln T - \ln \beta}{2}\right) \sim \mathcal{N}(0, 1)$.
(d) What is the hazard rate of $\ln T$?
(e) What is the variance of $\ln T$?

Man has often found it easier to sacrifice his own life rather than just learn the multiplication table!

W. Somerset Maugham

I submit, in the absence of evidence to the contrary, that non-exponential (i.e., non-Gaussian) laws do occur *in rerum naturâ* and that the 'ancient solitary reign' of the exponential (Gaussian) law of errors should come to an end.

Francis Edgeworth 1883

I am averse to students who gratuitously display the affectations and temperament of genius but show none of the required determination, self-discipline or dedication to hard work.

Edwin Hewitt

Very few men are wise by their own counsel; or learned by their own teaching. For he that was only taught by himself, had a fool for a master.

Ben Jonson

What constitutes a 'good lecture' is a qualification of the listener; to the average student a good lecture is one that provides as much entertainment as enlightenment, as displayed in the popular lectures in biology given by Isaac Asimov. However, some of the most esteemed lectures of all time have appealed only to a very limited audience, e.g., Newton's lectures at Cambridge and Feynman's lectures at Cal. Tech. Both R.A. Fisher (mathematical geneticist) and P.A.M. Dirac (physicist) were extraordinarily careful expositors, to put it politely. If either were asked a question during a presentation he would reread, with perhaps different emphasis, the same carefully crafted paragraph. After one such incident, a questioner said to Sir Ronald "That's confusing; would you say it more clearly ?" Fisher responded "Young man, if it could have been stated more clearly I would have done so."

John Rainwater

Government fostered gambling institutionalizes the idea that "windfall equals success" and so injures egalitarianism in society. Gambling, with its concomitant belief in luck, chance or fate, exacerbates fatalism in people and thereby diminishes their belief in the importance of the stern virtues, viz., industriousness, thrift, diligence and studiousness. It is no wonder thart skill-less gambling is growing in America at the time when the nations productivity, competitiveness and academic performance are decreasing.

Anon.

CHAPTER 3

Partitions and Selection

3.1. Binomial Coefficients and Sterling Numbers

The number of *combinations* (subsets) of size $k \le n$ that can be chosen (obtained) from a set of size n is denoted, for nonnegative integers k, n, by

$$\binom{n}{k} = \frac{n!}{k!(n-k)!} \quad \text{for } k = 0, 1, \ldots, n \qquad \text{where } n! = 1 \cdot 2 \cdots n. \tag{3.1}$$

These numbers $\binom{n}{k}$, read "n choose k," are also called *binomial coefficients* because they occur in the binomial expansion

$$(x+y)^n = \sum_{k=0}^{n} \binom{n}{k} x^k y^{n-k} \qquad \text{for all } x, y \in \Re.$$

The number of *permutations of n objects taken k at a time* (or ordered arrangements that can be formed) is often denoted by Pochhammer's symbol, $(n)_k$, as in Feller I, or as recommended by the Mathematical Association of America, namely, the kth factorial power of n as in (1.14)

$$n^{k\downarrow} := (n)_k := \prod_{i=1}^{k} (n+1-i), \quad \text{with} \quad k!\binom{n}{k} = (n)_k; \tag{3.2}$$

it is read "n permute k."
NB definitions of both $n^{k\downarrow}$ and $\binom{n}{k}$ can be easily extended for n real or complex.
Thus, we have the binomial expansion holding for all complex z

$$(1+t)^z = \sum_{k=0}^{\infty} \binom{z}{k} t^k \qquad \text{when } |t| < 1.$$

Note that the formulas in eqns (3.1) and (3.2) have obvious combinatoric interpretations for ordered sets (vectors) and unordered sets (combinations).

The expansions of the polynomial $x^{n\downarrow}$ in powers of x and that of x^n into a factorial polynomial are given, respectively, by

$$x^{n\downarrow} = \sum_{j=0}^{n} \left[{n \atop j} \right] \cdot x^j, \qquad \text{and} \qquad x^n = \sum_{j=0}^{n} \left\{ {n \atop j} \right\} \cdot x^{j\downarrow},$$

where the coefficients $\left[{n \atop j} \right]$ are called *Sterling numbers of the first kind* and $\left\{ {n \atop j} \right\}$ are called *Sterling numbers of the second kind.* These numbers first appeared in J.

Stirling's *Methodus Differentials* ... [1730]; they were prominent in the texts for Calculus of Finite Differences, c. 1860, e.g., G. Boole's and C. Jordan's (reprinted in [1960]), but their use has decreased with the advent of modern computing. For a unified interpretation with Gaussian coefficients see [55][2000].

The original definition of a Sterling-II number was

$$\left\{ {n \atop k} \right\} := \Delta^k \left(\tfrac{x^n}{k!}\right)\Big|_{x=0} = \frac{(-1)^k}{k!}\sum_{i=0}^{k}(-1)^i\binom{k}{i}i^n, \tag{3.3}$$

where here $\Delta f(x) := f(x+1) - f(x)$ and $\Delta^{m+1} f(x) := \Delta[\Delta^m f(x)]$ for $m = 1, 2 \ldots.$

Like the binomial coefficients, the Sterling-II numbers have a combinatorial interpretation; the integer $\left\{ {n \atop k} \right\}$, read "$n$ partition k," *counts the number of* partitions *(distinct, nonempty subsets) of a set of n distinct elements into k distinct non-empty subsets.* It is also the number of distinct allocations possible when throwing n distinquishable balls into k distinct cells leaving none of the cells empty, i.e., it is the number of ordered partitions.

In analogy with permutations and combinations we introduce $\{n\}_k$, read "n cover k," with the definitions of cover numbers given by

$$k!\left\{ {n \atop k} \right\} := \{n\}_k \quad \text{for } 1 \le k \le n \quad \text{with } \{0\}_0 := 1, \;\; \{n\}_0 := 0 \text{ for } n = 1, 2 \cdots. \tag{3.4}$$

NB that $\{n\}_k$ *counts the number of ways to arrange n indistinquishable balls in k enumerated nonempty cells.*

One can show that partition and cover numbers satisfy the recursion relations

$$\left\{ {n \atop k} \right\} = k\left\{ {n-1 \atop k} \right\} + \left\{ {n-1 \atop k-1} \right\} \quad \text{and} \quad \{n\}_k = k\{n-1\}_k + k\{n-1\}_{k-1}. \tag{3.5}$$

Unfortunately, there is yet no easy, self-contained formula to compute such a number as $5,487^{28\downarrow}$. A specifically crafted computer program would be required.

Exercise Set 3.A

1. Construct a table of Sterling-II (partition) numbers $\left\{ {n \atop k} \right\}$, for $2 \le k \le n \le 10$ for comparison with the familiar Pascal's triangle.

2. Show that definitions can be extended for all $x \in \Re$ by:

$$\binom{-x}{k} := (-1)^k\binom{x+k-1}{k} \quad \text{and} \quad \left\{ {x \atop k} \right\} := \sum_{j=0}^{k-1}\frac{(-1)^j(k-j)^x}{(k-j)!\,j!}.$$

3. Show that the partition and cover numbers are related by the factorial polynomial defined for real x by

$$x^n = \sum_{j=0}^{n} x^{j\downarrow}\cdot\left\{ {n \atop j} \right\} = \sum_{j=0}^{n}\binom{x}{j}\cdot\{n\}_j. \tag{3.6}$$

4. Using the expression in eqn (3.6), prove the recursion relations shown in eqn (3.5) for partition and cover numbers corresponding to the recursion used for Pascal's triangle.

5. Show directly from eqn (3.3), by "using the calculus of finite differences," that for $k \geq 1$,

$$\{n\}_k = \sum_{j=1}^{k}(-1)^{k-j}\binom{k}{j} j^n = \begin{cases} 0 & \text{for } n = 1, \ldots, k-1, \\ n! & \text{for } n = k, \\ > 0 & \text{for } n > k. \end{cases} \tag{3.7}$$

Hint: Let $\Delta f(x) := f(x+1) - f(x)$ and $Ef(x) = f(x+1)$. Then $\Delta = E - \mathfrak{t}$, where $\mathfrak{t}$ is here the identity function, so from eqn (3.3) we see $\{n\}_k = (E - \mathfrak{t})^k\,(x^n)|_{x=0}$. Hence, for example,

$$\{n\}_1 = 1, \quad \{n\}_2 = 2^n - 2, \quad \{n\}_3 = 3(1 - 2^n + 3^{n-1}), \cdots, \{n\}_n = n!.$$

6. Using eqn (3.3) prove $\{n\}_k = k\,(\{n-1\}_k + \{n-1\}_{k-1})$.

7. Show an analgous identity to that for Pascal's triangle, $\sum_{j=0}^{n}\binom{j}{m} = \binom{n+1}{m+1}$, viz.,

$$\left\{ {n+1 \atop k} \right\} = \sum_{j=0}^{n}\binom{n}{j}\left\{ {n-j \atop k-1} \right\} = \sum_{i=0}^{n}\binom{n}{i}\left\{ {i \atop k-1} \right\}.$$

Repeat this reduction to obtain a computational formula for $\left\{ {n \atop k} \right\}$ involving sums of multinomial weights for $\left\{ {n \atop k-2} \right\}$.

3.1.1. Lagrange Coefficients

Let us consider a vector of real positive numbers $\boldsymbol{\lambda}_n = (\lambda_1, \ldots, \lambda_n)$. We define the jth *Lagrange coefficient* as

$$\left\langle {\boldsymbol{\lambda}_n \atop j} \right\rangle = \prod_{i=1, i\neq j}^{n} \frac{\lambda_i}{\lambda_i - \lambda_j} \quad \text{for } j = 1, \ldots, n. \tag{3.8}$$

These quantities, resembling Lagrange interpolation coefficients, have some interesting properties. Obviously they are scale-invariant, i.e., dimensionless. But also they satisfy some surprizing identities, for example, the following:

$$\sum_{j=1}^{n}\left\langle {\boldsymbol{\lambda}_n \atop j} \right\rangle \lambda_j{}^r = \begin{cases} \sum_{i=1}^{n}\lambda_i^{-1} & \text{for } r = -1, \\ 1 & \text{for } r = 0, \\ 0 & \text{for } r = 1, \cdots, n-1, \\ (-1)^{n-1}\prod_{i=1}^{n}\lambda_i & \text{for } r = n. \end{cases} \tag{3.9}$$

These identities generalize some identities, given as exercises in Feller-I, p 65, [33], which involve the binomial coefficients. In particular the notable result

$$\sum_{j=1}^{n}\frac{(-1)^{j-1}}{j}\binom{n}{j} = \sum_{j=1}^{n}\frac{1}{j}, \tag{3.10}$$

which may be proved using induction or the hint supplied *loc. cit.*, namely "integrate the identity $\sum_{j=0}^{n-1}(1-t)^j = [1 - (1-t)^n] \cdot t^{-1}$." But a rabbit-out-of-the-hat leaves one curious to see if more than a rabbit can be obtained from the same hat.

From the definition in eqn (3.8), letting $\boldsymbol{n} = (1, 2, \ldots, n)$, we find $\left\langle {\boldsymbol{n} \atop j} \right\rangle = (-1)^{j-1}\binom{n}{j}$ for $j = 1, \cdots, n$, so not only does eqn (3.10) follow, but in addition from eqn (2.8) we see

$$\sum_{j=1}^{n}(-1)^{n-j}\binom{n}{j} j^r = \begin{cases} 0 \text{ for } r = 1, \cdots, n-1, \\ n! \text{ for } r = n. \end{cases}$$

Exercise Set 3.B

1. A discrete rv N is said to be *geometric* iff for some $0 < p < 1,\ q = 1 - p$ we have

$$\mathrm{P}[N = n] = qp^{n-1} \quad \text{for } n = 1, 2, \cdots. \tag{3.11}$$

   We write $N \sim \mathcal{G}(p)$.
   (a) Find $\mathrm{E}[N]$, $\mathrm{Var}[N]$, and the cdf $\mathrm{P}[N \leq n]$ for $n = 0, 1, 2, \cdots$.
   (b) Apply these results to the game of *Russian Roulette* to see how small a value of q would be acceptable. ('Russian Roulette' is any game in which there are repeated trials in which the pay-off for winning is trivial but losing is disastrous—such as passing in traffic on a two-lane road with obscured forward vision.)

2. A draws repeatedly, at random with replacement, from a half-deck of playing cards. A pays \$10 to B for each previously unobtained card but B pays \$3.50 to A when a drawn card is a duplicate. The game ends when all 26 cards have been recorded. Show that B has the advantage.

3. There is a problem in Feller I, see [33][p. 29], to show that with 23 people in a room the probability for two persons to have the same birthday (month and day) is greater than a half. This follows since,

$$\prod_{j=1}^{n}\left(1 - \frac{j}{365}\right) < \frac{1}{2} \quad \text{for } n = 22.$$

   This counterintuitive fact is so widely known, and misunderstood, that on a television talk show a guest asserted that it was likely that two persons in the audience had the same birthday. The host expressed his doubt and then asked, "How many people in this room have a birthday on April 19th?" and when no one raised a hand, he said triumphantly "See, I told you." Why didn't the odds work for that TV audience?
   (a) If there are 40 persons in a room what is the probability that at least two persons will have a birthday the same day?
   (b) If there are more than 60 persons in the room would you give 5 to 1 odds that at least two persons will have a birthday the same day?
   (c) An after-dinner speaker asserted to a his audience of 200 persons that not only were there two people in the room with the same birthday but he bet \$100 that he could get find a match by announcing the birthdays of only 5 persons! When the bet was accepted and a match occurred when the third birthday was announced, the wagerer refused to pay because he asserted the bet was rigged! What are the probabilities here? Was this an unusual occurrence?

4. If $N_i \sim \mathcal{G}(p_i)$ for $i = 1, 2$ with $N_1 \perp N_2$, show that

$$\mathrm{P}[N_1 + N_2 \leq n] = 1 - \frac{q_2}{q_2 - q_1} p_1^n - \frac{q_1}{q_1 - q_2} p_2^n \quad \text{for } n = 1, 2, \cdots.$$

5. Let $N_i \sim \mathcal{G}(p_i)$ for $i = 1, \cdots, k$ be independent rvs. Show that

$$\mathrm{P}\left[\sum_{i=1}^k N_i \le n\right] = \sum_{j=1}^k \left\langle {\boldsymbol{q}_k \atop j} \right\rangle (1 - p_j^n) \quad \text{for } n = 1, 2 \cdots, \tag{3.12}$$

where we define $\boldsymbol{q}_k = (q_1, \cdots, q_k)$ and $q_i = 1 - p_i$ for $i = 1, \ldots, k$ are all distinct with

$$\left\langle {\boldsymbol{q}_k \atop j} \right\rangle = \prod_{i=1, i \neq j}^k \frac{q_i}{q_i - q_j} \quad \text{for} \quad 1 \le j \le k.$$

(a) Check that for $j = 1, \cdots, k$

$$\left\langle {\boldsymbol{q}_{k+1} \atop j} \right\rangle = \left\langle {\boldsymbol{q}_k \atop j} \right\rangle \frac{q_{k+1}}{q_{k+1} - q_j} \quad \text{with} \left\langle {\boldsymbol{q}_{k+1} \atop 0} \right\rangle = 1, \quad \text{and} \quad \sum_{j=1}^{k+1} \left\langle {\boldsymbol{q}_{k+1} \atop j} \right\rangle = 1.$$

(b) Argue that for consistency we must have $\sum_{j=1}^k \left\langle {\boldsymbol{q}_k \atop j} \right\rangle p_j^n = 1$ for $n = 1, \cdots, k-1$ and then use eqn (3.9) to prove it.

(c) Let $W_k = \sum_{j=1}^k N_j$ and show using eqns (3.9) and (3.12) that

$$\mathrm{E}[W_k] = \sum_{j=1}^k \frac{1}{q_j} \quad \text{and} \quad \mathrm{Var}[W_k] = \sum_{j=1}^k \frac{p_j}{q_j^2}. \tag{3.13}$$

6. * One of the Markov Chain models for damage accumulation proposed in [9] starts at time $t = 0$ from a new-state 1 and progress through each damage-state until failure at state k. It has a transition matrix

$$\begin{pmatrix} p_1 & q_1 & 0 & 0 & \cdots & 0 & 0 \\ 0 & p_2 & q_2 & 0 & \cdots & 0 & 0 \\ 0 & 0 & p_3 & q_3 & \cdots & 0 & 0 \\ 0 & 0 & 0 & p_4 & \cdots & 0 & 0 \\ \cdot & \cdot & \cdot & \cdot & \cdots & \cdot & \cdot \\ \cdot & \cdot & \cdot & \cdot & \cdots & \cdot & \cdot \\ 0 & 0 & 0 & 0 & \cdots & p_{k-1} & q_{k-1} \\ 0 & 0 & 0 & 0 & \cdots & 0 & 1 \end{pmatrix}.$$

Here p_j is the probability of remaining in state j during one duty cycle, and $q_j = 1 - p_j$ is the probability of going from state j to state $j + 1$. This formulation p. 84 *loc. cit.* for the distribution of the time to failure can be seen by inspection to be equivalent with the distribution of $W_k = \sum_1^k N_j$, the waiting time for the sum of k independent geometric rvs. However, if the fact that $\sum_1^k \left\langle {\boldsymbol{q}_k \atop j} \right\rangle = 1$ is overlooked some complicated expressions ensue. What advantages can you see to a Markov Chain formulation?

3.2. Lotteries and Coupon Collecting

3.2.1. Lotteries

In the Western World lotteries began in Italy; the *Lotto de Firenze*, with a money prize, was instituted in 1530. Lotteries were in Genoa in the 1620s and in France by the 1750s. They were used in America to help finance both the Revolutionary and Civil wars. But scandals involving rigged winners or organizers who fled with sales and prize money, had become so frequent that every state in the U.S.A. had outlawed lotteries by 1893. All lotteries were prohibited by the Washington State Constitution of 1890. In other states, such as Oregon, only churches and other licensed charitable organizations could hold lotteries. But in 1964 the first modern lottery was held in New Hampshire and within 20 years they had spread to about 20 states, By the turn of the century the number of states instituting lotteries had doubled. Lotteries are now legal in all states except where vested gambling interests combine with the churches to prohibit it. Today there are national lotteries in China, Germany, England, France, Canada, Ireland, *et al.* What had happened?

The advent of modern computing machines had made strict accounting easier, cheating more difficult and reliable winner identification (for tax purposes) possible. State legislatures were informed by the vendors of ticket machines: "Lottery revenue will help pay for public improvements, education, health care for the elderly and aid to the indigent plus provide diversion and entertainment for the citizenry." But did it?

On Saturday May 30, 1992, the state of Texas initiated a lottery previously banned by state law for 147 years; it was the 35th state to do so. There were $22.7 million in ticket sales the first day in their scratch-game with a $300 million ticket sale possible in which only 500 tickets could win the top prize of $10,000. In fact there were only 30 winners of $10,000 and all the other 1,200 winners accumulated less than $200,000 in prize money; it was a take for the state of more than 16 million dollars, less their approximately 6 million dollars in set up costs. Part of the reason for this wild public enthusiasm was their complete ignorance of the odds.[1]

3.2.2. Coupon Collecting

The COUPON COLLECTING PROBLEM, e.g., see [82], is equivalent with the following:

> Balls in an urn, numbered 1 to m, are randomly drawn with replacement, and the number of each drawn ball is recorded. What is the number of draws, on average, which will be required to have recorded k different balls?

The CLASSICAL OCCUPANCY PROBLEM, is equivalent with the coupon collecting problem if you ask for more than the expectation,

> If we toss n (indistinquishable) balls at random into m enumerated (distinquishable) cells, what is the distribution of the number of occupied cells? What is the distribution of the number of tosses to occupy k different cells?

[1] This was illustrated in a cartoon strip where *Value-Priced Lottery Tickets* were on sale at half-price. These VP-tickets only reduced one's chance of winning by the infinitesimal amount of $1/13,987,000$!

The latter problem was first addressed and solved by Abraham De Moivre, in his seminal paper *De Mensura Sortis* [1712] and was later expanded in his treatise *The Doctrine of Chances* [1732], reprinted in [1756]. There "the classical occupancy distribution," namely, the probability of exactly k cells remaining empty when n indistinquisable balls are tossed into m distinquishable cells was first given. (He obtained it using only the inclusion-exclusion theorem!)

Let us deal first with the coupon collecting problem. Assume there are m distinct objects in the urn. Let N_j be the random number of draws, made with replacement, which are necessary to obtain the jth different object given that $(j-1)$ objects have been previously collected for $j = 1, 2, \ldots, m$. Clearly, with probability 1 we have $N_1 = 1$. The probability of drawing one of the same objects that had been previously obtained is the same on each successive draw. Thus, N_j, the waiting time (in draws) until the jth object is obtained, is a geometric random variable and we see $N_1 \perp N_2 \perp \cdots \perp N_k$. So we know by Exercise 3.B1 that for $k = 1, 2, \cdots \mathrm{P}[N_j = k] = q_j \cdot (p_j)^{k-1}$ where $p_j = (j-1)/m$ and $q_j = 1 - p_j$ and here $0^0 = 1$. In this case we define $S_n := \sum_{j=1}^{n} N_j$ to be the total number of draws to obtain $n \leq m$ different objects. Thus, by eqn (3.13) we find the expectation to be $\mathrm{E}N_j = \frac{1}{q_j} = \frac{m}{m+1-j}$ and so obtain the harmonic-sum solution for the coupon collecting problem. The expectation is easily computed using De Temple's half-integer approximation for the harmonic sum, see [30], namely, for any integer $m \geq 6$,

$$\sum_{j=1}^{m} \frac{1}{j} \doteq \ln(m + \frac{1}{2}) + \gamma \qquad \text{where } \gamma = 0.57721\cdots \tag{3.14}$$

is the Euler–Mascheroni constant.

Thus,

$$\mathrm{E}S_n = \sum_{j=1}^{n} \frac{1}{q_j} = \sum_{j=m+1}^{m} \frac{m}{j} \doteq m \ln\left(\frac{2m+1}{2n+1}\right), \tag{3.15}$$

with the approximation good for $1 \ll n \ll m$.

We now obtain, in general, the probability distribution first sought by De Moivre.

Theorem 1. *Let* $\{N_j : j = 1, \ldots, n\}$ *for* $n \geq 1$ *be independent rvs with* $N_j \sim \mathcal{G}(p_j)$, *where* $p_j + q_j = 1$. *Then, if the* q_j *are all distinct, with* $W_n = \sum_{j=1}^{n} N_j$, *we have for* $x = 1, 2, \cdots$,

$$P[W_n \geq x] = \sum_{j=1}^{n} \left\langle {\boldsymbol{q}_n \atop j} \right\rangle p_j^{x-1}, \quad P[W_n = x] = \sum_{j=1}^{n} \left\langle {\boldsymbol{q}_n \atop j} \right\rangle p_j^{x-1} q_j, \tag{3.16}$$

where $\boldsymbol{q}_n = (q_1, \cdots, q_n)$ *with*

$$\left\langle {\boldsymbol{q}_n \atop j} \right\rangle = \prod_{i=1, i \neq j}^{n} \frac{q_i}{q_i - q_j} \quad \textit{for } j = 1, 2, \cdots, n \quad \textit{and} \quad \sum_{j=1}^{n} \left\langle {\boldsymbol{q}_n \atop j} \right\rangle = 1. \tag{3.17}$$

PROOF. A proof was requested in Exer. 3.B5. Proceeding by induction let $\boldsymbol{q_n} = (q_1, \ldots, q_n)$ for $n = 1, 2, \cdots$ with $\left\langle {\boldsymbol{q}_1 \atop 1} \right\rangle = 1$. Then we have for any $n \geq 2$, and $j = 1, \ldots, n-1$

$$\left\langle {\boldsymbol{q}_n \atop j} \right\rangle = \left\langle {\boldsymbol{q}_{n-1} \atop j} \right\rangle \cdot \frac{q_n}{q_n - q_j} \quad \text{with} \quad \left\langle {\boldsymbol{q}_n \atop n} \right\rangle = \prod_{i=1}^{n-1} \frac{q_i}{q_i - q_n}. \tag{3.18}$$

Assume the proposition true for $n-1$. Then one convolves the geometric variables as follows:

$$\begin{aligned} \mathrm{P}[W_n \leq x-1] &= \sum_{y=1}^{x-1} \mathrm{P}[W_{n-1} \leq x-y] \cdot \mathrm{P}[N_n = y] \\ &= \sum_{y=1}^{x-1} \left(1 - \sum_{j=1}^{n-1} \left\langle {\boldsymbol{q}_{n-1} \atop j} \right\rangle p_j^{x-y-1}\right) q_n p_n^{y-1} \\ &= 1 - p_n^{x-1} - \sum_{j=1}^{n-1} \left\langle {\boldsymbol{q}_n \atop j} \right\rangle p_j^{x-1}[1 - (p_n/p_j)^{x-1}] \\ &= 1 - \sum_{j=1}^{n} \left\langle {\boldsymbol{q}_n \atop j} \right\rangle p_j^{x-1}. \end{aligned}$$

The proofs of the equations in (3.17) are trivial. □

Corollary 1. *If the q_i are not distinct the convolution can be obtained from eqn (3.16) merely by taking the appropriate limits.*

For example, say $q_1 = q_2$ with q_i distint for $i \geq 2$ then

$$\lim_{q_1 \to q_2} \sum_{j=1}^{n} \left\langle {\boldsymbol{q}_n \atop j} \right\rangle p_j^{x-1} = \left[\lim_{\epsilon \to 0} \frac{N(\epsilon)}{D(\epsilon)}\right] \prod_{i=3}^{n} \frac{q_i}{q_i - q_2} + \lim_{q_1 \to q_2} \sum_{j=3}^{n} \left\langle {\boldsymbol{q}_n \atop j} \right\rangle p_j^{x-1} \tag{3.19}$$

where $\epsilon = q_2 - q_1$ and we have

$$N(\epsilon) = q_2(p_2 + \epsilon)^{x-1} \prod_{i=3}^{n} (q_i - q_2) - (q_2 - \epsilon) p_2^{x-1} \prod_{i=3}^{n} (q_i - q_2 + \epsilon)$$

$$D(\epsilon) = \epsilon \prod_{i=3}^{n} (q_i - q_2 + \epsilon).$$

Both $N(\epsilon), D(\epsilon) \to 0$ as $\epsilon \to 0$ so we apply L'Hôpital's rule to obtain for the term in square brackets on the right in eqn (3.19) that

$$[\cdots] = (x-1) q_2 p_2^{x-2} + p_2^{x-1} \left(1 - \sum_{i=3}^{n} \frac{q_i}{q_i - q_2}\right).$$

The theorem above gives the distribution in the coupon-selection problem.

Corollary 2. *The pdf of* $S_n = \sum_1^n N_j$, *the total number of draws necessary to obtain* $n \leq m$ *different numbers when drawing at random with replacement from a set of* m *distinct numbers, is*

$$\begin{aligned} P[S_n = x] &= \frac{\binom{m-1}{n-1}}{m^{x-1}} \sum_{j=1}^{n-1} (-1)^{n-1-j} \binom{n-1}{j} j^{x-1} \\ &= \frac{m^{n\downarrow}}{m^x} \left\{ {x-1 \atop n-1} \right\} \quad \textit{for } x = n, n+1, \cdots. \end{aligned} \tag{3.20}$$

PROOF. If we let $q_j = (m+1-j)/m$ for $j = 1, \ldots, n$ then from eqn (3.17) we find, asas,

$$\left\langle {q_n \atop j} \right\rangle = (-1)^{n-j} \frac{j}{m+1-j} \binom{m}{n} \binom{n}{j} \quad \text{for } j = 1, \ldots, n, \tag{3.21}$$

so by substitution into eqn (3.16) and simplifying we obtain the first equality; then we recognize the formula from eqn (3.7) and simplify. □

This pdf provides surprisingly complicated identities, for example, by equating the moments of S_n to the formulas calculated in general. Consider the variance for the expected waiting time until n different numbers are obtained, namely, $\mathrm{E}[S_n^2] - \{\mathrm{E}[S_n]\}^2$.

Incidently this nonintuitive behavior of the (mathematical) expectation of the long waiting time (large number of draws) to obtain a prescribed coverage is the source of the claims by some observers, who record the winning combinations in the state lotteries, as well as the touts of certain winning strategies, that some winning numbers "stay hot" after they have once been chosen. A cursory examination of the frequency of winning numbers shows that many are chosen repeatedly before others, that have never been drawn, are chosen; see [99].

3.3. Occupancy and Allocations

3.3.1. Occupancy

If we toss n indistinquishable balls at random into m distinquishable (enumerated) cells and ask for the probability distribution of the number of occupied cells, we see this is, except for nomenclature, the same as the coupon collecting problem.

Suppose we want to find the distribution of K_n, the number of distinct objects that will have been obtained in n draws (with replacement) from an urn containing m distinct objects. By the duality principle we have the direct relationship

$$[K_n \leq k] = [S_{k+1} \geq n+1], \tag{3.22}$$

and so by substituting from eqn (3.21) into eqn (3.16) we have, asas, the cdf

$$\mathrm{P}[K_n \leq k] = \frac{\binom{m-1}{k}}{m^n} \sum_{i=1}^{k} \frac{(-1)^{k-i}}{m-i} \binom{k}{i} i^n \quad \text{for } k = 1, \cdots, m. \tag{3.23}$$

Since $\mathrm{P}[K_n = k] = \mathrm{P}[K_n \leq k] - \mathrm{P}[K_n \leq k-1]$ we then obtain, asas, for $k = 1, \cdots, m \wedge n := \min\{m, n\}$,

$$\mathrm{P}[K_n = k] = \frac{\binom{m}{k}}{m^n} \sum_{j=1}^{k} (-1)^{k-j} \binom{k}{j} j^n = \frac{\binom{m}{k}\{n\}_k}{m^n}. \tag{3.24}$$

Here again we recognize the combinatorial coefficient "n cover k" from eqn (3.7). This number counts the ways to arrange n indistinquishable balls into k distinguishable (enumerated) *nonempty* cells, with the convention $\{0\}_0 = 1$. We have seen these coefficients $\{n\}_k$ are related to the Sterling-II numbers, written $\left\{ {n \atop k} \right\}$, since $k! \left\{ {n \atop k} \right\} = \{n\}_k$.

We interpret eqn (3.24) as the probability of k cells being occupied if n indistinquishable balls are thrown, at random, into m distinquishable cells. This is now seen as a simple combinatoric formula; the numerator is the product of the number of ways we may select, from the m cells, those k that are to be occupied, viz., $\binom{m}{k}$ ways, times the number of ways that these k cells can be covered by n balls, viz., $\{n\}_k$. The denominator, m^n, counts all the ways n balls can be thrown at random into into m cells.

Exercise Set 3.C

1. Use the definition of equality of events to prove eqn (3.22).
2. Derive eqn (3.24) from eqn (3.23).
3. Consider the coupon collecting problem and let S_n be the total number of draws, with replacement, necessary to obtain n different numbers from a set of $m \geq n$ distinct numbers. Can you give a simple combinatorial argument that the correct formula in eqn (3.20) is

$$\mathrm{P}[S_n = x] = \frac{\binom{m}{n}\{x-1\}_{n-1}}{m^x} \quad \text{for } x = n, n+1, \cdots.$$

4. Using the set-combinatorical definition, prove the recursion relation for Sterling-II numbers,

$$\left\{ {n \atop j} \right\} = j \left\{ {n-1 \atop j} \right\} + \left\{ {n-1 \atop j-1} \right\} \quad \text{for } j = 1, 2, \cdots, n. \tag{3.25}$$

5. Using the approximate formula for the harmonic sum it is possible to count, approximately, how many quick-pick lottery tickets will have to be purchased in a state's "49 pick 6" lottery to expect that all the $m = \binom{49}{6} = 13,983,816$ different six-number combinations available for the players of *Lotto* (the most prominent game among the dozen or so offered to the public today) have been sold?

6. Calculate the reduction in the probabiity of winning when playing the *Power Ball* game instead of Lotto. The former requires the exact number has to be selected on the "power-ball" but the other, say 6, numbers can be obtained on any ball in any order.

Multiple Occupancy

If we toss n identical balls at random into m enumerated cells and let K_n be the number of occupied cells then $X_0 = m - K_n$ is the number of unoccupied cells, i.e., the number of cells with zero occupancy. Thus, we have its probability function given by

$$\begin{aligned} \mathrm{P}[X_0 = k] &= \frac{\binom{m}{k}\{n\}_{m-k}}{m^n} \\ &= \frac{m^{(m-k)\downarrow}\left\{ {n \atop m-k} \right\}}{m^n} \quad \text{for } k = 1, \cdots, m-1 \text{ when } m > 1. \end{aligned} \tag{3.26}$$

In physics this distribution is classical and when estimates of its probabilities are observed, one obtains what are called "Maxwell–Boltzman statistics" which are to be contrasted with other "statistics," obtainable from other theoretical distributions to be mentioned later.

From the above formulas we see there are events in the state lotteries where m is the total number of 6-number combinations out of 49 numbers specified and n is the number of Quick-pick tickets sold and both may be in the millions, for which no precise calculation of probabilities can be made! And so easily computed approximations are sought!

Let X_k be the number of cells which have *k-occupancy* for $k = 0, 1, \cdots, n$. Then clearly we must have the two restrictions

$$\sum_{k=0}^{n} X_k = m \qquad \text{and} \qquad \sum_{k=0}^{n} kX_k = n \tag{3.27}$$

Consider the indicator variable $X_{i,j} := 1$ iff the jth cell has i occupants and 0 otherwise. (Note $X_i = \sum_{j=1}^{m} X_{i,j}$.) Consider the number of balls in the first cell. After n tosses have been made this number will be a binomial rv since each toss has the same probability, namely, $1/m$, of going into the first cell. Thus, $X_{i,1} \sim \mathcal{B}(n, 1/m)$. By symmetry each cell is interchangeable so

$$\mathrm{E}[X_i] = m\mathrm{E}[X_{i,1}] = m\mathrm{P}[X_{i,1} = 1] = m\binom{n}{i}\frac{(m-1)^{n-i}}{m^n}.$$

It becomes possible by a somewhat more involved argument given by von Mises, see [81], to obtain the rth factorial moment of X_k,

$$\mathrm{E}[X_k^{r\downarrow}] = \frac{m^{r\downarrow} n!(m-r)^{n-rk}}{(n-rk)!(k!)^r m^n} \quad \text{whenever } r \le m, r \cdot k \le n. \tag{3.28}$$

Exercise Set 3.D

1. Show: $\mathrm{E}[X_0] = m\left(\frac{m-1}{m}\right)^n$ and, since $\mathrm{Var}[X_0] = \mathrm{E}[X_0^{2\downarrow}] + \mathrm{E}[X_0] - (\mathrm{E}[X_0])^2$, it follows that

$$\mathrm{Var}[X_0] = m(m-1)\left(1 - \frac{2}{m}\right)^n + m\left(1 - \frac{1}{m}\right)^n - m^2\left(1 - \frac{1}{m}\right)^{2n}.$$

2. Prove a theorem of R. von Mises [81] that if $m, n \to \infty$ so that $me^{-n/m} \to \lambda > 0$, then the distribution of X_0 becomes, in the limit, Poisson, namely, $\mathrm{P}[X_0 = k] - \frac{e^{-\lambda}\lambda^k}{k!} \to 0$. Consider the utility of this result for a lottery where $m \approx 1.2 \times 10^7$ and $n \approx 2.27 \times 10^5$. Can the exact probability be calculated?

3. Show that approximately $X_0 \sim \mathcal{B}in(m, p)$ with $p = e^{\nu}$, where $\nu = n/m$ for m large, since $\mathrm{E}[X_0] \approx me^{-\nu}$ and $\mathrm{Var}[X_0] \approx me^{-\nu}[1 - e^{-\nu}]$. Determine under what conditions the binomial approximation is better than the Poisson approximation and which should be used only when $n \ll m$.

4. If $\nu = n/m$ is fixed for n, m large then normalize X_k by setting $U_k = X_{,k}/m$, the fraction of cells having k-occupancy, and use eqn (3.28) to show that $\mathrm{E}[U_k] \longrightarrow \frac{e^{-\nu}\nu^k}{k!}$.

5. If for a given lottery prize of $\$D$ (which is shared if multiple winners occur) the participation m goes up exponentially with linearly increasing $\$D$, compute what size of lottery jackpot has the highest expected return.

6. R. von Mises, gave the joint (discrete) distribution of the number of cells having all occupancy numbers as

$$f_{X_0,\cdots,X_n}(x_0, \cdots, x_n) = \frac{m!n!}{m^n \prod_{j=0}^{n} (j!)^{x_j} (x_j!)} \quad \text{for all } x_j \geq 0,$$

where $\sum_{j=0}^{n} x_j = m$, $\sum_{j=0}^{n} jx_j = n$. Can you obtain this joint distribution by using conditional probabilities begining with X_0?

7. Show, using eqns (3.3) and (3.4), that the n-cover-k coefficients, $\{n\}_k$, satisfy the recursions

$$\{n\}_k = k\left(\{n-1\}_k + \{n-1\}_{k-1}\right), \qquad \{n\}_{k+1} = (k+1)\sum_{i=0}^{n-1} \binom{n-1}{i}\{i\}_k. \tag{3.29}$$

8. A metropolitan Police Department is under political pressure to avoid any action that can be misconstrued as "antigroupism," such as racial profiling or sport-car profiling (for stopping them, unecessarily as their drivers believe). Assume the population consists of r categories of persons in proportions $p_1 > \cdots > p_r$ with $\sum_1^r p_j = 1$; so each arrest is like drawing, with replacement, from an urn consisting of these r proportions of variously colored balls.
   (a) What is the the total number of arrests that must be made by each officer so as to be 95% sure (s)he had obtained at least one arrest in each category, when $p_r = .02$?
   (b) The arrest record of each officer is examined to see if his(her) observed percentage is significantly different from the known proportions, How would you set the limits of tolerable deviation so as to prevent invalid charges of racial profiling?
   (c) Certain categories of age, of employment, of education, of economic status do not have uniform likelihood of misprison and these may not be equally distributed within the group catgories above. What account should be made of these considerations?

9. The BAC company uses widgets, which it buys from ABC, under quality-control specifications. But BAC also has random inspections of each lot. If too high a number are found defective, the lot is returned. BAC suspects that ABC merely takes the returned shipment and repackages it with current production and sends it back. The reason for this supposition is that any widget that cannot be successfully installed by BAC workers is discarded, but the reason for unusability is recorded. BAC observed the higher frequency for failure during installation occurs in waves indicating that a lot having a higher proportion of defectives had passed inspection. Litigation may be impending and you are consulted about buttressing the argument with data and calculations. What model would you adopt? What statistics would you collect? Can you obtain their distribution? Would you ask about installation procedures? How would you counter the argument that random errors sometimes come in waves?

3.3.2. Allocations

Assume that n distinguishable cells are occupied by r indistinguishable balls placed there by some unknown natural mechanism. Let r_i be the number of occupants of cell i for $i = 1, \cdots, n$, i.e., it is the *occupancy number of cell* i. Then the cell's *occupancy vector* is the n-tuple $r_1, \ldots, r_n$, where $\sum_{i=1}^{n} r_i = r$.

(1) Let r balls be tossed independently and individually, each with probability $1/n$ of landing into each cell, then the Maxwell–Boltzman distribution for occupancy obtains, which is a special case of the multinomial distribution

$$\frac{r!}{r_1! \cdots r_n!} \frac{1}{n^r}, \quad \text{where } r_i \geq 0 \text{ for } i = 1, \ldots, n \text{ with } \sum_{i=1}^{n} r_i = r.$$

(2) Assume the n cells are the ordered spaces between $n+1$ vertical bars and the r balls are stars scattered among them. Thus, $| \star\star\star \,||\, \star\star |||| \star |$ denotes an allocation of 6 balls into 7 cells with occupancy numbers 3,0,2 0,0,0,1. We must start and end with a bar, however; the remaining stars and bars may be arranged in any order. (One may envisage this arrangement in a circle using n, bars but it takes more space.) By definition of the binomial (combinatorial coefficient) the total number of distinct allocations is

$$A(r, n) = \binom{n+r-1}{r} = \binom{n+r-1}{n-1}.$$

We use the notation from Feller's volume I, [33], for the "number of allocations."

By a similar argument the number of distinct allocations in the case when no cell must be left empty is

$$A_0(r, n) = \binom{n-1}{r-1}.$$

One obtains *Bose–Einstein* "statistics" whenever we have r indistinquishable particles allocated into n cells with each distinquishable arrangement having the same probability, namely,

$$1/A(r, n).$$

This is known in physics to be true of elementary particles in atomic arrangements that have a particular "intrinsic" spin.[2]

(3) To describe elementary particles with spin 1/2, i.e., electrons, neutrons, protons, it becomes necessary to assume no multiple occupancy can occur so each occupancy number is 0 or 1. This gives *Fermi–Dirac* statistics in which there are $\binom{n}{r}$ possible allocations each with the same probability.

Theorem 2. *We obtain Maxwell–Boltzman, Bose–Einstein, or Fermi–Dirac "statistics" from the occupancy distribution, with* $r_i \geq 0$ *for* $i = 1, \ldots, n$ *and* $\sum_{i=1}^{n} r_i = r$*, respectively, whenever*

$$P[R_1 = r_1, \cdots, R_n = r_n] = \begin{cases} \frac{r!}{r_1! \cdots r_n!} \frac{1}{n^r} \\ 1/A(r, n) \\ 1/\binom{n}{r} \qquad \textit{provided each} \quad r_i = 0 \textit{ or } 1. \end{cases} \tag{3.30}$$

The reason why spin of the particle makes this true is not completely understood.

3.4. Related Concepts

3.4.1. The Sum of Epstein Waiting Times

Let us consider the random waiting time until a fisherman catches his first fish. There is no "bonus" for waiting; if one waits an hour and catches nothing, then the waiting time remaining until a fish is caught is the same (in distribution) as when fishing started! Since the French word for fish is "Poisson" it is appropriately, but coincidentally, called a *Poisson process*. This "memoryless" feature implies that the random waiting time until the first "event," call it T_1, has an *Epstein* distribution

$$T_1 \sim \mathrm{Ep}(\lambda_1) \quad \text{iff} \quad \mathrm{P}[T_1 > t] = e^{-\lambda_1 t} \quad \text{for all } t > 0; \text{for some } \lambda_1 > 0.$$

Suppose we want the random waiting time until n different, independent events have occurred serially. Let $S_n = \sum_{i=1}^{n} T_i$, where $T_i \sim \mathrm{Ep}(\lambda_i)$ is the random waiting time for the ith kind of event for each $i = 1, \cdots, n$, with $\lambda_1, \cdots, \lambda_n$ all positive and distinct. Set $\boldsymbol{\lambda}_n = (\lambda_1, \ldots, \lambda_n)$; one can easily show, using induction, that the

[2] "Intrinsic spin" was a conceptual upheaval when introduced. It implied that each extant particle of matter possessed some fixed amount of spinning inertia, which was a multiple of a universal quantum, that did not depend upon any other physical attribute.

survival distribution of S_n is, in this case, given by

$$\bar{F}_n(t) := \mathrm{P}[S_n > t] = \sum_{j=1}^{n} \left\langle {\boldsymbol{\lambda}_n \atop j} \right\rangle e^{-\lambda_j t} \quad \text{for } t > 0, \tag{3.31}$$

where $\sum_{j=1}^{n} \left\langle {\boldsymbol{\lambda}_n \atop j} \right\rangle = 1$ and $\left\langle {\boldsymbol{\lambda}_n \atop j} \right\rangle$ are the same as that defined in eqn (3.16).

Since $\mathrm{E}T_i = 1/\lambda_i$ it is clear that

$$\mathrm{E}[S_n] = \int_0^{\infty} \bar{F}_{(k)}(t)\,dt = \sum_{i=1}^{n} \left\langle {\boldsymbol{\lambda}_n \atop j} \right\rangle \frac{1}{\lambda_j} = \sum_{i=1}^{n} \mathrm{E}T_i = \sum_{i=1}^{n} \frac{1}{\lambda_i}. \tag{3.32}$$

Using the last equality in eqn (3.32), where the $\left\langle {\boldsymbol{\lambda}_n \atop j} \right\rangle$ do not appear, another finesse is possible. This follows using the same method of reasoning used to obtain eqn (3.32). Straightforwardly we find identities for the coefficients in the distribution of the sum of Epstein (negative exponential) random variables, which have been rediscovered again and again; see, e.g., [94, 1965], p. 61 [56, 1970], p. 84 [9, 1985]. But more can be seen herein eqn (3.31), for example,

$$\mathrm{Var}[S_n] = \sum_{j=1}^{n} \frac{1}{\lambda_j^2} \quad \text{implies} \quad \sum_{i=1}^{n} \left\langle {\boldsymbol{\lambda}_n \atop j} \right\rangle \lambda_j^{-2} = \sum_{j=1}^{n} \frac{1}{\lambda_j^2} + \sum_{i>j=1}^{n} \frac{1}{\lambda_j \lambda_i},$$

which can now be verified directly using Lagrange coefficients.

NB the distribution of $S_n = \sum_{i=1}^{n} T_i$, when the T_i are iid $\mathrm{Ep}(\lambda)$ is easily found; it is instructive to compare eqn (3.31) when the λ_i are close in value by examining the limit as, say λ_n approaches λ_{n-1}, and repeating the steps in Corollary 1.

3.4.2. Interpolation and Numerical Integration

For any given set of points in the plane $\{(x_i, y_i) : i = 1, \ldots, n\}$, where w log $x_1 < \cdots < x_n$, the Lagrange interpolation polynomial of degree n, label it $P_n(t)$, is defined by

$$P_n(t) = \textstyle\sum_{j=1}^{n} y_j \prod_j^{(n)}(t), \quad \text{where} \quad \prod_j^{(n)}(t) := \left\langle {\boldsymbol{x}-t\boldsymbol{1} \atop j} \right\rangle \qquad \text{for all } x_1 \le t \le x_n, \tag{3.33}$$

using the vector $\boldsymbol{x} - t\boldsymbol{1} = (x_1 - t, \ldots, x_n - t)$ and the definition in eqn (3.8). NB that $P_n(x_j) = y_j$ for $j = 1, \ldots, n$ since, by definition,

$$\prod_j^{(n)}(t) = \prod_{i=1, i\neq j}^{n} \frac{x_i - t}{x_i - x_j} \quad \text{and} \quad \prod_j^{(n)}(x_k) = \begin{cases} 0 \text{ for } j \neq k, \\ 1 \text{ for } j = k. \end{cases} \tag{3.34}$$

Lagrange interpolation is no longer in favor; Kreyszig's authorative textbook, [57, p. 968] says

> The Lagrange polynomials are not practical in numerical work. Computations may become laborious and, more importantly, previous work is wasted in the transition to a polynomial of higher degree. However, they are of great interest in deriving other formulas. An important application of this type ... [is] the derivation of the famous Simpson's Rule of integration.

This asserted deficiency may be too strong and should be reexamined in those cases where a computational formula can be effected that retains previous work in the upgrading to a higher-degree polynomial. This can be done by interpolating, not between the points, but between the related lines or polynomials.

Theorem 3. *The Lagrange interpolation polynomials of degree n and $n+1$, based on the pairs $\{(x_i, y_i) : i = 1, \ldots, n+1\}$, are related by, writing $m_j = \frac{y_{n+1}-y_j}{x_{n+1}-x_j}$ for $j = 1, \ldots, n$,*

$$P_{n+1}(t) = \sum_{j=1}^{n} [y_j + m_j(t - x_j)] \prod\nolimits_{i=1}^{(n)}(t). \tag{3.35}$$

Proof. By definition we have $P_{n+1}(t) = \sum_{j=1}^{n+1} y_j \prod_j^{(n)}(t)$. But from eqns (3.18) and (3.34), we obtain for $j = 1, \cdots, n$

$$\prod\nolimits_j^{(n+1)}(t) = \prod\nolimits_j^{(n)}(t)\left(\frac{x_{n+1}-t}{x_{n+1}-x_j}\right), \quad \text{and} \quad \prod\nolimits_j^{(n+1)}(t) = 1 - \sum_{j=1}^{n} \prod\nolimits_j^{(n+1)}(t).$$

Thus, we have

$$\begin{aligned} P_{n+1}(t) &= \sum_{j=1}^{n} y_j \prod\nolimits_j^{(n)}(t)\left(\frac{x_{n+1}-t}{x_{n+1}-x_j}\right) \\ &\quad + y_{n+1}\left\{\sum_{j=1}^{n} \prod\nolimits_j^{(n)}(t) - \sum_{j=1}^{n} \prod\nolimits_j^{(n)}(t)\left(\frac{x_{n+1}-t}{x_{n+1}-x_j}\right)\right\}, \\ &= \sum_{j=1}^{n} \prod\nolimits_j^{(n)}(t)\left\{\frac{y_j(x_{n+1}-t)}{x_{n+1}-x_j} - \frac{y_{n+1}(x_j-t)}{x_{n+1}-x_j}\right\}, \end{aligned}$$

which can be seen to be equivalent with eqn (3.35). □

Corollary 3. *The relation between the Lagrange polynomial $P_n(x)$ based on (x_i, y_i) for $i = 0, \ldots, n$ and the augmented polynomial $P_{(n+m)}(x)$ based on (x_i, y_i) for $i = 0, \ldots, n+m$ is*

$$P_{(n+m)}(x) = \sum_{k=0}^{n} L_k^m(x) \prod\nolimits_k^{(n)}(x),$$

where, setting $L_k^0(x) = y_k$ for $k = 0, \ldots, n$, we define

$$L_k^j(x) = \frac{x_{n+1-j} - x}{x_{n+1-j} - x_k}\left[L_k^{j-1}(x) - L_{n+1-j}^{j-1}(x)\right] + L_{n+1-j}^{j-1}(x) \quad \text{for } j = 1, \ldots, n.$$

Theorem 4. *If the Lagrange polynomial, based on the points* (x_i, y_i) *for* $i = 0, \ldots, m,$ *where* $x_0 < x_1 < \cdots < x_m,$ *is* $P(x) = \sum_{k=0}^{m} y_k \prod_k^m(x)$ *for* $x_0 < x < x_m$*, then the integral is*

$$A_m = \int_{x_0}^{x_m} P(x)\,dx = \sum_{k=0}^{m} y_k B_k^m, \quad \text{where} \quad B_k^m = \int_{x_0}^{x_m} \prod\nolimits_k^{(m)}(x)\,dx.$$

Corollary 4. *In the case of equal-length subintervals; say* $x_i = x_0 + ih$ *for* $i = 0, \ldots, m$*, and set* $t = (x - x_0)/h$*, the remarkable formula is obtained*

$$B_k^m = h \int_0^m \binom{t}{k}\binom{m-t}{m-k}\,dt; \text{ thus } \sum_{k=0}^{m} B_k^m = hm \text{ and } B_k^m = B_{m-k}^m. \tag{3.36}$$

NB that for $m = 2$ we obtain $B_0^2 = h/3 = B_2^2$ and $B_1^2 = 4h/3$. And hence $A_2 = \frac{h}{3}[y_0 + 4y_1 + y_2]$, which by application to $2n$ equal subintervals gives the well known *Simpson's Rule* for numerical integration.

Suppose the entire interval of integration consists of $2n$ subintervals of which the first two are of equal length, as are the second two, and so on; then numerical integration is given by

$$\frac{1}{3}\left\{h_1 y_0 + h_{2n} y_{2n} + \sum_{i=1}^{n-1}(h_i + h_{i+1})y_{2i} + 4\sum_{i=1}^{n} h_i y_{2i-1}\right\}. \tag{3.37}$$

Exercise Set 3.E

1. Prove the results of Corollary 2 and eqn (3.36).
2. Verify for $m = 3$ that $B_0^3 = \frac{3h}{8} = B_3^3,\ B_1^3 = \frac{9h}{8} = B_2^3$, which yields another integration rule,

$$A_3 = \frac{3h}{8}\left[\sum_0^3 y_i + 3\sum_1^2 y_i\right].$$

   Also one finds for $m = 4$ that $B_0^4 = B_4^4 = \frac{14h}{45},\ B_1^4 = B_3^4 = \frac{64h}{45},\ B_2^4 = \frac{h}{15}$, which gives

$$A_4 = \frac{h}{15}\left[5(y_o + y_4) + 22(y_1 + y_3) + 8y_2 - \tfrac{1}{3}[y_0 + 2(y_1 + y_3) + y_4]\right].$$

   There are other formulas for A_m when $m \geq 5$.
3. For any vector $\boldsymbol{\lambda} = (\lambda_1, \ldots, \lambda_n)$ with $\lambda_i \in \Re$, prove algebraically that the identities in eqn (3.9) hold along with these:

$$\sum_{i=1}^{n}\left\langle{\lambda_n \atop j}\right\rangle \lambda_j^r = \begin{cases} \sum_{i=1}^{n}\lambda_i^{-1} & \text{if } r = -1, \\ \sum_{i=1}^{n}\lambda_i^{-2} + \sum_{i>j=1}^{n}(\lambda_i\lambda_j)^{-1} & \text{if } r = -2. \end{cases}$$

4. Show that for all $\lambda_j, \lambda \in \Re$,

$$\sum_{j=1}^{n}\left\langle{\lambda_n \atop j}\right\rangle \frac{\lambda_j}{\lambda_j - \lambda} = \prod_{j=1}^{n}\frac{\lambda_j}{\lambda_j - \lambda}.$$

5. Are the identities in Exer. 3 and 4 above new if $\lambda_i = i$ for $i = 1, \ldots, n$?

6. How may one interpret these identities above when $\boldsymbol{\lambda} = \boldsymbol{\nu} - t\mathbf{1}$?

7. If a component survival function was of the family $\bar{F}(x, \alpha, \beta)$ for which the mean and variance were known, explain how you could use eqn (3.37) to determine the parameters α and β.

8. Previously in "intelligence tests" it was common to ask such questions as: "Given the numbers y_1, y_2, y_3, y_4, what is y_5?" This answer was "correct" if it agreed with the "generating function" $g(5) = y_5$, which the test makers had chosen.
   If one utilizes Lagrange polynomials, viz., $L(x) = \sum_{j=1}^{4} y_j \prod_j^{(4)}(x)$, when the test specifies $y_i = i$ for $i = 1, \ldots, 4$, will $L(5) = 5$? Does $L(251) = 251$? What about $y_i = ai^2 + b$ for given constants a and b? Refer to eqn (3.34).

9. The *Fibonacci* series $\{f_n\}_{n=0}^{\infty}$, which arises often in natural science, is defined by $f_0 = 0$, $f_1 = 1$ with $f_n = f_{n-2} + f_{n-1}$ for $n \geq 2$.
   (a) Show that $\sum_{n=0}^{\infty} f_n 2^{-n+1} = 1$.
   (b) Let N be the discrete rv defined by $P[N = n] = f_n 2^{1-n}$ for $n = 1, 2 \cdots$. Show the mean and variance of N are, respectively, the integers 2 and 11.

10. Let Albert toss a fair coin until he gets three heads or three tails in a row. He wins if he gets three heads first. If A is the number of the toss when Al gets his third successive heads, show $P[A = n] = 2^{-n} f_{n-2}$ for $n = 3, 4 \cdots$. What are the mean and variance of A?

11. The *Fibonacci polynomial*, $\wp_n(x)$ satisfies the recursion given by $\wp_n(x) = x\wp_{n-1}(x) + \wp_{n-2}(x)$. Show that $\wp_n(x)$ is the coefficient of t^n in the expansion of $t/(1 - xt - t^2)$. What are $\wp_0(x)$ and $\wp_1(x)$?

12. *Splines* are polynomials (usually cubic), used in *industrial design* to obtain smooth curves, which are fitted at various points across an interval but with a matching specified slope at one end. How does the use of Lagrange interpolation compare with this?

A state lottery is a public subsidy of intelligence. The lottery yields public income that is calculated to lighten the tax burden of us prudent abstainers at the expense of the benighted mass of wishful thinkers.

Prof. W. V. Quine

No one ever went broke in America underestimating the intelligence of the American public.

Henry L. Mencken

CHAPTER 4

Coherent Systems

4.1. Functional Representation

Consider a system that for its functioning requires that each of its components must function, e.g., the current passes iff all components are operational or a chain holds iff all n links are not broken. Such a system is represented pictorially by the diagram

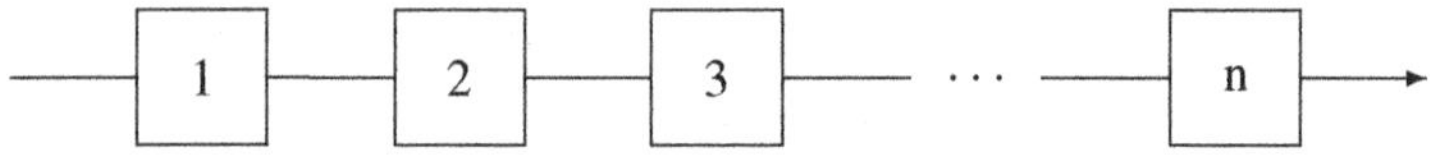

Let us indicate the state of the ith component with the binary variable

$$x_i = \begin{cases} 1 & \text{if } i\text{th component works,} \\ 0 & \text{if } i\text{th component fails,} \end{cases} \qquad \text{for } i = 1, \cdots, n;$$

these variables are binary, i.e., they assume only two values, because they indicate *dichotomic* states, which are contrasting, mutually exclusive such as operative-inoperative, normal-anomalous, good-bad, or sweet-sour, depending upon the application.

The operation of the system with these components is also indicated by such a binary (dichotomic) function

$$\varphi(x_1, \cdots, x_n) = \begin{cases} 1 & \text{if the system works} \\ 0 & \text{if the system fails.} \end{cases}$$

The type of system described above is called a *series* system and its operation is indicated by the performance function

$$\varphi(x_1, \cdots, x_n) = \prod_{i=1}^{n} x_i .$$

Another common system is called a *parallel* system, which functions iff any one of its components funtions. It is represented by the diagram:

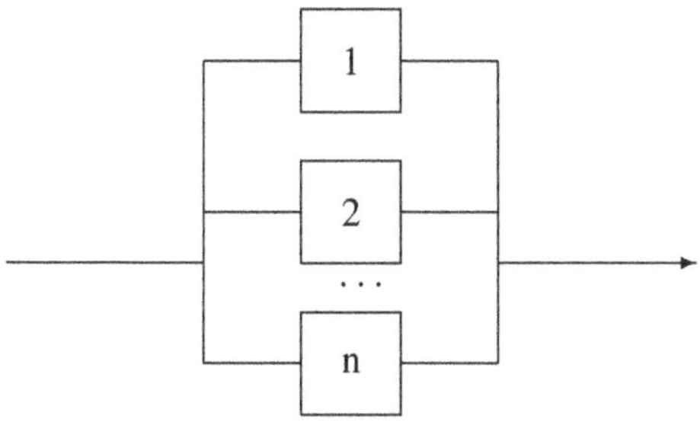

and its operation is indicated by the performance function

$$\varphi(x_1, \cdots, x_n) = \coprod_{i=1}^{n} x_i := 1 - \prod_{i=1}^{n}(1 - x_i),$$

where here the relationship between the "product" and "co-product" symbols is defined.

Note that for two components we have only two possible structures, namely,

series	$\kappa(x_1, x_2) = x_1 x_2 := x_1 \sqcap x_2,$
parallel	$\rho(x_1, x_2) = x_1 + x_2 - x_1 x_2 := x_1 \sqcup x_2.$

One can construct more elaborate structures by the process of composition, i.e., a series-parallel system

$$\varphi(x_1, x_2, x_3, x_4) = \kappa[\rho(x_1, x_2), \rho(x_3, x_4)] = \rho(x_1, x_2) \cdot \rho(x_3, x_4).$$

But not all systems of interest can be represented as series-parallel or parallel-series or even as a combination of such systems. What is the class of structure functions of interest in reliability theory? A starting point is to examine the following class:

An indicator function φ of n indicator variables is a *coherent system* iff it is monotone, i.e.,

$$\varphi(x_1, \cdots, x_n) \leq \varphi(y_1, \cdots, y_n) \quad \text{whenever} \quad x_1 \leq y_1, \cdots, x_n \leq y_n,$$

but not identically constant since we must have

$$\varphi(0, \cdots, 0) = 0, \quad \text{and} \quad \varphi(1, \cdots, 1) = 1.$$

The mathematical specification of φ, of course, must correspond to the actual operation and nonoperation of the system it represents.

A system is said to be a *k-out-of-n (majority) system* iff its performance function is, for some $1 \leq k \leq n$, given by the indicator function

$$\varphi(x_1, \cdots, x_n) = \mathrm{I}(\textstyle\sum_{i=1}^{n} x_i \geq k) := \begin{cases} 1 & \text{if } \sum_1^n x_i \geq k, \\ 0 & \text{if } \sum_1^n x_i < k. \end{cases}$$

Clearly a 1-out-of-n system is a parallel system and an n-out-of-n system is a series system. But what about other values of k?

Consider a 2-out-of-3 system. It can be represented as follows:

$$\begin{aligned}\varphi(x_1, x_2, x_3) &= x_1x_2 \sqcup x_1x_3 \sqcup x_2x_3\\ &= x_1x_2x_3 + x_1x_2(1 - x_3) + x_1x_3(1 - x_2) + x_2x_3(1 - x_1).\end{aligned}$$

Clearly such representations, if written out explicitly, can become quite lengthy for high orders of n.

Now consider the home theater system described in the diagram below:

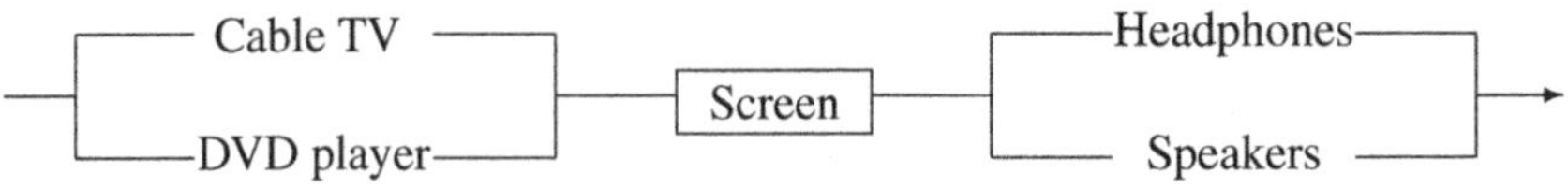

With the preceding notation we can write

$$\varphi(x_1, x_2, \cdots, x_5) = (x_1 \sqcup x_2)x_3(x_4 \sqcup x_5)$$

and maintain a compact notation for the structure function.

Note that for every coherent system φ there exists a *dual* system ψ defined by

$$\psi(x_1, \cdots, x_n) = 1 - \varphi(1 - x_1, \cdots, 1 - x_n);$$

thus, series and parallel systems are dual.

All components within a system should be relevant. We would say the first component is *irrelevant* to the system φ iff

$$\varphi(1, x_2, \cdots, x_n) \equiv \varphi(0, x_2, \cdots, x_n) \qquad \text{for all values of } x_2, \cdots, x_n.$$

That is, the first component is irrelevant whenever its state does not affect the operation of the system. NB this does not mean irrelevant components should not be considered in reliability studies since being irrelevant to operation may be relevant to maintenance or the safety of operators.

Let us, for the present, assume that all components are relevant in the systems analyzed. We now come to the problem of representing the performance function when the system consists of an enumeration of all possible states of the components and the corresponding state of the system.

Consider the following enumeration:

$\varphi(\boldsymbol{y})$	1	1	1	0	0	0	0	0
y_1	1	1	1	0	1	0	0	0
y_2	1	1	0	1	0	1	0	0
y_3	1	0	1	1	0	0	1	0

How can we express $\varphi(x_1, \cdots, x_n)$ mathematically? Note the following identity; here we set $\boldsymbol{x} = (x_1, \cdots, x_n)$,

$$\varphi(\boldsymbol{x}) = x_1\varphi(1, x_2, \cdots, x_n) + (1 - x_1)\varphi(0, x_2, \cdots, x_n) \quad \text{for } x_1 = 0, 1. \quad (4.1)$$

This is called the *pivotal reduction*. Continuing its implementation, one obtains the representation of φ in terms of its enumerated values, viz.,

$$\varphi(\boldsymbol{x}) = \sum_{y_1=0}^{1} \cdots \sum_{y_n=0}^{1} \varphi(\boldsymbol{y}) \prod_{j=1}^{n} x_j^{y_j}(1-x_j)^{1-y_j}, \tag{4.2}$$

where $\boldsymbol{y} = (y_1, \ldots, y_n)$, and we make the convention that $0^0 = x_j^0 = (1-x_j)^0 = 1$ for any j.

If we were to utilize this formula for the system whose states are enumerated above we would find

$$\begin{aligned}\varphi(\boldsymbol{x}) &= x_1x_2x_3 + x_1x_2(1-x_3) + x_1x_3(1-x_2)\\ &= x_1(x_2 + x_3 - x_2x_3) = x_1(x_2 \sqcup x_3).\end{aligned}$$

One immediately realizes that one need *only list those vectors that indicate the state of the components for which the system functions.* This is a substantial savings since in a structure of n components there are 2^n possible state vectors. Many modern systems have 300,000 components and listing all possible state vectors would be tedious. But this enumeration, which usually affects a reduction by more than a half since components often fail in more ways than they work, is still not sufficient for practical purposes.

A rule of thumb, long used by designers, is that redundancy among the components is superior to redundancy among systems. Using the theory of coherent systems one can prove that

Theorem 1. *Let φ be any coherent system of order n. Then*

$$\varphi(x_1 \sqcup y_1, \ldots, x_n \sqcup y_n) \geq \varphi(\boldsymbol{x}) \sqcup \varphi(\boldsymbol{y}),$$

where $x \sqcup y = 1 - (1-x)(1-y)$ for binary variates x and y. Equality holds for all state vectors $\boldsymbol{x}$ and $\boldsymbol{y}$ iff the system φ is parallel, i.e.,

$$\varphi(\boldsymbol{x}) := \sqcup x_i = 1 - \sqcap(1-x_i).$$

Let us now consider a set of n components. We shall, without loss of generality, identify the set of components with their enumeration. Thus, the set of components is $C = \{1, 2, \cdots, n\}$. For a given state vector of component operation, say $\boldsymbol{x} = (x_1, \cdots, x_n)$, we define the set of indices (components) that are zero (not operating) and unity (operating), respectively, with the notation

$$C_0(\boldsymbol{x}) = \{i \in C \;:\; x_i = 0\}\,,\;\; C_1(\boldsymbol{x}) = \{i \in C \;:\; x_i = 1\}.$$

Let φ be a given coherent system. Then a *path (vector)* of φ is a state vector $\boldsymbol{x}$ such that $\varphi(\boldsymbol{x}) = 1$. The corresponding *path set* is $C_1(\boldsymbol{x})$.

A *minimal path (vector)* is a state vector $\boldsymbol{z}$ such that $\varphi(\boldsymbol{z}) = 1$ but

$$\boldsymbol{y} \prec \boldsymbol{z} \text{ implies } \varphi(\boldsymbol{y}) = 0.$$

Here $\boldsymbol{y} \prec \boldsymbol{z}$ means $y_i \leq z_i$ for $i = 1, \cdots, n$ with $y_i < z_i$ for some $i = 1, \cdots, n$. Correspondingly, the set $C_1(\boldsymbol{z})$ is called a *minimal path set*.

In an entirely analogous fashion we can formulate the concepts of *cut vector, cut set, minimal cut vector*, and *minimal cut set*.

Consider the Wheatstone-bridge system.

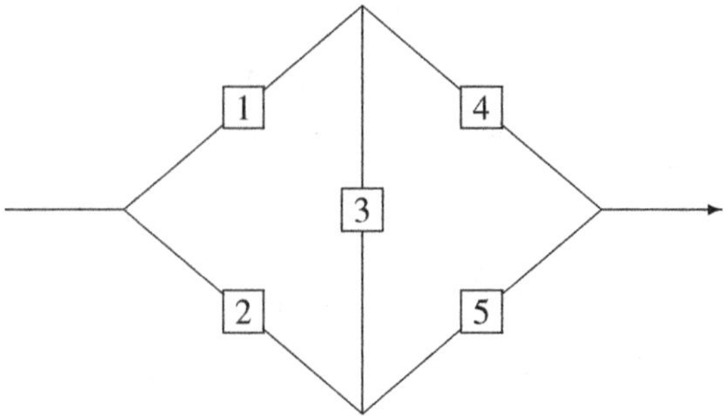

The min-paths (minimal path sets) are

$$\mathrm{P}_1 = \{1, 4\}, \quad \mathrm{P}_2 = \{2, 5\}, \quad \mathrm{P}_3 = \{1, 3, 5\}, \quad \mathrm{P}_4 = \{2, 3, 4\}.$$

The min-cuts (minimal cut sets) are

$$\mathrm{K}_1 = \{1, 2\}, \quad \mathrm{K}_2 = \{4, 5\}, \quad \mathrm{K}_3 = \{1, 3, 5\}, \quad \mathrm{K}_4 = \{2, 3, 4\}.$$

Suppose we have the jth min-path set, say, P_j, of φ given. Then we can form the corresponding series subsystem

$$\rho_j(\boldsymbol{x}) = \prod_{i \in \mathrm{P}_j} x_i;$$

ρ_j is the jth min-path series function. Suppose there are altogether $j = 1, \cdots, p$ such min-path sets from φ. By definition, $\varphi(\boldsymbol{x}) = 1$ iff at least one *min-path series subsystem* is functioning; hence,

$$\varphi(\boldsymbol{x}) = \coprod_{j=1}^{p} \rho_i(\boldsymbol{x}) = 1 - \prod_{j=1}^{p} [1 - \rho_j(\boldsymbol{x})].$$

Similarly we argue from the enumeration of the min-cuts that there is a *min-cut parallel subsystem*

$$\kappa_j(\boldsymbol{x}) = \coprod_{i \in \mathrm{K}_j} x_i \qquad \text{for } j = 1, \cdots, k$$

and the corresponding structural representation in terms of these min-cut subsystems. Since all these cuts must be operational, we have

$$\varphi(\boldsymbol{x}) = \prod_{j=1}^{k} \kappa_j(\boldsymbol{x}).$$

Let us consider the min-path representation of the bridge structure:

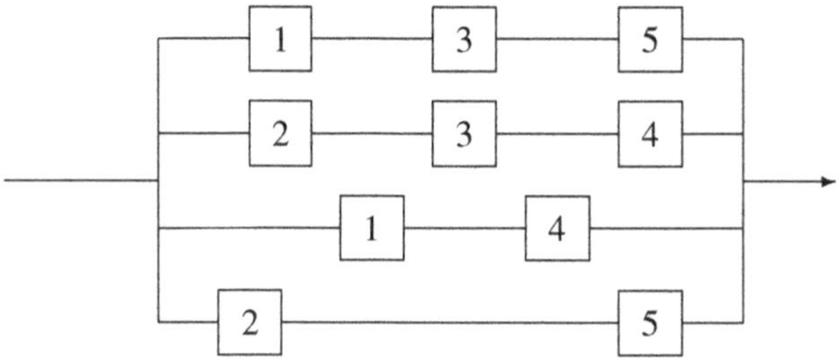

Thus, we have the succinct expression

$$\varphi(\boldsymbol{x}) = \coprod_{i=1}^{4} \rho_i(\boldsymbol{x}),$$

where

$$\begin{aligned}\rho_1(\boldsymbol{x}) &= x_1 x_3 x_5,\\ \rho_2(\boldsymbol{x}) &= x_2 x_3 x_4,\\ \rho_3(\boldsymbol{x}) &= x_1 x_4,\\ \rho_4(\boldsymbol{x}) &= x_2 x_5.\end{aligned}$$

Now let us obtain the min-cut representation of this same system from its min-cut diagram.

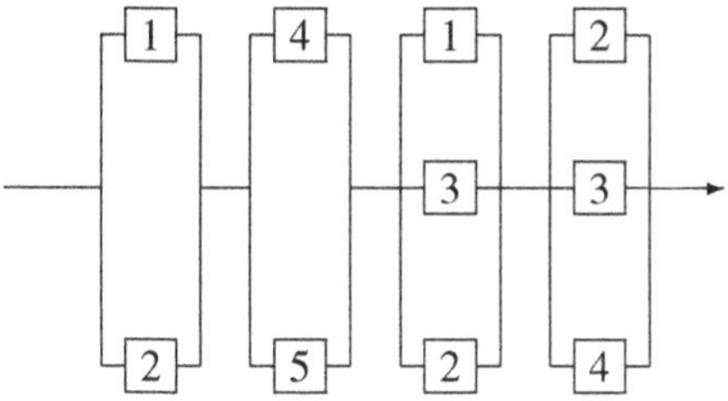

Thus, we have

$$\varphi(\boldsymbol{x}) = \prod_{j=1}^{4} \kappa_i(\boldsymbol{x}),$$

where

$$\begin{aligned}\kappa_1(\boldsymbol{x}) &= 1 - (1 - x_2)(1 - x_2),\\ \kappa_2(\boldsymbol{x}) &= 1 - (1 - x_4)(1 - x_5),\\ \kappa_3(\boldsymbol{x}) &= 1 - (1 - x_1)(1 - x_3)(1 - x_5),\\ \kappa_4(\boldsymbol{x}) &= 1 - (1 - x_2)(1 - x_3)(1 - x_4).\end{aligned}$$

Notice that some of the components are not relevant to some min-cut or min-path subsystems.

4.2. Event-Tree Depiction

There are alternative visual methods of representing coherent systems, called *event trees*, instead of the functional representation which is often more useful for analysis. When only the mechanisms by which system failure occurs are considered, rather than of success, the tree is called a *fault tree*.

The basic events are represented by circles and squares (or diamonds in some depictions).

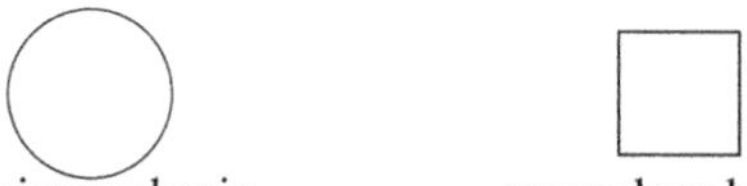

primary basic secondary basic

A system or subsystem of importance is represented by a rectangle.

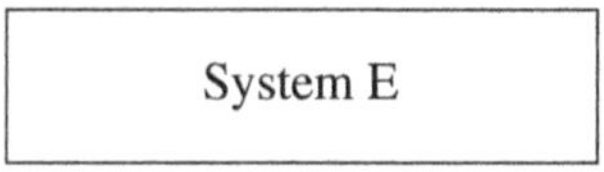

Below each major event is an AND Gate or an OR gate.

AND gate OR gate

The output from an AND gate occurs iff all input events occur. The output from an OR gate occurs iff at least one of the input events occurs. Consider two branches of such a tree: the AND and OR gates.

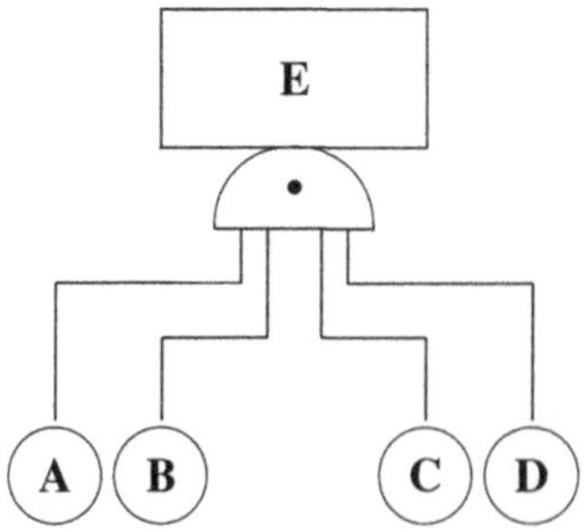

$E = A \sqcap B \sqcap C \sqcap D$

The AND Gate

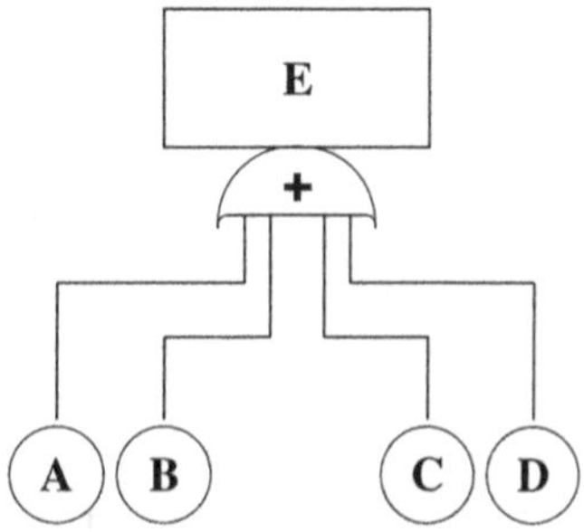

$E = A \sqcup B \sqcup C \sqcup D$

The OR Gate

These correspond to series and parallel arrangements of subsystems; the left-hand diagram says, "the top event E is caused by the simultaneous events A, B, C, and D occurring." The right-hand diagram says "the top event E is caused by A or B or C or D."

In general, event trees serve three purposes:

1. An event tree aids in the systematical determination of possible causes of failure. When properly used, an event tree often leads to the discovery of concatenations of events leading to the top event, which otherwise might not have been recognized as possible causes of failure of the system being analyzed.
2. The event tree serves as an easily comprehended summary of the system design that can be used to disclose weak points in design and how they might contribute to undesirable events and, more importantly, what changes in design can alleviate them.
3. The event tree provides an easily understood, efficient and convenient method for the computation of the probability of the top event, whether it is success or failure.

However, it should not be inferred that event trees can be used to show that all conceivable courses have been considered. Usually only a fraction of the modes of failure that are ultimately observed in service have been previously considered, and as in most system failures in modern society, many are due to human carelessness, inattention, or stupidity.

We conclude that there is a one-to-one correspondence between a fault (event) tree and its coherent system representation. To evaluate the reliability performance of a complex equipment it seems conceptually easier for engineers to construct a fault (event) tree. But for additional analysis and optimal reliability redesign, as well as for confidence bounds on service life, it becomes necessary to obtain a coherent system representation.

4.2.1. Associated Random Variables

We now consider a type of dependence between rvs that arises from a consideration of coherent systems. We say the components of the rv $\boldsymbol{X}$ are *associated* (or the set or vector is *associated*) iff for any pair of coherent systems φ_1 and φ_2 we have the covariance nonnegative, i.e.,

$$\mathrm{Cov}[\varphi_1(\boldsymbol{X}), \varphi_2(\boldsymbol{X})] \geq 0.$$

This leads to the generalization:

Definition 1. The rv $\boldsymbol{X} = (X_1, \ldots, X_n)$ is *associated*, i.e., is arv, iff

$$\mathrm{Cov}[\Gamma(\boldsymbol{X}), \Delta(\boldsymbol{X})] \geq 0,$$

for all increasing binary functions Γ and Δ.

NB the restriction that Γ and Δ be binary is not a loss of generality since this implies, under the assumption, the conclusion is true for all increasing functions Γ and Δ for which the covariance exists.

Tidpat the following conclusions can be reached.

1. Any single random variable is associated.
2. Any subset of an associated rv is associated.
3. Independent rvs are *a fortiori* associated.
4. If $\boldsymbol{X}_1$ and $\boldsymbol{X}_2$ are each separately arv, when $\boldsymbol{X}_1$ is statistically independent of $\boldsymbol{X}_2$, (we write $\boldsymbol{X}_1 \perp \boldsymbol{X}_2$,) then the combined vector $(\boldsymbol{X}_1, \boldsymbol{X}_2)$ is associated.
5. Increasing functions of arvs are associated.

We also have the following result.

Theorem 2. *If $\boldsymbol{X}$ is a binary arv, then $(1 - X_1, \cdots, 1 - X_n)$ is also an arv.*

Exercise Set 4.A

1. Prove all five properties of association.
2. Prove that if X and Y are binary (indicator) rvs, then $\text{cov}(X, Y) \geq 0$ implies X, Y are associated. Is this true for any rvs?
3. Let T_1 and T_2 be rvs with finite expectations and covariance. Then
$$\text{Cov}[T_1, T_2] = \int_{-\infty}^{\infty} \int_{-\infty}^{\infty} \text{Cov}[\text{I}(T_1 > s), \text{I}(T_2 > t)\, ds\, dt.$$
4. Using Exercise 3, show that if $\boldsymbol{T}$ is associated and f_1 and f_2 are increasing functions, then $f_1(\boldsymbol{T})$ and $f_2(\boldsymbol{T})$ are associated.
5. Prove that if (T_1, T_2) is an associated rv, then $\text{cov}(T_1, T_2) \geq 0$, whenever the covariance exists.
6. If $\boldsymbol{X}$ is an rv of independent states of n components, show that for any coherent system the min paths $\rho_1(\boldsymbol{X}), \ldots, \rho_p(\boldsymbol{X})$ are associated. Is this true for the set of min cuts?
7. If (X, Y) are jointly normal rvs, then X, Y are associated iff $\text{cov}(X, Y) \geq 0$.
8. If $\boldsymbol{X}$ is associated, then $a\boldsymbol{X}$ is associated for any $a \in \Re$.
9. If $\boldsymbol{X} = (X_1, \ldots, X_n)$ is a set of iid rvs, then the corresponding set of order observations, $X_{1:n} \leq X_{2:n} \leq \cdots \leq X_{n:n}$, is associated.

Exercise Set 4.B

Problems for Discussion

(a) What circumstances do you think led to the decision to construct a nuclear generating plant in California virtually over an earthquake fault line?
(b) Why do you think that the ACRS, an advisory committee to the Nuclear Regulatory Commission, recommended that off-site power be employed for use in the control center of each nuclear generating plant when on-site power, being generated by the plant itself, would be cheaper and more convenient?

(c) What, do you think was the reason that, up until 1946, it was the practice at the Boeing company that the designers of each airplane be aboard during the first takeoff? Is this a good idea in your opinion?
(d) Does it follow that high-performance aircraft, with advanced avionics and "smart" weaponry, are always tactically more desirable than lower-performance aircraft? Why? How is high-performance related to reliability?
(e) The Volvo company designed a braking system which, for safety reasons, nullified the braking capacity of the diagonally opposite wheel if any brake cylinder lost fluid. This would cause a large symmetric reduction in braking capability but could prevent the skidding that would result if half as much braking capability were deployed on one side as the other. Obtain the coherent system that would describe this performance.

4.3. Evaluation of Reliability

If $\boldsymbol{X} = (X_1, \ldots, X_n)$ is a set of Bernoulli variates indicating the state of the n components for a coherent system φ, then the system has reliability

$$R = \Pr[\varphi(\boldsymbol{X}) = 1] = \mathrm{E}\varphi(\boldsymbol{X}). \tag{4.3}$$

If all the X_i are stochastically independent, then the system reliability is completely determined by the component reliabilities, say $p_i = \mathrm{E}X_i$ for $i = 1, \ldots, n$, and we can write

$$R(\boldsymbol{p}) = \mathrm{E}\varphi(\boldsymbol{X}), \quad \text{where } \boldsymbol{p} = \mathrm{E}(\boldsymbol{X}). \tag{4.4}$$

For example, if $\varphi(\boldsymbol{X}) = \prod_{i=1}^n X_i$, then $R(\boldsymbol{p}) = \prod_{i=1}^n p_i$, and also if $\varphi(\boldsymbol{X}) = \mathrm{I}(\sum_{i=1}^n X_i \geq k)$ we find, setting $q_i = 1 - p_i$, that

$$R(\boldsymbol{p}) = \sum_{j=k}^{n} \sum_{i=1}^{\binom{n}{j}} \prod_{\ell \in A_{j,i}} p_\ell \prod_{\ell \notin A_{j,i}} q_\ell,$$

where $A_{j,i}$ is the ith subset of size j selected from the n components $\{1, \ldots, n\}$, of which there are $\binom{n}{j}$. What is the functional nature of the reliability $R(\boldsymbol{p})$? We can see by pivotal reduction from eqn (4.1) that

(a) $R(\boldsymbol{p})$ is multilinear in $p_1, \ldots, p_n$.
(b) $R(\boldsymbol{p})$ is strictly increasing in each p_i for $0 < p_i < 1$.

We also see if all component reliabilties are the same, i.e., $p_i = p$ for $i = 1, \ldots, n$ that for series and parallel systems, respectively, their reliabilties are

$$R_S(\boldsymbol{p}) = p^n \quad \text{and} \quad R_P(\boldsymbol{p}) = 1 - (1 - p)^n,$$

and for a k-out-of-n system, the reliability is

$$R(\boldsymbol{p}) = \sum_{j=k}^{n} \binom{n}{j} p^j (1 - p)^{n-j}.$$

Under the assumption of component independence, we obtain the probabilistic version of the designer's rule of thumb as stated in Theorem 1.

Corollary 1. *The reliability of any coherent system satisfies the inequality*

$$R(\boldsymbol{p} \sqcup \boldsymbol{p}') \geq R(\boldsymbol{p}) \sqcup R(\boldsymbol{p}').$$

This result shows that redundancy at the component level yields higher reliability than redundancy at the subsystem level.

Since exact representations of the reliability can be obtained for series and parallel systems of independent components by merely substituting the indicator variable for performance by the reliability of that component it is often incorrectly assumed that this operation will hold for all coherent systems. But it does not.

Suppose we have a system-state representation given by

$$\varphi(x_1, x_2, x_3) = x_1 x_3 (x_2 \sqcup x_3),$$

with component reliabilities $\mathrm{E}X_i = p_i$ for $i = 1, 2, 3$. We find asas

$$\mathrm{E}\varphi(X_1, X_2, X_3) = p_1(p_2 p_3 + p_3 - p_2) \neq p_1 p_3 (p_2 \sqcup p_3).$$

So how do we obtain the exact reliability? If we utilize eqn (4.2), we see

$$\mathrm{E}\varphi(\boldsymbol{X}) = \sum_{all\, \boldsymbol{x}} \varphi(\boldsymbol{x}) \prod_{i=1}^{n} p_i^{x_i} (1 - p_i)^{1 - x_i}.$$

But the large order of most complex structures, with each component's reliability near unity, makes this method practically infeasible for computation in most realistic situations.

Let us consider system performance as being a binary random variable, say Φ, which is determined through its coherent system φ using its min-path representation and its p min paths, i.e.,

$$\Phi = \varphi(X_1, \cdots, X_n) = \coprod_{j=1}^{p} \rho_j(\boldsymbol{X}),$$

where each component's performance is an independent Bernoulli variate. Let the reliability of the ith component be

$$\Pr[X_i = 1] = r_i \qquad \text{for } i = 1, \cdots, n.$$

Denote the performance event of the jth min path by

$$\mathrm{E}_j = [\rho_j(\boldsymbol{X}) = 1] = \bigcap_{i \in \mathrm{P}_j} [X_i = 1] \qquad \text{for } j = 1, \cdots, p.$$

Then the reliability of the system, i.e., the probability that it is operational, is

$$\Pr[\Phi = 1] = \Pr\left[\bigcup_{j=1}^{p} \mathrm{E}_j\right],$$

where

$$\Pr(\mathrm{E}_j) = \prod_{i \in P_j} r_i.$$

Let us consider again the Wheatstone bridge's min paths:

$$\begin{aligned}
\rho_1(\boldsymbol{x}) &= x_1x_3x_5 & \Pr(\mathrm{E}_1) &= r_1r_3r_5\\
\rho_2(\boldsymbol{x}) &= x_2x_3x_4 & \Pr(\mathrm{E}_2) &= r_2r_3r_4\\
\rho_3(\boldsymbol{x}) &= x_1x_4 & \Pr(\mathrm{E}_3) &= r_1r_4\\
\rho_4(\boldsymbol{x}) &= x_2x_5 & \Pr(\mathrm{E}_4) &= r_2r_5
\end{aligned}$$

It will have reliability R, where

$$R = \Pr[\mathrm{E}_1 \cup \mathrm{E}_2 \cup \mathrm{E}_3 \cup \mathrm{E}_4].$$

This may be evaluated using the inclusion-exclusion theorem:

$$R = \sum_i \Pr(\mathrm{E}_i) - \sum_{i>j} \Pr(\mathrm{E}_i\mathrm{E}_j) + \sum_{i>j>k} \Pr(\mathrm{E}_i\mathrm{E}_j\mathrm{E}_k) - \cdots \pm \Pr(\bigcap \mathrm{E}_i).$$

Utilizing the independence to write, for example,

$$\Pr(\mathrm{E}_1\mathrm{E}_2\mathrm{E}_4) = \prod_{i\in P_1\cap P_2\cap P_4} r_i = \prod_{i\neq 3} r_i,$$

and cancelling out terms, we obtain the lengthy representation

$$\begin{aligned}
R = {} & r_1r_3r_5 + r_2r_3r_4 + r_1r_4 + r_2r_5 - r_1r_3r_4r_5 - r_1r_2r_3r_5\\
& - r_1r_2r_3r_4 - r_2r_3r_4r_5 - r_1r_2r_4r_5 + 2r_1r_2r_3r_4r_5.
\end{aligned}$$

If we knew that all $r_i \geq r$, then a lower bound for the reliability function would be (why?) $R(r) \geq 2r^3 + 2r^2 - 5r^4 + 2r^5$. But this method of expansion is good only in theory. In practice nearly all component reliabilities are so near unity that arithmetic operations are subject to round-off error and the computed reliability may be imprecise. Better to compute the unreliability. For example in the case above, letting $r = 1 - \epsilon$, we find $1 - R(1-\epsilon) \leq 4\epsilon^2 - 2\epsilon^3 - 3\epsilon^4 + 2\epsilon^5$ and there is no first-order term. Often there is no second or third. Exact computation requires great precision.

4.3.1. System Life

Suppose that components within a system each have a random operational life (length) when in service; say that the rv $T_i > 0$ is the time to failure of the ith component. Then the indicator function

$$X_i(t) = \mathrm{I}(T_i > t) = \begin{cases} 1 & \text{if } T_i > t,\\ 0 & \text{if } T_i \leq t, \end{cases}$$

when T_i has hazard function H_i, has expectation

$$\mathrm{E}X_i(t) = \Pr[T_i > t] = e^{-H_i(t)} \qquad \text{for } t > 0.$$

The *performance process* of the system is indicated by the process

$$\Phi(t) = \varphi(X_1(t), \cdots, X_n(t)) \quad \text{at any time } t > 0,$$

and the reliability is given by $R(t) = \mathrm{E}\Phi(t)$ for $t > 0$. An analytic expression for this can be evaluated by the methods described previously.

If $T = \sup\{t : \Phi(t) = 1\}$ denotes the life length of the system Φ, then $\Pr[T > t] = R(t)$ so that the expected life of the system is, by a previous exercise, the integral of the reliability

$$\mathrm{E}T = \int_0^\infty R(t)\,dt.$$

For example, if the life length of each component is exponential so that the component hazard functions are $H_i(t) = \lambda_i t$ for $i = 1, \cdots, n$, and the structure is, say,

$$\varphi(x_1, \cdots, x_n) = (x_1 + x_2 - x_1 x_2) \prod_3^n x_i,$$

then the reliability becomes

$$R(t) = \left[e^{-\lambda_1 t} + e^{-\lambda_2 t} - e^{-(\lambda_1+\lambda_2)t}\right] e^{-t\sum_3^n \lambda_i},$$

and consequently the mean life is

$$\mathrm{E}T = \int_0^\infty R(t)\,dt = \frac{1}{\lambda_1 + \sum_3^n \lambda_i} + \frac{1}{\sum_2^n \lambda_i} - \frac{1}{\sum_1^n \lambda_i}.$$

Exercise Set 4.C

1. Suppose that the hazard rate for a rv $T > 0$ is defined for some $a > 0$ by $h(t) = \lambda \mathrm{I}(t \geq a)$ for all $t > 0$.
   (a) Find the density associated with the time of failure, T
   (b) Find the density of $T_1 + \cdots + T_m$ where the $T_i \sim T$ are iid.

2. Suppose the failure time of an integrated circuit has the following density:

$$f(t; a, r) = \frac{(r+1)a^{r+1}}{(a+t)^{r+2}} \quad \text{for } t > 0.$$

   (a) For what values of a and r is $f(t; a, r)$ a *bona fide* density?
   (b) Obtain an expression for the hazard function and identify the class of distributions to which it belongs.

3. Suppose that k different sources each manufacture the same device, which has an exponential life, but each source generates a different hazard rate for its product and each source supplies a different fraction of the inventory. Assume the ith manufacturer has supplied a proportion p_i of the components with hazard rate λ_i for $i = 1, \cdots, k$.
   (a) If a device is selected at random from the entire population, what is the density of the time of failure of the one selected?
   (b) Obtain an expression for the reliability and hazard function of the selected device.

4. If the length of life of a satellite is exponentially distributed with an expected life of 5 years and three such satellites are now in service, one of age 2 years, one of age 4 years and one just launched, what is the probability that at least two will still be in service after two years?

5. If component i, with lifelength T_i, has hazard function $H_i(t)$ for $i = 1, 2, 3, 4$ and T_1, T_2, T_3, T_4 are monotonely associated rvs, find a lower bound for the reliability of the system below in terms of the H_i.

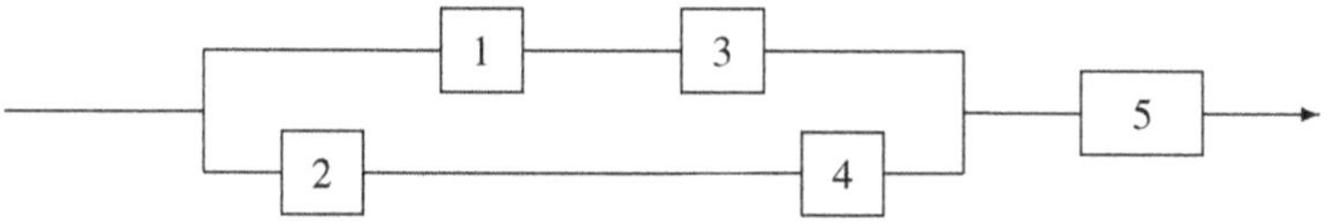

6. Suppose that n independently functioning components are connected in a series system. Assume that the logarithm of the time-to-failure for each component is normally distributed with expectation of 10 hours and standard deviation of 2 hours.
   (a) If $n = 4$, what is probability that the system will be functioning after 10 years of operation?
   (b) If the probability of failure during the first 5 years cannot exceed 0.01, what is the largest value that n can assume?

7. Suppose that the sdf of a life, in months, is known and differentiable. In carrying out preventative maintenance, the company wants to decide how many months after installation such a component should be replaced to minimize the expected cost per component. If there is a restriction that the shortest practicable elapsed time between installation and replacement is E months, find an expression for the optimum number of months for replacement, when the cost per component with life L being replaced after m months is given, in two cases, by
   (a) $C(L, m) = a|L - m|$ for some $a > 0$.
   (b) $C(L, m) = a1(L < m) + b(L - m)^+$ for some $0 < a < b$.

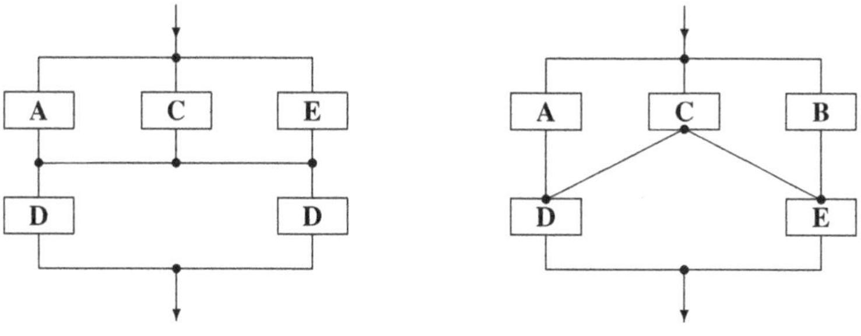

(a) **(b)**

8. Consider the components **A,B,C,D,E** as indicated above:
   (a) Assuming independence of all five components, obtain an expression for the reliability of the system in each of the two cases (a) and (b).
   (b) Derive the reliability in the case that A and D and B and E are associated.

9. The following is a specific example of the use of fault-tree analysis to determine the reliability of a subsystem of a nuclear generating plant, which is illustrated in Figures 4.1 and 4.2. The top event is the rupture of a high-temperature reactor. The oxident and fuel are continuously reacted at high temperature to form the desired products. Because of

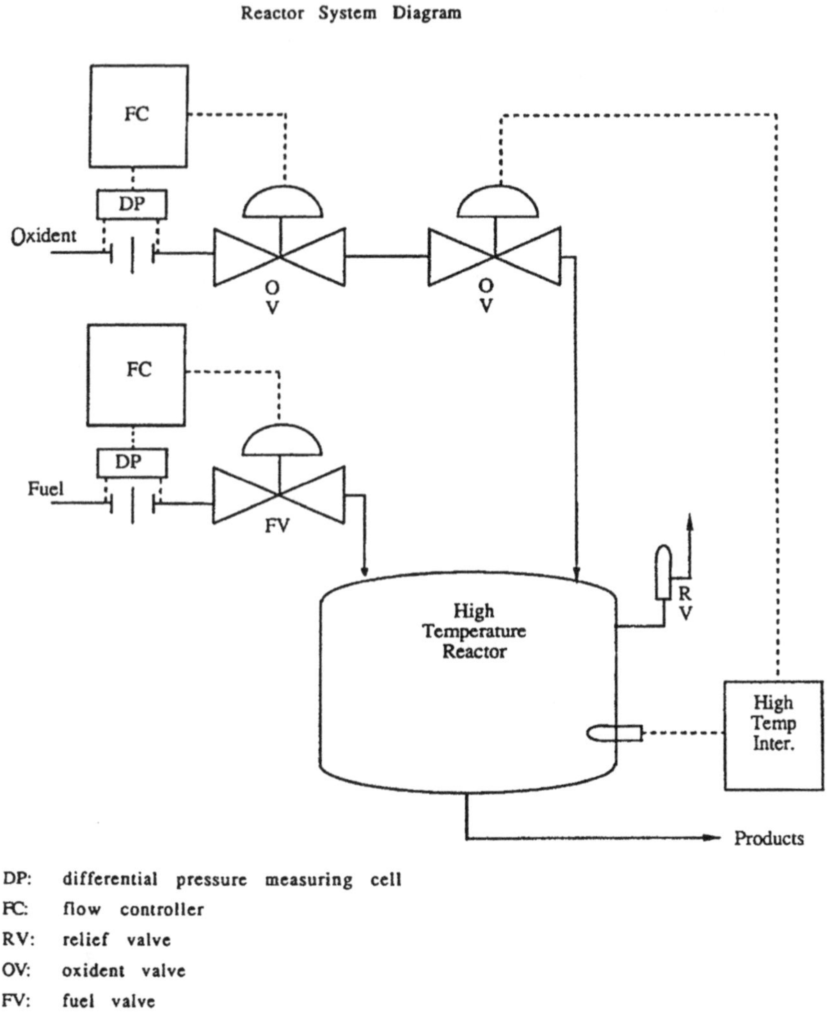

Figure 4.1. Reactor System Diagram.

the heat of reaction, the rate must be carefully controlled by maintaining the necessary oxident-to-fuel ratio. If this ratio becomes too high, the reaction can 'run away' causing the pressure to rise precipitiously, rupturing the vessel. Consequently, the oxident and fuel feeds to the reactor are precisely metered in the proper ratio. To safe-guard the reactor in the event of a run-away reaction, a safety relief valve is provided. In addition a high-temperature interlock-valve system is installed to protect against further rupture.

(a) Write out the coherent system for these components using the fault tree as given in Figure 4.2 and determine the minimum reliability of the components so that

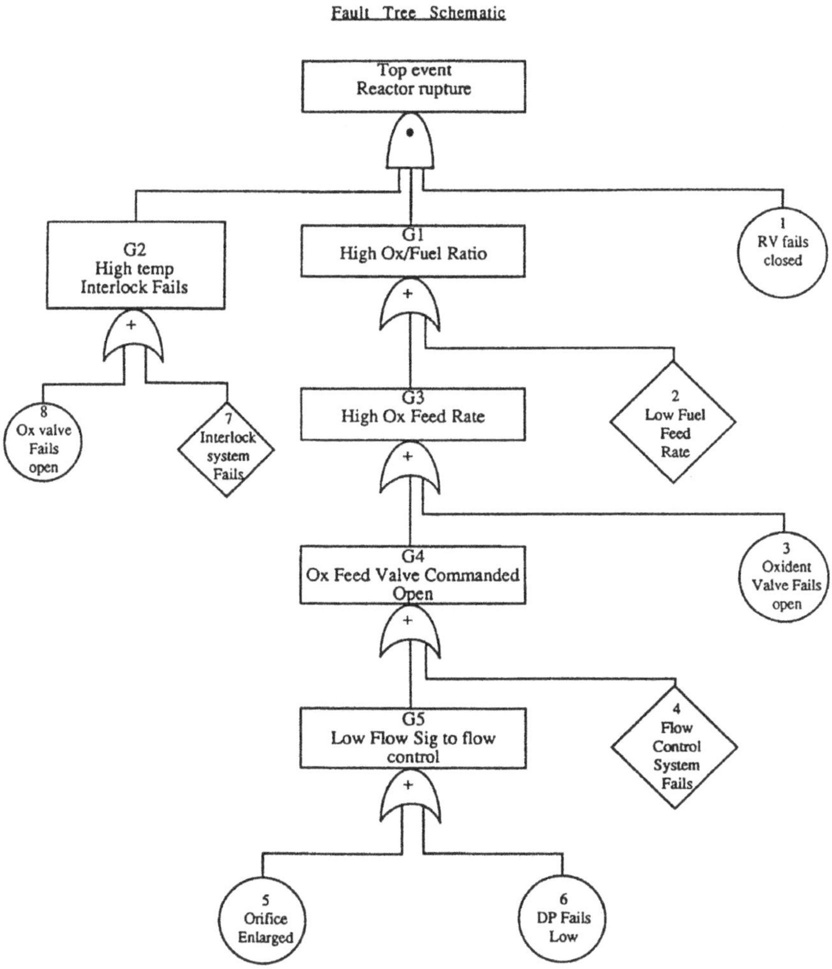

Figure 4.2. Fault-tree schematic.

the top event has an unreliability less than 10^{-6} per unit of time, assuming all the components have exponential service lives.

(b) Criticise the universal assumption of constant hazard rates for every component, as made in part (a), and indicate a more realistic approach.

(c) Construct a model with one or more critical components subject to wear-out.

10. Consider the possibility of constructing small fail-safe nuclear power plants near metropolitan areas, thus reducing the power losses and expense of constructing long-distance power transmission lines. This would facilitate the utilization of the waste heat of the reactor for domicile heating, and the smaller size would lessen its desirability as a target for terrorism or sabotage. Do a calculation of the entailed risk (expected

loss) over a 35 year period for such design using the reliability of a reactor located near Seattle WA (population 1.5×10^6 within a 20 mile radius) and one which supplies power to the NW power grid from Hanford WA (population 15×10^4 within 20 miles). Remember the area contaminated in an accident is proportional to the generating power of the plant.
(a) Why do you think such systems have not been considered?
(b) Would public concern for their own safety be a factor? Do you think such fear could be alleviated by publishing the safety record of the operating history of the nuclear-powered submarines and aircraft carriers? Why?

11. Fail-safe structures have long been utilized in the design of airframes. This capacity for safety has shifted the dominant failure mode from structural integrity to maintenance and inspection-adequacy during operations. Since airlines are unregulated, may not attempts to increase profitability for the stockholders tend to reduce or eliminate the often 'needless' inspections? Prepare a presentation to managerial staff on this problem by choosing a life-distribution for a component in each unit of a fleet of 250 units. Assume each unit has a mean life of 25,000 hours and the component replacemnt cost is 10^{-3} that of the unit cost, but an in-service failure of the component will cost 300 times the cost of unit. (Remember the variance of unit-life is not known and your job-future depends upon the adequacy of the answer.)

4.4. Use of Association to Bound Reliability

We shall use the abreviation *arv* to stand for an *associated random vector* or *a set of associated random variables.*

Theorem 3. *If* $\mathbf{X}$ *is an arv of binary variates then:*

$$(i) \qquad \Pr[\textstyle\prod_{i=1}^{n} X_i = 1] \geq \prod_{i=1}^{n} \Pr[X_i = 1],$$

and

$$(ii) \qquad \Pr[\textstyle\coprod_{i=1}^{n} X_i = 1] \leq \coprod_{i=1}^{n} \Pr[X_i = 1].$$

PROOF. Since X_1 and $\prod_{i=2}^{n} X_i$ are arv's, we have by definition

$$\mathrm{Cov}(X_1, \textstyle\prod_{i=2}^{n} X_i) = \mathrm{E}[X_1 \cdot \prod_{i=2}^{n} X_i] - \mathrm{E}[X_1]\mathrm{E}[\prod_{i=2}^{n} X_i] \geq 0.$$

If we repeat this argument n times, we obtain the inequality $\mathrm{E}[\prod_{i=1}^{n} X_i] \geq \prod_{i=1}^{n} \mathrm{E}[X_i]$, which is equivalent with (i). To prove (ii) we see it is equivalent with $\mathrm{E}[\prod_{i=1}^{n}(1 - X_i)] \geq \prod_{i=1}^{n} \mathrm{E}[1 - X_i]$. Set $Y_i = 1 - X_i$ for $i = 1, \cdots, n$; thus $\mathbf{Y}$ is also arv by Theorem 2. So to show $\mathrm{E}[\prod_{i=1}^{n} Y_i] \geq \prod_{i=1}^{n} \mathrm{E}[Y_i]$, we have successively by definition of association

$$\mathrm{E}[\textstyle\prod_{i=1}^{n} Y_i] \geq \mathrm{E}[Y_1]\mathrm{E}[\prod_{i=2}^{n} Y_i] \geq \mathrm{E}[Y_1]\mathrm{E}[Y_1]\mathrm{E}[\prod_{i=3}^{n} Y_i] \geq \cdots \geq \prod_{i=1}^{n} \mathrm{E}[Y_i]. \quad \square$$

Corollary 2. *If* $\boldsymbol{T}$ *is any arv, then bounds on its distribution and survival are*

$$\Pr\{\bigcap_{i=1}^{n}[T_i > t_i]\} \geq \prod_{i=1}^{n} \Pr[T_i > t_i] \tag{4.5}$$

and

$$\Pr\{\bigcap_{i=1}^{n}[T_i \leq t_i]\} \geq \prod_{i=1}^{n} \Pr[T_i \leq t_i]. \tag{4.6}$$

Proof. Let $X_i = \mathrm{I}(T_i > t_i)$ for $i = 1, \ldots n$; they are arv by the properties of association. We then apply (i) of the preceding theorem to obtain (4.5). By (ii) we must show

$$\Pr\{\coprod_{i=1}^{n} \mathrm{I}(T_i > t_i) = 1\} \leq 1 - \prod_{i=1}^{n}[1 - \Pr(T_i > t_i)],$$

and the result followes by the calculus of probability. □

We also have immediately the solution

Corollary 3. *If* $\boldsymbol{T}$ *is any arv, then*

$$\Pr[\min_{i=1}^{n} T_i > t] \geq \prod_{i=1}^{n} \Pr[T_i > t] \tag{4.7}$$

and

$$\Pr[\max_{i=1}^{n} T_i > t] \leq \coprod_{i=1}^{n} \Pr[T_i > t]. \tag{4.8}$$

We know, for any independent components, that the series and parallel systems have, respectively, the lowest and highest performances among all coherent systems. We now claim that this is true also for component lives that are arv.

Theorem 4. *Let* φ *be a coherent system with* $\boldsymbol{X}$ *an arv of indicator variates with reliabilities* $E[\boldsymbol{X}] = \boldsymbol{p} \coloneqq (p_1, \ldots, p_n)$. *Then*

$$\prod_{i=1}^{n} p_i \leq \Pr[\varphi(\boldsymbol{X}) = 1] \leq \coprod_{i=1}^{n} p_i.$$

Proof. Since $\prod X_i \leq \varphi(\boldsymbol{X}) \leq \coprod X_i$, by taking expectations we obtain

$$\mathrm{E}\prod_{i=1}^{n} X_i \leq \mathrm{E}[\varphi(\boldsymbol{X})] \leq \mathrm{E}\coprod_{i=1}^{n} X_i,$$

and thus by the inequality (4.5) we have $\mathrm{E}\prod X_i \geq \prod \mathrm{E}X_i$, the result follows. □

Theorem 5. *If* φ *is a coherent system with* $\rho_1(\boldsymbol{x}), \ldots, \rho_p(\boldsymbol{x})$ *as the min-path (series) subsystems and* $\kappa_1(\boldsymbol{x}), \ldots, \kappa_k(\boldsymbol{x})$ *as the min-cut (parallel) subsystems when* $\boldsymbol{X}$ *is an arv of indicator variates with reliabilities* $E[\boldsymbol{X}] = \boldsymbol{p}$, *then*

$$\prod_{j=1}^{k} \Pr[\kappa_j(\boldsymbol{X}) = 1] \leq \Pr[\varphi(\boldsymbol{X}) = 1] \leq \coprod_{i=1}^{p} \Pr[\rho_i(\boldsymbol{X}) = 1].$$

Proof. All the $\kappa_j(\boldsymbol{x})$ for $j = 1, \ldots, k$ are increasing functions of $\boldsymbol{x}$ and since $\boldsymbol{X}$ is an arv, it follows that $\{\kappa_1(\boldsymbol{X}), \ldots, \kappa_k(\boldsymbol{X})\}$ is an arv. Thus, using the preceding results, we have

$$\mathrm{E}\varphi(\boldsymbol{X}) = \Pr\left[\prod_{j=1}^{k} \kappa_j(\boldsymbol{X}) = 1\right] \geq \prod_{j=1}^{k} \Pr\left[\kappa_j(\boldsymbol{X}) = 1\right],$$

establishing the lower bound. The upper bound is obtained similarly. □

If we assume that the components are independent, we obtain the following result.

Corollary 4. *If φ is a coherent system having independent components $\boldsymbol{X}$ with reliabilities $E[\boldsymbol{X}] = \boldsymbol{p}$, then we have upper and lower bounds on the reliability given by*

$$\ell_\varphi(\boldsymbol{p}) \leq \hbar(\boldsymbol{p}) \leq u_\varphi(\boldsymbol{p}),$$

where $\hbar(\boldsymbol{p}) = E\varphi[\boldsymbol{X}]$ with

$$\ell_\varphi(\boldsymbol{p}) \equiv \prod_{j=1}^{k} \coprod_{i \in K_j} p_i, \quad \text{and} \quad u_\varphi(\boldsymbol{p}) \equiv \coprod_{j=1}^{p} \prod_{i \in P_j} p_i.$$

Proof. This follows directly from the preceding result. □

We also have the following:

Remark 1. *For any coherent structure φ, we see that*

(a) $\ell_\varphi(\boldsymbol{p})$ and $u_\varphi(\boldsymbol{p})$ are increasing functions.
(b) $\ell_\varphi(\boldsymbol{p}) < \hbar(\boldsymbol{p}) < u_\varphi(\boldsymbol{p})$ for $0 < p_i < 1$ for $i = 1, \cdots, n$,

when not all the min-paths and the min-cuts are disjoint.

We now give the min-max bounds on reliability for coherent systems.

Theorem 6. *Let φ be a coherent system with min-paths $P_1, \ldots, P_p$ and min-cuts $K_1, \ldots, K_k$. Let the components $\boldsymbol{X}$ have reliabilities $E\boldsymbol{X} = \boldsymbol{p}$, with $E\varphi(\boldsymbol{X}) = \hbar(\boldsymbol{p})$. Then the following bounds always obtain:*

$$\max_{1 \leq j \leq p} \Pr[\min_{i \in P_j} X_i = 1] \leq E\varphi(\boldsymbol{X}) \leq \min_{1 \leq j \leq k} \Pr[\max_{i \in K_j} X_i = 1]. \tag{4.9}$$

If the components are associated, the bounds can be explicitly given as

$$\max_{1 \leq j \leq p} \prod_{i \in P_j} p_i \leq \hbar(\boldsymbol{p}) \leq \min_{1 \leq j \leq k} \coprod_{i \in K_j} p_i. \tag{4.10}$$

PROOF. By the representations given previously, we have

$$\varphi(\boldsymbol{X}) = \max_{1\le j\le p}\min_{i\in P_j} X_i = \min_{1\le j\le k}\max_{i\in K_j} X_i$$

Thus, we have for all $1 \le s \le p$ and $1 \le t \le k$

$$\min_{i\in P_s} X_i \le \varphi(\boldsymbol{X}) \le \max_{i\in K_t} X_i.$$

Inequality (4.9) now follows by maximizing over s and minimizing over t.

The inequality (4.10) follows from inequality (4.9) and Theorem 3. □

4.5. Shape of the Reliability Function

We now examine the behavior of the reliability of coherent systems.

Lemma 1. *Let $\varphi(\boldsymbol{X})$ be an operational coherent system. Then*

$$\mathrm{Cov}\left(\varphi(\boldsymbol{X}), \textstyle\sum_{i=1}^n X_i\right) \ge \mathrm{Var}(\varphi(\boldsymbol{X})).$$

PROOF. It is sufficient to prove that

$$\mathrm{Cov}\left(\varphi(\boldsymbol{X}), \textstyle\sum_{i=1}^n X_i - \varphi(\boldsymbol{X})\right) \ge 0.$$

But this is true since both arguments are increasing functions of $\boldsymbol{X}$. □

Lemma 2. *Let $\varphi(\boldsymbol{X})$ be an operational coherent system with $E\varphi(\boldsymbol{X}) = \hbar(\boldsymbol{p})$. Then*

$$\mathrm{Cov}\left(\varphi(\boldsymbol{X}), \textstyle\sum_{i=1}^n X_i\right) = \sum_{i=1}^n p_i q_i \frac{\partial \hbar(\boldsymbol{p})}{\partial p_i} \quad \textit{where } q_i = 1 - p_i.$$

PROOF. Let $\boldsymbol{\delta}_i = (\delta_{1,i}, \ldots, \delta_{n,i})$, where $\delta_{i,j} := \mathrm{I}(i = j)$ is Kronecker's delta, and we write $\bar{\boldsymbol{\delta}}_i = (1 - \delta_{1,i}, \ldots, 1 - \delta_{n,i})$. We note that

$$\boldsymbol{\delta}_j \vee \boldsymbol{x} = (x_1, \cdots, x_{j-1}, 1, x_j, \cdots, x_n), \tag{4.11}$$

and

$$\bar{\boldsymbol{\delta}}_j \wedge \boldsymbol{x} = (x_1, \cdots, x_{j-1}, 0, x_j, \cdots, x_n). \tag{4.12}$$

Then

$$\begin{aligned}
\mathrm{Cov}(\varphi(\boldsymbol{X}), X_i) &= \mathrm{E}[\varphi(\boldsymbol{X})X_i] - \mathrm{E}[\varphi(\boldsymbol{X})]\mathrm{E}X_i \\
&= p_i\hbar(\boldsymbol{\delta}_i \wedge \boldsymbol{p}) - p_i\hbar(\boldsymbol{p}) \\
&= p_i\hbar(\boldsymbol{\delta}_i \wedge \boldsymbol{p}) - p_i\left[p_i\hbar(\boldsymbol{\delta}_i \wedge \boldsymbol{p}) + q_i\hbar(\bar{\boldsymbol{\delta}}_i \vee \boldsymbol{p})\right] \\
&= p_i q_i\left[\hbar(\boldsymbol{\delta}_i \wedge \boldsymbol{p}) - \hbar(\bar{\boldsymbol{\delta}}_i \vee \boldsymbol{p})\right] = p_i q_i \frac{\partial \hbar(\boldsymbol{p})}{\partial p_i},
\end{aligned}$$

from which the result is obvious. □

We also have the next result.

Theorem 7. *If $\hbar(\boldsymbol{p})$ is the reliability of a coherent system of order $n \geq 2$, then*

(a) $\sum_{i=1}^{n} p_i q_i \frac{\partial \hbar(\boldsymbol{p})}{\partial p_i} \geq \hbar(\boldsymbol{p})[1 - \hbar(\boldsymbol{p})]$ *for* $\mathbf{0} < p < 1$.
If all $p_i \equiv p$ *and we set* $h(p) = \hbar(p, \ldots, p)$ *then*

(b) $pq \frac{\mathrm{d}h(p)}{\mathrm{d}p} > h(p)[1 - h(p)]$ *for* $0 < p < 1$.

(c) If $h(p) = p_0$ *for some* $p_0 \in (0, 1)$, *then* $\operatorname{sgn}[h(p) - p] = \operatorname{sgn}[p - p_0]$.

(d) If φ *has no path or cut sets of size 1, then there exists a fixed point* $0 < p_0 < 1$, *for which* $h(p_0) = p_0$.

Proof. Since $\varphi(\boldsymbol{X})$ is a Bernoulli variate, we know that $\operatorname{Var}[\varphi(\boldsymbol{X})] = \hbar(\boldsymbol{p})[1 - \hbar(\boldsymbol{p})]$, then using the two lemmas above we obtain part (a). The other conclusions follow by taking all component reliabilities equal. □

If we examine the reliability of a k-out-of-n system, when all component reliabilities are equal, we have

$$h(p) = \mathrm{E}\left\{\mathrm{I}(\textstyle\sum X_i \geq k)\right\} = \sum_{i=k}^{n} \binom{n}{i} p^i (1-p)^{n-i}$$

$$= 1 - k\binom{n}{k} \int_p^1 t^{k-1}(1-t)^{n-k}\, dt.$$

For a 6-out-of-10 system we see the shape given in Figure 4.3.

For virtually all practical instances when all component's reliabilities are equal, one sees that the graph of the function $h(\cdot)$ is "S-shaped." But what exactly does "S-shaped" mean? Does it mean that the radius of curvature changes in a specific way or that $h(\cdot)$ is initially convex increasing and then becomes concave increasing? We make the following:

Definition 2. A differentiable increasing function h mapping (0, 1) onto (0, 1) is S-shaped iff

$$\frac{h(p)}{ph'(p)} \quad \text{is increasing, with range } \left(1, \tfrac{1}{h'(1)}\right).$$

We now have some immediate consequences.

Remark 2. *If h is S-shaped, there exists a unique fixed point $0 < p_0 < 1$ such that $h(p_0) = p_0$.*

Remark 3. *If h is S-shaped, then the dual reliability, $h_D(p) = 1 - h(1 - p)$, is also S-shaped.*

Remark 4. *If two functions, say h_1, h_2, are S-shaped, then the composite function $h_1[h_2(p)]$ is S-shaped.*

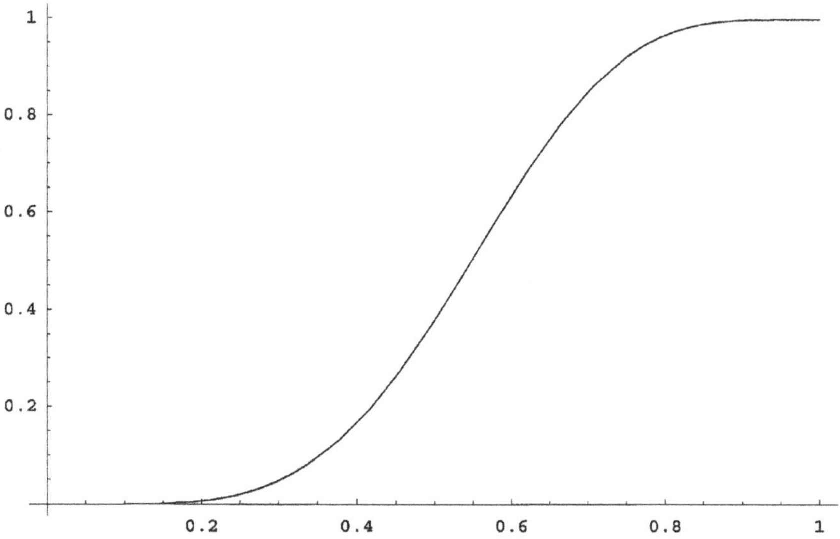

Figure 4.3. S-shaped reliability function of 6-out-of-10 system.

Remark 5. *If h is S-shaped with fixed point p_0, then setting $h_n(p) = h[h_{n-1}(p)]$ for $n = 1, 2, \cdots$ with $h_0(p) = p$, we find*

$$\lim_{n\to\infty} h_n(p) = I(p > p_0) \quad \textit{for all } p \neq p_0.$$

From the remarks above, we find the following.

Theorem 8. *If h is S-shaped, then the function Υ_h is also increasing, where*

$$\Upsilon_h(p) = \frac{h(p)}{p} \cdot \frac{1-p}{1-h(p)} \quad \textit{for} \quad p \in (0, 1).$$

NB that h need not be differentiable for Υ_h to be increasing. The behavior of Υ_h as measured by a suitable functional can be used to determine the degree of S-shapedness.

But usually the system reliability is a function of a vector of component reliabilities, what does S-shapedness mean in the multivariate case? This last definition is what is needed.

Definition. If $\hbar(\boldsymbol{p})$ is a multivariate reliability, it is S-shaped iff

$$\Upsilon_{\hbar}(\boldsymbol{p}) = \frac{\hbar(\boldsymbol{p})}{\prod p_i} \cdot \frac{\prod(1-p_i)}{1-\hbar(\boldsymbol{p})}$$

is monotone increasing in $\boldsymbol{p}$.

4.6. Diagnostics and Importance of System Components

4.6.1. Importance

How can one assess the relative importance of components in the performance of the system? Let us fix our attention on the jth component and determine how well the system φ will operate with that component operational or how it will operate with that component failed. Consider the state of the remaining components when the jth component makes the difference whether the system operates or not. Using the notation of eqns (4.11) and (4.12) we have,

$$\varphi(\boldsymbol{\delta}_j \vee \boldsymbol{x}) - \varphi(\bar{\boldsymbol{\delta}}_j \wedge \boldsymbol{x}) = 1; \tag{4.13}$$

the jth component is more important to operation when this holds than when the difference is 0. If eqn (4.13) holds, then $\boldsymbol{\delta}_j \vee \boldsymbol{x}$ is a critical path vector for j and $C_1(\boldsymbol{\delta}_j \vee \boldsymbol{x})$ is the critical path set for j. We set

$$n_\varphi(j) = \sum_{\{\boldsymbol{x}: x_j=1\}} \varphi(\boldsymbol{\delta}_j \vee \boldsymbol{x}) - \varphi(\bar{\boldsymbol{\delta}}_j \wedge \boldsymbol{x})$$

to denote the total number of critical path sets for j. Then we take as the measure of *system importance* of the jth component in φ as

$$I_\varphi(j) = \frac{n_\varphi(j)}{2^{n-1}}. \tag{4.14}$$

Thus, we can, for any system φ of order n, rank the importance of the components, namely $I_\varphi(1), \ldots, I_\varphi(n)$, as to its functioning.

If the components have different reliabilities, we may want to determine the *reliability importance* of a component to the system's operation. This is defined, when $\mathrm{E}\varphi(\boldsymbol{X}) = \hbar(\boldsymbol{p})$, as

$$I_\hbar(j) = \frac{\partial \hbar(\boldsymbol{p})}{\partial p_j} = \mathrm{E}[\varphi(\boldsymbol{\delta}_j \vee \boldsymbol{x}) - \varphi(\bar{\boldsymbol{\delta}}_j \wedge \boldsymbol{x})]. \tag{4.15}$$

If all the reliabilities are functions of time t, which occurs when, say $X_i = \mathrm{I}(T_i > t)$, with T_i the service life of the ith component, then the component reliabilities are $p_j(t)$ for $1 \le j \le n$ and all $t > 0$. Thus, the rate of decrease of reliability over time is given by

$$\frac{\partial \hbar[\boldsymbol{p}(t)]}{\partial t} = \sum_{j=1}^{n} I_\hbar(j) \frac{dp_j(t)}{dt}.$$

4.6.2. Diagnostics Using Reliability

The mathematical insight that comes from the knowledge of the min-path representation of a coherent system can also be used to help design an *expert diagnostic system*, necessarily done in conjunction with a computer for large fail-safe systems, to determine the order in which components should be examined and repaired to reestablish, with minimum expected cost, the operation of the system.

Theorem 9. *Given $\hbar$ as the reliability function of a coherent system with the ith component having life $T_i \sim F_i$ for $i = 1, \ldots, n$, then the probability that component i caused the system failure, given that the system failed at time $t > 0$, is given by*

$$\frac{\Delta_i \hbar[\bar{\boldsymbol{F}}(t)] f_i(t)}{\sum_{j=1}^{n} \Delta_j \hbar[\bar{\boldsymbol{F}}(t)] f_j(t)},$$

where $f_j = F_j^{'}$ and $\bar{\boldsymbol{F}}(t) = [\bar{F}_1(t), \ldots, \bar{F}_1(t)]$.

PROOF. We merely note that

$$\Delta_j \hbar[\bar{\boldsymbol{F}}(t)] = \Pr\{\varphi[\delta_i \vee \boldsymbol{X}(t) - \varphi[\bar{\delta}_i \wedge \boldsymbol{X}(t)]\}$$

is the probability that at time $t > 0$ the system is functioning iff the ith component is functioning. That is the system is functioning if component i is functioning but failed otherwise. Thus the numerator (times dt) is the probability that component i causes system failure in time $(t, t + dt)$ while the denominator is the corresponding probability of system failure in the same interval. □

Using the theorem of total probability we have

Theorem 10. *The probability that the ith component causes system failure during $(0, t)$, given that system failure occurs in time $(0, t)$ is*

$$\frac{\int_0^t \Delta_i \hbar[\bar{\boldsymbol{F}}(u)] \, d\, F_i(u)}{\sum_{j=1}^{n} \int_0^t \Delta_j \hbar[\bar{\boldsymbol{F}}(u)] \, d\, F_i(u)}.$$

If we let $t \to \infty$ in the above, then we obtain the importance of the ith component in the system which has reliability function $\hbar$, namely,

$$I_\hbar(i) = \int_0^\infty \Delta_i \hbar[\bar{\boldsymbol{F}}(u)] \, d\, F_i(u).$$

The importance function has the following probabilistic properties, under mild conditions,

$$0 < I_\hbar(i) < 1, \quad \text{and} \quad \sum_{i=1}^{n} I_\hbar(i) = 1.$$

Exercise Set 4.D

1. Discuss why the contractors for nuclear power generation, whose principal goal is making a profit for the stockholders, are often at swordpoints with the governmental Advisory Committee on Reactor Safe-Guards whose primary responsibility is the health and safety of the citizenry.

2. List all the reliability concepts that you think would be useful to discuss from either side in such presentations as above. What do you think the sticking points would be?

4.7. Hazard Rates and Pólya Frequency Functions

One of the types of functions most intimately connected with the behavior of reliability functions is the class of frequency functions introduced by George Pólya. We shall denote the class of *Pólya frequency functions of order 2* by writing, merely, PF_2.

Definition 3. Any real-valued function, say h defined on $\Re$, is PF_2 iff both (a) and (b) hold, where

(a) $h(x) \geq 0$,
(b) for any two pairs of points in $\Re$, say $x_1 < x_2$ and $y_1 < y_2$, the determinant of the 2×2 matrix D is nonnegative, where $D = (d_{i,j})$ with $d_{i,j} = h(x_i - y_j)$. Equivalently one may substitute for (b) either:
(b′) $\ln h$ is concave on $\Re$, or
(b″) for each fixed $\delta > 0$, the ratio $\frac{h(x+\delta)}{h(x)} \in\downarrow$ in x for each x in an interval contained in the support of h.

We now have the immediate result.

Theorem 11. *Let $X \sim F$, where $X > 0$. Then F is* IHR *iff $\bar{F}$ is* PF_2.

The proof is an exercise.

The extent of the range for which the hazard rate is increasing determines the optimality of length of burn-in. This is shown here.

Remark 6. *If $h \in\uparrow$ on (τ, ∞), then burn-in should not exceed time τ.*

Proof. Note that

$$\mathrm{E}[T|T > \tau] = \int_0^\infty \exp\{-\int_0^t h(x+\tau)\,dx\}\,dt.$$

If $\tau < \tau_1 < \tau_2$, we must have $h(x+\tau_1) \leq h(x+\tau_2)$ for all $x \geq 0$ so that by integration

$$\int_0^t h(x+\tau_1)\,dx \leq \int_0^t h(x+\tau_2)\,dx.$$

Hence

$$\begin{aligned}
E[T|T > \tau_1] &= \int_0^\infty \exp\{-\int_0^t h(x+\tau_1)\,dx\}\,dt \\
&\geq \int_0^\infty \exp\{-\int_0^t h(x+\tau_2)\,dx\}\,dt = E[T|T > \tau_2]
\end{aligned}$$

Thus, the expected remaining life is decreased with a burn-in extending within the region (τ, ∞) on which h is increasing. □

We also note that

$$E[T|T > \tau] = \int_0^\infty \frac{\bar{F}(t+\tau)}{\bar{F}(\tau)}\,dt. \tag{4.16}$$

Thus, if $\bar{F}(t+\tau)/\bar{F}(\tau)$ is, for fixed $t > 0$, increasing in τ for some initial interval $0 < \tau < \tau_0$, then for that interval we must have $E[T|T > \tau]$ increasing for the same interval.

4.8. Closure Properties

Let T_i for $i = 1, \ldots, n$ be the life lengths of n independent components with $X_i(t) = I(T_i > t)$ indicators of their survival time for any $t > 0$. If φ is a coherent system of order n, define its life length $T \sim F_\varphi$ by the sdf

$$\bar{F}_\varphi(t) = \Pr\{\varphi[X_1(s), \cdots, X_n(s)] = 1 \text{ for } 0 \leq s < t\}.$$

Let $E[X_i(t)] = \bar{F}_i(t)$. Then $\bar{F}_\varphi(t) = \hbar_\varphi[\bar{F}_1(t), \ldots, \bar{F}_n(t)]$, where $\hbar_\varphi$ is the reliability function for the coherent φ.

We now consider the question whether the life of the system, T, will necessarily be IHR if each of the lives of the n components, T_i for $i = 1, \ldots, n$ is IHR?

NB a coherent structure of two parallel components each with constant hazard rate, label them λ_1, λ_2, has reliability

$$\bar{F}(t) = 1 - (1 - e^{-\lambda_1 t})(1 - e^{-\lambda_2 t}) \quad \text{for } t > 0.$$

Its hazard rate becomes

$$h(t) = \frac{\lambda_1 e^{-\lambda_1 t} + \lambda_2 e^{-\lambda_2 t} - (\lambda_1 + \lambda_2)e^{-(\lambda_1+\lambda_2)t}}{e^{-\lambda_1 t} + e^{-\lambda_2 t} - e^{-(\lambda_1+\lambda_2)t}},$$

from which we see that $h(0) = 0$ and $h(\infty) = \min(\lambda_1, \lambda_2)$. One can show that

$$h'(t) \geq 0 \quad \text{iff} \quad \lambda_1^2 e^{-\lambda_2 t} + \lambda_2^2 e^{-\lambda_1 t} \geq (\lambda_1 - \lambda_2)^2.$$

Thus, for certain values of the λ_i the hazard rate is initially increasing but becomes terminally decreasing.

We note that despite the intuitive appeal of the IHR classification, if one places components with IHR lives in coherent systems we do not necessarily obtain a system with an IHR life. But since one manufacturer's systems are often another manufacturer's components, it is of interest to consider what class of life lengths would contain all the IHR distributions and yet be closed under the formation of coherent systems. This consideration necessitates another

Definition 4. Let $g : \Re_+ \to \Re$. We say that g is *star-shaped* on $\Re_+$ iff $g(x)/x \in\uparrow$.

Let f be a convex function such that $f(0) = 0$; then for $0 < \alpha < 1$ we have

$$\alpha f(x) + (1-\alpha)f(y) \geq f(\alpha x + (1-\alpha)y) \quad \text{for } x, y \in \Re_+.$$

If we set $y = 0$, then $\alpha f(x) \geq f(\alpha x)$. For we see $f(x)/x \geq f(\alpha x)/(\alpha x)$ any $x > 0$. Therefore, f is star-shaped. This argument goes backward. Thus,

Theorem 12. *A function g is* star-shaped *on $\Re_+$ iff $g(\alpha x) \leq \alpha g(x)$ for any $0 < \alpha < 1$ and all $x \in \Re_+$.*

We relax hazard rate behavior by introducing IHRA (DHRA), which means the *hazard rate increases (decreases) on average*, respectively. Let $T \sim F$ have hazard $H = -\ln(1-F)$ then:

(i) $F \in$ IHR iff H is convex increasing, i.e., $H' = h > 0, h' > 0$.
(ii) $F \in$ IHRA iff $H(t)/t \in\uparrow$ for $t > 0$.
(iii) $F \in$ DHRA iff $-H(t)/t \in\uparrow$ for $t > 0$.

Remark 7. *$F \in$*IHRA *iff $H(t) = -\ln[1-F(t)]$ is* star-shaped *for $t > 0$.*

Let F be IHRA; then $-\ln[\bar{F}(t)]/t \in\uparrow$. Therefore, $\ln[\bar{F}(t)^{1/t}] \in\downarrow$. Thus, we see

$$\exp\{-H(t)/t\} = \bar{F}(t)^{1/t} \in\downarrow \quad \text{iff } F \text{ is IHRA.}$$

Therefore, for $0 < \alpha < 1$, we have

$$[\bar{F}(\alpha t)]^{1/\alpha t} \geq [\bar{F}(t)]^{1/t} \quad \text{iff} \quad [\bar{F}(\alpha t)]^{1/\alpha} \geq \bar{F}(t).$$

Thus, we have

Theorem 13. *F is* IHRA *iff $\bar{F}(\alpha t) \geq [\bar{F}(t)]^{\alpha}$.*

There also follows, from the definition of star-shaped, the result

Theorem 14. *F is* IHRA *iff for each $\lambda > 0$ the difference $\bar{F}(t) - e^{-\lambda t}$ has at most one change of sign, which must occur from $+$ to $-$, if at all.*

Lemma 3. *If $0 \leq \alpha, \lambda \leq 1$ and $0 \leq x \leq y$, then*

$$(\lambda y)^{\alpha} + (1-\lambda^{\alpha})x^{\alpha} \geq [(\lambda y) + (1-\lambda)x]^{\alpha}.$$

PROOF. Since $f(x) = x^\alpha$ is concave increasing in x, for any $0 \le \alpha \le 1$, it follows that

$$f(u_1 + \delta) - f(u_1) \ge f(u_2 + \delta) - f(u_2) \quad \text{for} \quad u_1 \le u_2.$$

Now take $\delta = \lambda(y - x),\ u_1 = \lambda x,\ u_2 = x.$ □

Recall that a system φ is *monotonic* iff, except for irrelevant components, it is coherent.

Theorem 15. *Let $\hbar(\boldsymbol{p})$ be the reliability of a monotone system. Then we have*

$$\hbar(\boldsymbol{p}^\alpha) \ge [\hbar(\boldsymbol{p})]^\alpha \qquad \textit{for } 0 < \alpha \le 1, \tag{4.17}$$

where $\boldsymbol{p}^\alpha = (p_1^\alpha, \cdots, p_n^\alpha)$.

PROOF. Use induction on the order of the system and the preceding lemma. □

We can now state the central closure theorem for coherent systems.

Theorem 16. *If each independent component of a coherent system has an* IHRA *life distribution, then the system itself has an* IHRA *life distribution.*

PROOF. Let F be a cdf of system life and F_i the cdf of component life for $i = 1, \ldots, n$. Then for $0 < \alpha \le 1$ we have

$$\bar{F}(\alpha t) = \hbar[\bar{F}_1(\alpha t), \ldots, \bar{F}_n(\alpha t)],$$

and since each F_i is IHRA, then $\bar{F}_i(\alpha t) \ge [\bar{F}_i(\alpha t)]^\alpha$ for $i = 1, \ldots, n$. By monotonicity of each argument we see

$$\bar{F}(\alpha t) \ge \hbar\{[\bar{F}_1(\alpha t)]^\alpha, \ldots, [\bar{F}_n(\alpha t)]^\alpha\} \ge \{\hbar[\bar{F}_1(\alpha t), \ldots, \bar{F}_n(\alpha t)]\}^\alpha = [\bar{F}(\alpha t)]^\alpha,$$

where the second inequality follows by (4.17) above. Thus, we see that $[\bar{F}(\alpha t)] \ge [\bar{F}(\alpha t)]^\alpha$ from which we conclude that T is IHRA. □

NB a coherent system of IHR components has an IHRA life.

4.8.1. Further Closure Properties

Let $\mathcal{Q}$ denote some class of life distributions; we define the extensions of that class as follows:

- $\mathcal{Q}^{CS}$ is the closure of the set of life distributions from $\mathcal{Q}$ under the formation of coherent systems.
- $\mathcal{Q}^{LD}$ is the closure of the set of life distributions from $\mathcal{Q}$ under all limits in distribution obtained by using elements from $\mathcal{Q}$.

Consider the IHRA class, the class of Epstein distribitions, which we denote by EP, and the degenerate distributions denoted by

$$DG = \bigcup_{t>0} \{\varepsilon_t : \varepsilon_t(x) = \mathrm{I}(x \leq t) \text{ for } x > 0\}.$$

We state without proof

Lemma 4. *These inclusions hold:*

$$\{\text{IHRA}\} \subset \{\text{EP,DG}\}^{CS,LD}, \qquad \{\text{DG}\} \subset \{\text{EP}\}^{CS,LD}, \qquad \mathcal{Q}^{LD,CS} \subset \mathcal{Q}^{CS,LD}.$$

From this we can obtain the important

Theorem 17. $\{\text{IHRA}\}^{CS} = \{\text{IHRA}\} = \{\text{EP}\}^{CS,LD}$.

There are additional closure results having to do with the convolution of distributions, which govern the addition of life lengths.

Theorem 18.

If F_1 *and* F_2 *are* IHR, *then* $F_1 * F_2$ *is* IHR.
If F_1 *and* F_2 *are* IHRA, *then* $F_1 * F_2$ *is* IHRA.
If F_1 *and* F_2 *are* DHR, *then* $F_1 * F_2$ *is necessarily neither* DFR *nor* DHRA.

But there is another important closure property that concerns the class of distributions with decreasing hazard rate. We first define a mixture of distributions: let $F(\cdot : \theta)$ be a distribution for each $\theta \in \Theta$ and G a distribution on $\Theta \subset \Re^m$, then the *mixture* is

$$F(x) = \int_\Theta F(x;\theta)\, dG(\theta).$$

Theorem 19. *If* $F(\cdot;\alpha)$ *is* DHR *(or* DHRA*) for each* $\alpha \in A$ *then* F *is* DHR *(or correspondingly* DHRA*).*

Suppose that an electronic device has a life $T > 0$ and $T \sim \text{EP}(\lambda)$ for some $\lambda > 0$. Let $\Lambda \sim G$ where we have

$$G(\lambda) = \frac{(\lambda \wedge b - a)^+}{b - a} \quad \text{for } \lambda > 0; 0 \leq a < b,$$

i.e., G is a uniform cdf on (a, b). Then we find the cdf to be

$$F(t) = \int_a^b \int_0^t \lambda e^{-\lambda s}\, ds\, dG(\lambda),$$

and upon simplifying we obtain the sdf as

$$\bar{F}(t) = \frac{e^{-at} - e^{-bt}}{t(b-a)} \quad \text{for } 0 < t < \infty,$$

which we can be sure is itself DHR from the mixture theorem but is not obvious otherwise.

There have been other nonparametric classifications of distributions proposed for hazard functions but none has proved to be as interesting or as mathematically rich as the IHR and IHRA classifications. One of the most intuitively appealing is the "bath-tub", shaped hazard rate. Several parametric families have been proposed, but the right definition seems not to have been found. Two candidates are "convex hazard rate" (i.e., H^{-1} is S-shaped) and "initially decreasing and terminally increasing hazard rates." But parametric families with these properties are not easily found.

To become more applicable in engineering the concept of *coherent structures* has to be extended to include the structural strength of each component, as determined by finite-element analysis for the strength of the elements or the re-allocation of load after the failure of one element. It thus becomes more realistic to maximize the reliability of such structures subject to constraints in weight and/or cost, given a specified performance level.

The utility of coherent systems is not widely recognized although some application has been made outside engineering reliability, e.g. in group interaction in sociology but not, seemingly, in computer "virus" protection. The term *irreducible complexity* is used by the (creationist) Discovery Institute of Seattle to describe what we call a series system; "a single system composed of several $\cdots$ parts that contribute to the basic function, where the removal of any one part causes the system to cease functioning." The eye is mentioned as such a system. They make the nugatory claim that evolution by natural selection is powerless to have developed such features in animals.

> Nothing in biology makes sense except in the light of evolution.
>
> Theodosius Dobzhansky

Exercise Set 4.D

1. Using §4.5, prove a version of the famous result of John von Neumann that arbitrarily good relays could be constructed from arbitrarily crummy ones.

2. A person whom you just met tells you he has two children and you learn subsequently that one is a boy. Show that the probability the second child is a boy is not one-half.

3. On a TV show a valuable prize is behind one of three screens. The contestant chooses one screen. Then the emcee removes one of the two remaining screens and shows the prize is not behind it. He then asks the contestant if he wishes to change his initial selection to the other screen. Can the contestant improve his chances of winning by doing so?

   This problem appeared in *Parade* Magazine; Marilyn vos Savant answered "yes" but she received many letters, some from from Ph.D's, who said "no." Show that "yes" is correct as stated. However, if the emcee does not know where the prize is located and when he opens a curtain with the prize behind it the contestant gets nothing, then it follows that even knowing the strategy above the contestant cannot improve his(her) initial chances of winning beyond one-third. (The expected expense is what concerns the sponsors.)

Some concepts are deeper than others. Anyone who thinks he understands quantum mechanics doesn't understand it.

Niels Bohr

Rational thinking imposes limitations on one's perception, but expands one's comprehension of reality.

John Nash

The power of instruction is seldom of much efficacy except in those happy dispositions where it is almost superfluous.

Edward Gibbon

Call no man a scholar who does not rejoice when one of his own errors is discovered, *by someone else.*

Donald Foster

Expertise is too often characterized by an inability to accept new points of view, stemming from the depths of its preoccupation with its own conclusions. Witness the testimony:

That 'atom bomb' is the biggest damn fool thing we have ever done ⋯ that bomb will never go off, and I speak as an expert on explosives.

Admiral William D. Leahy, Chief of Naval Operations, World War II

You know the most amazing thing happened to me tonight as I was coming here to this lecture. You won't believe what happened when I passed through the parking lot. I saw a car with the license plate ARW 357. Can you imagine? Of all the millions of license plates in this state, what was the chance I would see that particular one? Amazing!

Richard P. Feynman

The creative principle resides in mathematics. In a certain sense, therefore, I hold it true that pure thought can grasp reality, just as the ancients dreamed.

Albert Einstein

Diogenes, with his lantern, might have found one honest man because:

In Art the mass of people no longer seek consolation and exultation, but those who are refined, rich, unoccupied... seek what is new, strange, original, extravagant, scandalous. I myself, since Cubism and before, have satisfied these masters and critics with all the changing oddities which have passed through my head, and the less they understood me, the more they admired me. By amusing myself with these games, with all the absurdities, puzzles. rebuses, arabesques, I became famous and that very quickly. And fame for a painter means sales, gains, fortunes, riches. And today, as you know, I am rich. But when I am alone with myself, I have not the courage to think of myself as an artist in the great and ancient sense of the term. Giotto, Titian, Rembrant were great painters. I am only a public entertainer who has understood his times and exploited them as best I could... Mine is a bitter confession, more painful than it may appear, but it has the merit of being sincere.

Pablo Picasso

CHAPTER 5

Applicable Life Distributions

Often a number of parametric distributions can be used to summarize a given sample of life-length data. Sometimes several of them can do it quite well. For example, if we take the Data-Set VII in Chapter 9 (101 observations of the fatigue-life of aluminum coupons) we find there are several unimodal, skewed to the left, two-parameter life distributions that will fit it adequately *in the region of central tendency.* These include the Galton, Weibull, Gamma, and fatigue-life distributions; certainly there are others. How does one decide which of these distributions is most appropriate? In certain instances it makes little difference which of these families of distributions is adopted for use. But if the life of airframe components, made of the same material as that tested, must be predicted under many different loading conditions, all at some fraction of the maximum stress applied during the test, great differences arise among the families in their realistic predictive capability when the service-life is extrapolated from test data.

Obtaining fatigue-life data at unrealistically high stress levels is necessitated by having to complete the testing within a small fraction of the design life. After all, time is money. This is called an *accelerated test* since the stress level is beyond that encountered in service. What is desired is a method to calculate a safe-life for critical components when the maximum stress in service is, say, one-hundredth of that imposed in the test. That is, we must have a statistical model in which the parameters of the life distribution are constructs of the physical factors, such as the stress regime and the type of material (both of which are known to be of primary importance) so that if these physical factors are changed the appropriate modifications to the distribution of service life are possible, with valid predictions over the range of applicable service-life conditions. This is especially true whenever public health and safety are at risk.

5.1. The Gaussian or Normal Distribution

Under what conditions should the normal distribution be used? It is applied so universally and so uncritically that, simultaneously, it is the most used, and misused, distribution in statistics. The Central Limit Theorem (the limit theorem which is central to so much of statistical theory) is given by the classical Lindeberg–Feller normal convergence criterion.

Theorem 1. *If $X_i \sim F_i$, with mean μ_i and finite variance σ_i^2, are independent for $i = 1, 2, \ldots$, then setting*

$$S_n = \sum_{i=1}^{n}(X_i - \mu_i) \quad \textit{and} \quad s_n^2 = \sum_{i=1}^{n} \sigma_i^2,$$

we have, as $n \longrightarrow \infty$,

$$\frac{S_n}{s_n} \rightsquigarrow N(0, 1) \quad \textit{and} \quad \max_{i \le n} \frac{\sigma_i}{s_n} \to 0$$

if and only if, for every $\epsilon > 0$,

$$g_n(\epsilon) = \frac{1}{s_n^2} \sum_{i=1}^{n} \int_{|x-\mu_i| \ge \epsilon s_n} x^2 \, dF_i(x) \to 0.$$

Other versions suppress the assumption of independence between summands substituting concepts such as the *uniform asympotic negligibility*, by Loéve [62, 1955], but they, being asymptotic in nature, are difficult to verify in practice.

Thus, in any situation where the rv of interest is thought to be the result of a number of factors, no one of which is of predominant importance, but all of which are additive in contribution then the initial choice for a distribution should, in the absence of any contrary information, likely be Gaussian.

There is a belief, sometimes expressed by those using large statistical analysis software-packages, that if one's data do not fit the analytical model, then the data should merely be transformed so that they do fit. This is sometimes supported by quoting the

Lemma 1. *If $X \sim F$ and $Y \sim G$ are both continuous variates, then there exists a transformation ψ such that $X \sim \psi(Y)$.*

Thus, in theory, all iid data sets from a continuous distribution can be transformed to normality. However, if one knew the transformation by which to accomplish this exactly one would not need the data in the first place. On the other hand, if one examines the nature of the mechanism (process) by which the data are generated one can often, based on scientific understanding, derive a distribution which will behave in closer accordance with the phenomenon generating the data than one can find by merely fitting constants, using statistical estimation procedures, to an *ad hoc* distribution. Moreover, the derivation of such a distribution, in accord with the underlying physical principles involving known concepts, lends credibility to any extrapolation which may be necessary for applications under conditions different than those first tested.

One such general argument was presented in Cramér's classic text. [28, 1951].

> Suppose that n random additive (independent) impulses are recieved, say $X_1, \ldots, X_n$. Let Y_n denote the magnitude of the response after n impulses. If we assume that the incremental change in the response is proportional to the size of the impulse and

inversly proportional to some function of the state of the response (a reasonable assumption in many biological situations) then, for some unknown constant c, we have the change in the response expressed by $\Delta Y_n = c \cdot X_n/q(Y_{n-1})$. We then may write

$$c\sum_{i=1}^{n} X_i = \sum_{i=1}^{n} \Delta Y_i \cdot q(Y_{i-1}) \approx \int_0^{Y_n} q(y)\,dy := Q(Y_n),$$

where the transformation Q is the integral. Then the distribution of $Q(Y_n)$ tends to normality as $n \to \infty$.

Thus, we have a legitimate derivation of a transformation of the variate Y_n, namely $Q(Y_n)$, which will be approximately (at least asymptotically) normal.

5.2. Epstein's Distribution

The Epstein distribution has the "lack of memory" property, namely, when $T \sim \mathrm{Ep}(\lambda)$, then

$$\Pr[T > t + x | T > t] = \Pr[T > x] \quad \text{for all } t, x \geq 0. \tag{5.1}$$

What this means, when T denotes the life of some component, is that the distribution of life remaining in a component in use, of any age, is the same as a new one! Such a component may fail but it *does not wear out!* Examples of components the life of which may be described by the Epstein law are electrical fuses, jeweled bearings, liquid crystals, transistors, and most importantly, integrated circuits; all may fail from external causes but not from aging.

Are there other distributions with this memory-less property? No, Epstein's law is the unique distribution with the non-aging property! Let us define the hazard and set $H(t) = -\ln\{\Pr[T > t]\}$. From eqn (5.1) it follows that

$$\Pr[T > t + x] = \Pr[T > x]\Pr[T > t],$$

which is equivalent with the equation $H(t + x) = H(t) + H(x)$. But this equation is known (under the mild condition of measurability) to have a unique nontrivial solution, namely, $H(t) = at$ for some constant a.

We can show this result from an elementary probability argument as follows: Assume $T \sim G$. Then the non-aging property can be written as $\bar{G}(t + x) = \bar{G}(t)\bar{G}(x)$ for all $t, x \geq 0$. Let $t = x$ and this becomes $\bar{G}(2t) = [\bar{G}(t)]^2$. Repeating this argument we obtain $\bar{G}(nt) = [\bar{G}(t)]^n$, which we may rewrite as $\bar{G}(t) = [\bar{G}(t/n)]^n$. Since we must have $0 \leq \bar{G}(1) \leq 1$ we can set $\bar{G}(1) = e^{-\lambda}$ for some $\lambda \geq 0$. We now must consider two cases.

Case I: If $\bar{G}(1) = 1$, then from the equation above $\bar{G}(n) = 1$ as $n \to \infty$ and so $\bar{G}(\infty) \neq 0$. This is a contradiction.

Case II: If $\bar{G}(1) = 0$, then from the other equation $\bar{G}(1/n) = 0$ as $n \to \infty$ and therefore $\bar{G}(0) \neq 1$. Another contradiction.

So we must have $\bar{G}(1/m) = e^{-\lambda/m}$ and $\bar{G}(n/m) = e^{-n\lambda/m}$ for any positive integers n, m. Thus we have for some $\lambda > 0$,

$$\bar{G}(y) = e^{-\lambda y} \quad \text{for all rational} \quad y > 0.$$

By continuity this is true for all $y \geq 0$ and hence

$$G(y) = 1 - e^{-\lambda y} \quad \text{for all} \quad y \geq 0.$$

5.2.1. The Erlang-k Distribution

An engineer named Erlang at the Bell Telephone Laboratories, early in the twentieth century, found that the length of telephone conversations seemed to be memory-less. (Apparently the conversation continues until the "accident" occurs when neither party can think of anything to add worthy of response to the latest sentence.) The question then is what is the length of k such calls?

Theorem 2. *Let $Y_i, \ldots, Y_k$ be iid with $Y_1 \sim \mathrm{Ep}(\lambda)$ then $Z = \sum_{i=1}^{k} Y_i$ has the* Erlang-*k distribution with density*

$$g_k(z) = \frac{\lambda^k z^{k-1}}{(k-1)!} e^{-\lambda z} \quad \textit{for} \;\; z > 0 \; \textit{and} \;\; k = 1, 2, \ldots. \tag{5.2}$$

Proof. The proof is by induction on k. For $k = 1$ we have $g_1(z) = \lambda e^{-\lambda z}$, which is an exponential distribution. Then assume eqn (5.2) is true for $k \in \aleph$. Then we must show the representation to be true for $k + 1$. We have by the convolution formula

$$\begin{aligned} g_{k+1}(z) &= \int_0^z g_k(z-y)\lambda e^{-\lambda y}\, dy, \\ &= \int_0^z \frac{\lambda^k (z-y)^{k-1}}{(k-1)!} e^{-\lambda(z-y)} \lambda e^{-\lambda y}\, dy, \\ &= \frac{\lambda^{k+1} e^{-\lambda z}}{(k-1)!} \int_0^z (z-y)^{k-1}\, dy. \end{aligned}$$

Hence by induction the proof is complete. □

5.3. The Galton and Fatigue-Life Distributions

5.3.1. The Log-Normal Distribution

Under what conditions do we use the log-normal distribution? Principally, whenever we have an rv that is the result of a large number of (nonnegative) multiplicative impulses, say $X_1, X_2, \ldots$. Under any circumstances, for which the central limit theorem holds for $Y_i = \ln X_i$ for $i = 1, 2 \ldots$, then there exist constants a_n and b_n such that

$$\prod_{i=1}^{n} \left(\frac{X_i}{b_n}\right)^{a_n} \rightsquigarrow \text{log-Normality}, \tag{5.3}$$

so the Galton distribution of eqn (2.31) would apply.

5.3.2. The Fatigue-Life Distribution

Let $X_i > 0$ with $X_i \sim F$ be iid variates for $i = 1, 2, \ldots$ with

$$EX_i = \mu, \quad \text{Var}(X_i) = \sigma^2 \quad \text{for } i = 1, 2 \ldots.$$

Here X_i denotes the (random, nonnegative) incremental damage sustained by a component during its ith duty cycle and $S_n = \sum_{i=1}^{n} X_i$ denotes the cumulative damage after n cycles. Assume that for failure to occur, the cumulative damage must exceed or equal a critical level, say ω. Let N_ω be the smallest (random) number of duty cycles at which cumulative damage first exceeds that critical level, i.e., it is the waiting time, in cycles, until failure. The failure event is

$$[N_\omega = n] = [S_{n-1} < \omega, S_n \geq \omega]$$

and by duality,

$$\Pr[N_\omega \leq n] = \Pr[S_n \geq \omega] = \Pr\left[\frac{S_n - n\mu}{\sqrt{n}\sigma} \geq \frac{\omega - n\mu}{\sqrt{n}\sigma}\right].$$

If we set $Z_n = (n\mu - S_n)/\sqrt{n}\sigma$ then we have

$$\begin{aligned}\Pr[N_\omega \leq n] &= \Pr\left[Z_n \leq \left(\frac{\sqrt{n}\mu}{\sigma} - \frac{\omega}{\sqrt{n}\sigma}\right)\right] \\ &= \Pr\left[Z_n \leq \frac{1}{\alpha}\left(\sqrt{\frac{n}{\beta}} - \sqrt{\frac{\beta}{n}}\right)\right] = \Pr[Z_n \leq \frac{1}{\alpha}\xi(n/\beta)],\end{aligned}$$

where we define

$$\alpha = \frac{\sigma}{\sqrt{\mu\omega}}, \quad \text{and} \quad \beta = \frac{\omega}{\mu}, \tag{5.4}$$

and the function $\xi(x) = \sqrt{x} - \frac{1}{\sqrt{x}}$ for $x > 0$, was given in eqn (2.32). Intuitively Z_n will, by the CLT, converge to normality relatively quickly, say before n reaches 10^2 cycles, but $\beta = \omega/\mu$, the nominal number of cycles to failure will in practice be of the order of 10^6 cycles. Hence we have given for all values of n of practical interest, a heuristic derivation of the distribution used in the calculation of metal fatigue [18, 1969].

$$\Pr[N_\omega \leq n] \doteq \Phi[\frac{1}{\alpha}\xi(n/\beta)].$$

Consider the limiting condition when the expected incremental damage per cycle becomes vanishingly small while the critical level of cumulative damage becomes very large. We can realize this by replacing μ by $\mu/\sqrt{m}$ and ω by $\omega\sqrt{m}$, for m large; then in eqn (5.4) α remains unchanged while β becomes βm.

Now if we fix the ratio $t = n/m$ and denote by $[y]$ the greatest integer not exceeding $y > 0$, we obtain asas

$$\Pr[N_{\omega\sqrt{m}} < tm] = \Pr[Z_{[tm]} \leq \xi(t/\beta)/\alpha].$$

Thus, by the central limit theorem, with $t = n/m$ fixed at any rational number

$$\lim_{n,m\to\infty} \Pr[N_{\omega\sqrt{m}}/m < t] = \Phi[\xi(t/\beta)/\alpha].$$

Thus, if we pass to the continuous analogue and count time in the same units as β we say, $T \sim \mathcal{FL}(\alpha, \beta)$ for some $\alpha, \beta > 0$ iff, for ξ as given in eqn (2.32),

$$\Pr[T \leq t] = \Phi\left[\frac{1}{\alpha}\xi\left(\frac{t}{\beta}\right)\right] \quad \text{for} \quad t > 0.$$

5.4. Discovery and Rediscovery

Shortly after the turn of the 20th century, Erwin Schrödinger, in his derivation of the distribution of the first passage time for a particle moving along a line subject to Brownian motion, assumed the particle moves with an expected velocity of v per unit time, so if time $t > 0$ is fixed, the random distance D over which the particle travels is a normal random variable with pdf

$$f_D(d;t) = \frac{1}{\sqrt{2\pi\delta t}} d \exp[-(d - vt)^2/2\delta t] \quad \text{for } d \in \Re, \tag{5.5}$$

where δ here is the *diffusion-rate* constant. (Note a negative distance traveled here means travel in the reverse direction.)

From this Schrödinger showed that the (random) time T, to cover a fixed distance d, has the density

$$f_T(t;d) = \frac{d}{\sqrt{2\pi\delta t^3}} \exp[-(d - vt)^2/2\delta t] \quad \text{for } t > 0. \tag{5.6}$$

During WWII, Abraham Wald studied the distribution of the random number of independent stochastic summands, which may assume both positive and negative values, to first exceed a fixed bound ω. Consider the sequence $X_1, X_2, \ldots$ of iid variates with $E[X] = \mu > 0$ and $\mathrm{Var}(X) = \sigma^2 > 0$; what is the limiting distribution of the random variable N_ω, when $\omega > 0$ is defined by

$$[N_\omega = n] = [S_1 < \omega, S_2 < \omega, \ldots, S_{n-1} < \omega, S_n \geq \omega]?$$

NB that X_i need not be non-negative and so the stopping event is different than the preceeding case for non-negative increments. Wald, *loc. cit.*, proved his lemma, namely $\mathrm{E}N_\omega \cdot \mu = \omega$ and that

$$\lim_{EN_\omega \to \infty} \Pr[N_\omega \omega \le y\mu] = F_Y(y),$$

where

$$F'_Y(y) = f_Y(y) = \sqrt{\frac{\nu}{2\pi}} e^{\nu} y^{-3/2} \exp[-\frac{\nu}{2}(y + y^{-1})], \qquad \text{for } y > 0,$$

with

$$\nu = \omega\mu/\sigma^2.$$

Our standard notation for the *Wald distribution*; is $Y \sim \mathcal{W}(1/\sqrt{\nu}, 1)$.

Exercise Set 5.A

1. If $X \sim F$ and $Y \sim G$ with F and G continuous, known distributions, what is the transformation ψ such that $\psi(X) \sim Y$? (Hint: use the probability integral transform.) What are the difficulties of transforming a data-set so that it becomes an iid exponential dataset to which the simple techniques of the exponential can be applied?
2. Show that the parameterization of Wald's distribution can be transformed to those of Schrödinger and Tweedie.
3. Show that the cumulant generating function, $C(t)$, of the inverse-Gaussian distribution, as given in eqn (2.36), is

   $$C(t) = \frac{\lambda}{\mu}\left[1 - \sqrt{1 - 2\mu^2 t/\lambda}\right],$$

   and use it to compute the mean and variance.
4. Use the cumulant generating function derived above to show that the distribution of $\bar{X}_n$, the sample mean of n iid observations from an inverse-Gaussian distribution, has the inverse-Gaussian distribution and find its mean and variance.
5. Note eqn (5.5) can be obtained by a single multiplication, namely, $df_D(d;t) = tf_T(t;d)$, from eqn (5.6). Why can this not be deduced from the simple relation $d = vt$?
6. Using the notation of eqn (4.14), prove that

   $$\frac{N_\omega - (\omega/\mu)}{\sigma\sqrt{\omega}\mu^{-3/2}} = \frac{N_\omega - \beta}{\alpha\beta} \rightsquigarrow Z \text{ as } \omega \to \infty.$$

   This result was called a central limit theorem for renewal theory by Billingsley, [17].
7. Give an interpretation of the ratio $N_{[\omega\sqrt{m}]}/m$ used in the derivation of the $\mathcal{F}L$-distribution.

5.5. Extreme Value Theory and Association

5.5.1. Gumbel's Theory

Let $T_i \sim F$ be iid for $i = 1, 2, \ldots, n$ and set $T_{1,n} = \min(T_1, \ldots, T_n)$. Then

$$\Pr[T_{1,n} > x] = [\bar{F}(x)]^n \quad \text{for any} \quad n \in \aleph.$$

Can we find two sequences of constants (a_n, b_n) by which to standardize $T_{1,n}$ so that $(T_{1,n} - b_n)/a_n$ has a limiting distribution as $n \to \infty$? Now

$$\Pr\left[\frac{T_{1,n} - b_n}{a_n} > x\right] = \Pr[T_{1,n} > a_n x + b_n] = \left[\bar{F}(a_n x + b_n)\right]^n.$$

We propose a general method of finding a minimum extreme-value distribution by using the transformed variate $U_n = nF(T_{1,n})$. We note that

$$\Pr[U_n > y] = \Pr[nF(T_{1,n}) > y] = \Pr[T_{1,n} > F^{-1}(y/n)] = \{\bar{F}[F^{-1}(y/n)]\}^n.$$

If the distribution F can be easily inverted, it may be possible to find a_n and b_n and thus the limiting distribution as $n \to \infty$.

Example 1. If T is uniform (0, 1) then $F(x) = x$ for $0 < x < 1$ and

$$\Pr[U_n > y] = \left(1 - \frac{y}{n}\right)^n \to e^{-y}.$$

Here we chose $a_n = 1/n$ and $b_n = 0$.

Example 2. Let $\bar{F}(x) = e^{-x^\alpha}$ for $x > 0; \alpha > 0$. Thus we see

$$\bar{F}^n(a_n x + b_n) = \exp[-n(a_n x + b_n)^\alpha] \equiv e^{-x^\alpha} \quad \text{for all} \quad n \in \aleph,$$

by choosing $a_n = n^{-1/\alpha}$, $b_n = 0$ for $n = 1, 2, \ldots$.

We now state two important results of Gumbel, see [42].

Theorem 3. *If $T_i \sim F$ are iid for $i = 1, 2, \ldots$ and we set $T_{1,n} = \min(T_1, \ldots, T_n)$, then on $\Re_+$, $\Re_-$ and $\Re$ the three possible limiting distributions, in standard form, for $T_{1,n}$ as $n \to \infty$ are*

$$F_1(x) = 1 - e^{-x^\alpha} \quad \textit{for} \quad x > 0; \alpha > 0. \tag{5.7}$$

and its reflection about zero, viz.,

$$F_2(x) = e^{-(-x)^\alpha} \quad \textit{for} \quad x < 0; \alpha < 0, \tag{5.8}$$

and

$$F_3(x) = 1 - e^{-e^x} \quad \textit{for} \quad -\infty < x < \infty. \tag{5.9}$$

The distribution F_1 is now called a *Weibull distribution.* Weibull himself called it "Gumbel's (first form of the) asymptotic distribution of extreme smallest values."

The form F_2 is of less interest in reliability but it is called a *negative Weibull*. The third, F_3, is now called *Gumbel's distribution of minima.*

Theorem 4. *If $T_i \sim F$ are iid for $i = 1, 2, \ldots$ and we set $T_{n,n} = \max(T_1, \ldots, T_n)$ then on $\Re_+$, $\Re_-$ and $\Re$ the three possible limiting distributions, in standard form, for $T_{n,n}$ as $n \to \infty$ are*

$$G_1(x) = e^{-(x)^{-\alpha}} \quad \textit{for } x > 0; \alpha > 0, \tag{5.10}$$

$$G_2(x) = e^{-(-x)^{\alpha}} \quad \textit{for } x < 0; \alpha > 0, \tag{5.11}$$

$$G_3(x) = e^{-e^{-x}} \quad \textit{for } -\infty < x < \infty. \tag{5.12}$$

Today G_3 is called *Gumbel's law of maxima.* Gumbel's laws have been used successfully to describe and predict extreme loads of all kinds. Gumbel's maximum describes highest flood levels, highest temperatures and greatest stress or strain, while Gumbel's minimum describes the lowest temperatures, pressures, rainfall, snowpack, etc. However, *the necessary assumption is that the basic variates generating extremes were iid so when this assumption fails it is not surprising that certain predictions fail, as they might under global change.*

Of course, these results can often be used to predict the extreme load on structures and systems, i.e., the maximum gust, maximum snowload, maximum traffic density, maximum powerdraw, etc. The most successful applications are made when these distributions are parameterized in such a way that they can be easily estimated or the parameters related to determinable physical values. One says that a rv $T \sim G$ has Gumbel's maximum distribution, and we write $T \sim \mathcal{G}\max(\alpha, \beta)$ whenever

$$G(t) = \exp[-e^{-\alpha(t-\beta)}] \quad \text{for } t \in \Re; \alpha > 0, \beta \in \Re. \tag{5.13}$$

In this case we have

$$ET = \beta + \frac{\gamma}{\alpha}, \quad \mathrm{Var}(T) = \frac{\pi^2}{6}. \tag{5.14}$$

Here the third constant, after α, β, is $\gamma = .57721\,56649\,01532\,86061 \cdots$ which is called the *Euler-Mascheroni* constant, and its algebraic nature is unknown. The nature and value of π are well known.

A rv $T \sim G$ has Gumbel's minimum distribution, and we write $T \sim \mathcal{G}\min(\alpha, \beta)$ whenever the survival distribution has the form

$$\bar{G}(t) = \exp[-e^{\alpha(t-\beta)}] \quad \text{for } t \in \Re; \alpha > 0, \beta \in \Re. \tag{5.15}$$

Correspondingly, in this case we have

$$ET = \beta - \frac{\gamma}{\alpha}, \quad \mathrm{Var}(T) = \frac{\pi^2}{6}. \tag{5.16}$$

Again, γ denotes the Euler-Mascheroni constant.

5.5.2. Maximum Loads and Association

We shall now consider the asymptotic distribution of maxima and its applications. But we shall leave the corresponding and analgous study of minima for the student. In order to fix ideas let us consider the average daily discharge, in appropriate units, of a given river at a certain location on a given day. Let us assume that the day of the year is selected at random. This is an rv with a known marginal distribution, say $X \sim G$. Let us consider $(X_1, \ldots, X_n)$ as the successive n-day record of such discharge. However this is not an iid sample.

The hourly fluctuation in flow is caused by diurnal temperature variation and recent rainfall. The *average daily discharge* is obtained by averaging over a 24 hour period. Assume that there exists an $a > 0$ such that

$$\bar{G}(x)e^{ax} \to 0 \quad \text{as } x \to \infty.$$

This means only that extremely large values of the daily flow cannot be too likely!

Let $Y = \max(X_1, \ldots, X_{365})$ be the *yearly flood.* Now reasonably the Y_i are increasing functions of such variables of weather as rainfall, spring temperature. Moreover, other factors affecting the river drainage basin such as the depth of the winter snowpack, the hectares of forest-clearcut areas, the acreage of farmland cultivated and in summerfallow, the acres of pavement and black-top plus other factors, might all well be considered.

But note well the $X_1, X_2, \ldots$ are not independent since successive days have values which are clearly related: accordingly we assume that for any integer $n \in \aleph$ the set $\mathbf{X} = (X_1, \ldots, X_n)$ is arv, i.e.,

$$\text{Cov}[\Gamma(\boldsymbol{X}), \Delta(\boldsymbol{X})] \geq 0 \quad \text{for all binary increasing functions } \Gamma, \Delta.$$

We recall Corollary 1 of Section (4.4):

Theorem 5. *If $T_1, \ldots, T_n$ are arv's (not necessarily binary), then*

$$\Pr[T_1 > t_1, \ldots, T_n > t_n] \geq \prod_{i=1}^{n} \Pr[T_i > t_i]$$

and

$$\Pr[T_1 \leq t_1, \ldots, T_n \leq t_n] \geq \prod_{i=1}^{n} \Pr[T_i \leq t_i].$$

Consider the two rvs $\hat{Y}_n = \max(\hat{X}_1, \ldots, \hat{X}_n)$ and $Y_n = \max(X_1, \ldots, X_n)$ where the $\hat{X}_i$ are iid and distributed by G and X_i are arv each with marginal distribution G, then it follows that

$$\Pr[Y_n \leq x] \geq \Pr[\hat{Y}_n \leq x] = G^n(x),$$

and so

$$\bar{F}_{Y_n}(x) \leq \bar{F}_{\hat{Y}_n}(x) = 1 - [1 - \bar{G}(x)]^n \quad \text{iff } Y_n \preceq \hat{Y}_n.$$

Recall $X \preceq Y$ is read "X is stochastically less than or equal to Y" and means

$$X \preceq Y \text{ iff } \bar{F}_X \leq \bar{F}_Y \text{ iff } F_X \geq F_Y \text{ iff } H_X \geq H_Y.$$

From the preceding results of extreme value theory, for real sequences $\{a_n\}$, $\{b_n\}$, we have that

$$\mathcal{L}\left(\frac{\hat{Y}_n - b_n}{a_n}\right) \rightsquigarrow F_G \quad \text{where } F_G(x) = \exp\{-e^{-\alpha(x-\beta)}\} \quad \text{for } x \in \Re.$$

Since $Y_n \preceq \hat{Y}_n$, the extreme value distribution is conservative. We now apply our reasoning concerning maximum floods to load fluctuations within a reoccurring series of duty cycles.

Theorem 6. *Under the assumption of association of load fluctuations within successive duty cycles, the maximum load over any given period of time will be stochastically less than under the assumption of independence of the load fluctuations when both have the same marginal distribution.*

Two questions similar to those concerning returning floods still remain:

- What is the distribution of the largest critical load fluctuation over a period of m duty cycles?
- How often does a critical load return with a stress of x or greater ?

To answer the first question we note that $\max_{i=1}^m Y_i \preceq \max_{i=1}^m \hat{Y}_i$ so that $\Pr[\max_{i=1}^m Y_i \leq x] \geq \exp\{-me^{-\alpha(x-\beta)}\}$ for $x \in \Re$ for some constants α, β.

To answer the second question we note that the (expected) number of observations in an iid sequence until a variate exceeds the value x for the first time is called the (expected) *return period.* What is the minimum n until $[Y_n > x]$ is observed for $n = 1, 2, \cdots$? Consider the indicator variables $\mathrm{I}(Y_i > x)$ for $i = 1, 2, \ldots$. They are Bernoulli rv's. Let N_x be the smallest index such that $\mathrm{I}(Y_i > x) = 1$, with $\hat{N}_x$ correspondingly defined. Then by duality

$$\Pr[N_x > n] = \Pr[Y_1 \leq x, \ldots, Y_n \leq x] \geq F^n(x) = \Pr[\hat{N}_x > n].$$

The result is again that extreme value theory is conservative since $\hat{N}_x \preceq N_x$, i.e., the return period is stochastically less under independence than under association.

NB If N is geometric, i.e., $\Pr[N = n] = pq^{n-1}$ for $n = 1, 2, \ldots$ then $\mathrm{E}N = p^{-1}$ and $\mathrm{Var}(N) = qp^{-2}$, so that it follows

$$\Pr[\hat{N}_x = n] = \bar{F}(x) \cdot [F(x)]^{n-1} \quad \text{for } n = 1, 2, \ldots$$

and so we see $\mathrm{E}\hat{N}_x = 1/\bar{F}(x) \leq \mathrm{E}N_x$.

Thus *the expected return for a period for an extreme load of magnitude exceeding x is* $\tau(x) = 1/\bar{F}(x)$. How could one determine the variance of the return period of a flood of magnitude exceeding x?

In practice, the yearly maximum meteorological conditions, earthquake, snow load, rainfall, or flood are often taken to have a *Gumbel maximum law;* an assumption not contradicted by experience *as long as environmental conditions remain the same.* This has the distribution

$$F(t) = \exp\{-e^{-\alpha(t-\beta)}\} \quad \text{for } t \in \Re \;\; \alpha > 0, \beta \in \Re. \tag{5.17}$$

Theorem 7. *If F is Gumbel's maximum law as given in eqn (5.17), then the period for the expected return for a deviation larger than x is*

$$\tau(x) = 1/\bar{F}(x) \asymp e^{\alpha(x-\beta)} \quad \textit{for } x \textit{ large.} \tag{5.18}$$

Proof. We need merely note that

$$\lim_{x\to\infty} \frac{e^{-\alpha(x-\beta)}}{1 - \exp\{-e^{-\alpha(x-\beta)}\}} = \lim_{y\to\infty} \frac{y^{-1}}{1 - e^{-1/y}} = 1.$$

□

Of course, it is folly to apply this result when the underlying distribution is changing over time, for example, when the yearly run-off of a river changes with agriculture or logging practice as well as due to increased urban sprawl, with its entailed increase of paved area. This theorem does not apply to any river which floods due to drainage from such areas, under such changing conditions.

Exercise Set 5.B

1. What distribution would you initially choose in order to make a prediction of the extreme fluctuation in
   (a) winter snow load, for safety in roof construction.
   (b) lowest winter temperature, for safety in water hookup.
   (c) maximum summer temperature, for installation of either air conditioning or reflective windows.
   (d) hardest freeze-thaw cycle for determining expense of highway construction.
   (e) wind-load, for danger of tree-fall to electric power lines.
   (f) maximum daily traffic for maximum load on a bridge.
   (g) heaviest truck, exceeding legal weight-in-motion bounds.
2. What records would you request to substantiate your assumptions in Exercise 1? Where would you obtain such records?

"Acts of God" and Other Extreme Overloads

The continual reoccurrence of accidents caused by human carelessness or extreme fluctuations in weather also must be of concern. We will speak of them as "quakes" and discuss earthquakes. The maximum annual earthquake is often assumed to have a magnitude, for some $\alpha, \beta > 0$, which is given by the cdf

$$G(x) = \exp\{-\beta e^{-\alpha x}\} \quad \text{for } x \in \Re, \tag{5.19}$$

where x is in *Richter's Scale.* The most probable (modal) value is $(\ln \alpha)/\beta$.

The expected return period in years for large quakes of magnitude x in Richter's scale would be given asymptotically, from eqn (5.19), by

$$\tau(x) \asymp e^{\alpha x}/\beta.$$

In a repetition of n years the annual largest earthquake with a magnitude of at least x is expected to occur $\phi(x)$ times. We see $\phi(x) = n/\tau(x) \asymp n\beta e^{-\alpha x}$ using the binomial assumption. This is known as *Richter's equation* when it is written as

$$\ln \phi(x) \asymp \ln(n\beta) - \alpha x. \tag{5.20}$$

Another question of interest is that of finding the distribution of the maximum annual quake over a period of m years. If we assume that the yearly maximum, say X, has Gumbel's maximum distribution G as expressed in eqn (5.17), then $Y_m = \max_{i=1}^{m} X_i \sim G_m$ where by extreme-value theory we have

$$G_m(y) = [G(y)]^m = \exp\{-e^{-\alpha(y-\beta_m)}\},$$

with $\beta_m = \beta + (\ln m)/\alpha$ as the most probable value of the worst quake in m years. Correspondingly the the expected value of the worst quake in m years is $\mathrm{E}Y_m = \beta_m + (\gamma/\alpha)$ where $\gamma = .57721\cdots$ is the Euler–Mascheroni constant.

Exercise Set 5.C

1. Assume, at a given locality, that the daily increments of increased strain along an earthquake fault (caused by movement of the tectonic plates) is a rv, say X_i for day $i = 1, 2, \ldots$. A quake occurs when for the first time (the smallest value of n) we have

$$\sum_{i=1}^{n} X_i > W,$$

   where W is the critical threshhold for strain resistance within the fault.
   (a) Make reasonable distributional assumptions about W and X_i for $i = 1, 2, \ldots$.
   (b) Do you think the waiting times would form an iid sequence? Why?
   (c) What would be the family of distributions of the waiting time between quakes?
   (d) Discuss how you would relate, as far as possible, the parameters of these distributions to the physically measurable (determinable) geological parameters of the region.
   (e) What influence would data from the geological history of the region have upon the model being constructed?

2. What are the underlying assumptions when one assumes a Gumbel extreme-value distribution for maximum annual load. How does one make certain they are verifiable?

3. In addition to the impersonal acts of God there are also the intentional, personal acts of "Old Nick", in various personifications, which cause both economic loss and personal suffering. This extends from acts of terrorism to the launching of computer-code viruses. How can reliability theory play a role in such situations?

The great difference between pure mathematics and applied mathematics is the nature of the problems that arise. One famous problem in an applied mathematics textbook of the last century merely said "A pile of coal is burning."

Anon.

In the protection of the public weal and safety one must design nuclear facilities not only to withstand the anticipated loads during their duty cycles but also consider the consequences of unanticipated episodic overloads caused by operator error (ignorance) or by acts of God, both being inevitable during their service life. $\cdots$ In addition there is the security issue involving sabotage, the risk of which must also be considered.

Advisory Commission on Nuclear Safeguards
United States Nuclear Regulatory Commission

Theoretical scientists, inching away from the safe and known, skirting the point of no return, confront nature with a free invention of the intellect. They strip the discovery down and wire it into place in the form of mathematical models or other abstractions that define the perceived relation exactly. The now naked idea is scrutinized with as much coldness and outward lack of pity as the naturally warm human heart can muster. They put it to use, devising experiments or field observations to test its claims. By the rules of scientific procedure it is then either discarded or temporarily sustained. Either way, the central theory encompassing it grows. If the abstractions survive, they generate new knowledge from which further exploratory trips of the mind can be planned. Through the repeated alternation between flights of the imagination and the accretion of hard data, a mutual agreement on the workings of the world is written, in the form of natural law.

Edward O. Wilson

CHAPTER 6

Philosophy, Science, and Sense

6.1. Likelihood without Priors

Mathematics is not one of the natural sciences; by itself it is (only) a consistent, logical construct. It is a set of definitions, syllogisms, theorems, and tautologies in which every result is true merely by definition of the terms involved. Nevertheless that "Fermat's last theorem," viz., "there is no solution in integers for $x^n + y^n = z^n$ for any $n > 2$," is true is considerably more profound (and harder to prove) in number theory than would be the assertion in sociology that "all bachelors are unmarried." While the methods of proof are different, the reason for the truth of both statements is exactly the same.

A (natural) science is defined as an ordered pair. The first component of the pair is a set of experimental laboratory procedures involving determinations of specific physical quantities, call them parameters; the second element of the pair, called the *theory*, is a set of theoretical mathematical definitions and theorems, and the parameters in these formulas are identified with those in the experimental procedures. Thus, the mathematical equations form an acceptably correct theory, when their parameters, having been empirically determined and all variables identified with measurable physical quantities, allow sufficiently accurate verification and prediction as well as scientific explanation of the phenomena.

Science deals only with ostensibly meaningful terms and concepts, those which can be verified through the senses or can be logically derived from and related to them. A statement (hypothesis) is *scientific* (otherwise perhaps worthy of poetic consideration) only if it is theoretically both *verifiable* and *falsefiable.* Verifiable means that possible experimental procedures are known whereby the statement can be demonstrated, empirically, to be true. Falsifiable means there are experiments logically possible which could refute the statement. Often these are the same. Moreover, any single discrepant outcome means the theory is false, or a mistake in experimental measurement has been made. A single concordant outcome supports a theory or shows experimental precautions were not carefully observed, such as was the case in "cold fusion." Thus, advance of science depends upon a continuing repetition of corrobarative experiment and the concurrence of predictive results.

Of course, the accuracy of all scientific predictions can be checked only within the limits of verification then possible using current technology. Hence scientific explanation becomes a series of improving approximations to reality, consistent with the theory proposed. As a consequence, scientific hypotheses are not so frequently overthrown (refuted) as they are improved and generalized. Einstein's theory did not disprove Newtonian mechanics as much as it merely showed it was not universally applicable, and it provided some limits to its applicability. Indeed, Newtonian mechanics holds neither at the atomic scale nor at the galactic scale.

Statistics, which is principally concerned with the proper treatment of data, is not a subset of mathematics but is an applied discipline. For example, the central limit theorem, as it is jocularly stated, is believed by mathematical statisticians because they think it has been empirically verified and it is accepted by experimentalists because they think it has been theoretically proved. Of course, its centrality is due to bolstering from both sides.

Assume that we have an experiment, say ς, whose outcome depends upon a state of nature, $\omega \in \Omega$. Unfortunately, not all the possible complex states of nature in Ω can be determined any more than can the exact particular ω resulting in a given measurement. Thus, we have $X_1 = \varsigma(\omega_1), \cdots, X_n = \varsigma(\omega_n)$ as the successive, perhaps almost deterministic, outcomes of n experimental trials, except the states $\omega_1, \ldots, \omega_n$ cannot, perhaps, be precisely or completely determined.

A chaotic function is one for which its functional values may be arbitrarily far apart even though the corresponding arguments may be arbitrarily close together. An experiment ς is often chaotic in exactly this sense and in such circumstances we regard its outcome as being a *random variable.* Consider the outcome "heads" or "tails" of a tossed coin which is the result of its initial rotational velocity, initial momentum, trajectory, and the elasticity of the impact surface. Thus, if we regard (ω_1, ω_2) as points in the (rotational velocity $\times$ momentum)-plane, we see from a generic depiction of Joseph Keller's fair coin rotating about a diameter, Figure 6.1, that even though initial rotational velocity and momentum of a toss are very close, the outcome can be different.

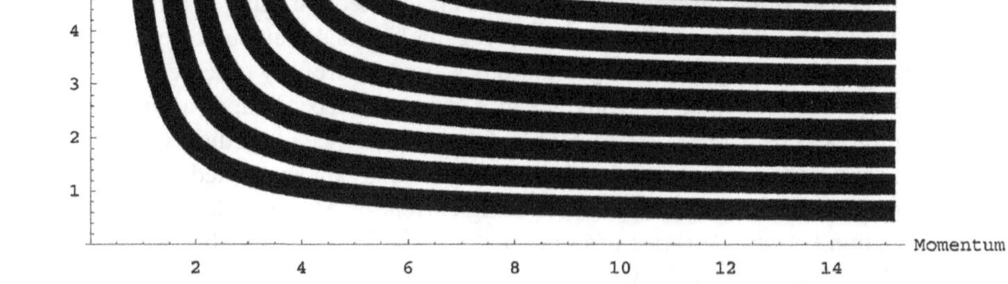

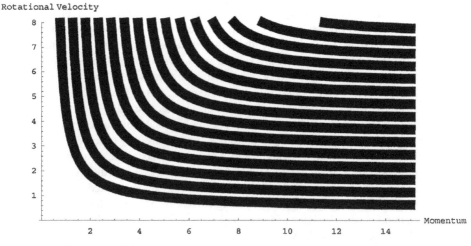

Figure 6.1. Outcome as a function of momentum and rotational velocity.

A good distributional model says that a random variable $X = \varsigma(\omega)$ has a distribution in a specified family, $X \sim F(\cdot, \theta)$, for some $\theta \in \Theta$ whenever the parameter space Θ can be related in some cognitively meaningful (scientific) fashion to our knowledge of the states of nature in Ω.

EXAMPLE 1. The fatigue life of metal, i.e., when failure occurs between 10^2 and 10^6 cycles, has been empirically found, by metallurgists, to be well described by a distribution with two parameters, say (α, β). The shape parameter α is mainly controlled by the type of metal since it is influenced by the ductility, fracture toughness, the heat treatment, and direction of rolling. The scale parameter β is related to the "melt of the metal" (quality control during forming) and the strains imposed by the duty cycle through stress intensity factors influenced by the geometry of the structural component.

Some commonly accepted statistical methods ignore scientific knowledge about the phenomenon in question when estimates of the parameters of service life are made by utilizing only the times of failure of a small set of nominally identical components and disregarding other factors engineers know are significant. For example, if it is known from metallurgy that the coefficient of variation in the distribution of metallic fatigue-life depends upon which metal is being used, should not the proper information of the type of metal be included in any estimate of the shape parameter?

But the question is how should such "engineering knowledge" be used? One answer is to form an *a priori* (prior) density on the parameter space Θ which is then combined with life-data using Bayes' theorem, to produce new parameter distributions *a posteriori* (posterior) which produce more "precise" estimates (usually smaller variance). Justification for this procedure is precisely what is at issue here. The validity of the results of a Bayesian analysis depends both on the model and the prior. Since this prior density is made-up from *personal opinion* (called "engineering judgment" though often independent of it), how can one determine even if it is ostensibly meaningful? How can one distinguish, in Bayesian predictions made using a posterior distribution, between the incorporation of valid or specious data (or even pure data manipulation) except to wait for such occurrences as managerial opinions which occasioned 'The Challenger Disaster'? One must regard such managerial decisions, which disregard engineering data and caution, as having been influenced by personal *prior* information. If all relevant previous test-data have been retained in a data bank, then past experience is available for utilization in combined likelihoods for any given situation. Thus, all former testing could theoretically be utilized (perhaps requiring extensive computation) along with proper knowledge of the effect on the parameter space of life under different test and service conditions.

In the case presented one must make use of relationships between the statistical parameters inferred from their metallurgy.

NB to regard the pair $\theta|X$ as a conditional random variable is a markedly different assumption than calculating the distribution of the estimate $\hat{\theta}(\boldsymbol{X})$ based on data $\boldsymbol{X}$. One might compare the engineering example with the situation in genetics where the influence of inheritance is known. For example, if X is the height of the son and Y, Z are the heights, respectively, of the father and mother, then $X|(Y, Z)$ makes objective sense since the inheritance mechanism is well understood and it provides a better predictive model, with smaller variance, than would just X.

6.2. Likelihood for Complete Samples

Suppose that $\boldsymbol{X} = (X_1, \ldots, X_n)$ represents a sample with a joint density from a parametric family:

$$f_{\mathrm{X}}(\boldsymbol{x}; \theta) \quad \boldsymbol{x} \in \mathcal{X} \subseteq \Re^n \quad \text{for some } \theta \in \Omega.$$

We want to find the value $\hat{\theta} \in \Omega$ which makes most probable *a posteriori* our having observed that set of data that has been observed. The procedure is: given any observation vector $\boldsymbol{x}$, find the value $\hat{\theta}(\boldsymbol{x})$ which maximizes the likelihood $\mathcal{L}(\theta|\boldsymbol{x})$ [or equivalently the log-likelihood $L(\theta|\boldsymbol{x})$], which we define, respectively, by

$$\mathcal{L}(\theta|\boldsymbol{x}) = f_{\mathrm{X}}(\boldsymbol{x}; \theta) \qquad \text{and} \qquad L(\theta|\boldsymbol{x}) = \ln f_{\mathrm{X}}(\boldsymbol{x}; \theta) \quad \text{for } \theta \in \Omega.$$

We define the mle (maximum likelihood estimate), based on $\boldsymbol{x}$, as the function on $\mathcal{X}$, call it $\hat{\theta}(\boldsymbol{x})$, and defined by

$$L(\hat{\theta}(\boldsymbol{x})|\boldsymbol{x}) = \sup_{\theta \in \Omega} L(\theta|\boldsymbol{x}).$$

Note that the function $\hat{\theta}(\cdot)$ may not be a single-valued function on $\mathcal{X}$, i.e., the mle value $\hat{\theta}(\boldsymbol{x})$ may not always be unique for every $\boldsymbol{x}$.

Example 2. Consider an iid sample from the Wald distribution with parameters α and β unknown, see eqn (2.39) in a preceding section, the joint density for a complete sample is

$$g(\boldsymbol{y}; \alpha, \beta) = \prod_{i=1}^{n} \left[\frac{\sqrt{\beta}}{\alpha} y_i^{-3/2} \varphi[\xi(y_i/\beta)/\alpha] \right],$$

from which it follows that the log-likelihood is, except for constants independent of the parameters,

$$L(\alpha, \beta|\boldsymbol{x}) \cong \sum_{i=1}^{n} \left[\frac{1}{2} \ln \beta - \ln \alpha - \frac{1}{2\alpha^2} \xi^2(y_i/\beta) \right].$$

From this asas we have the stationary equations

$$\frac{\partial L}{\partial \alpha} = \sum_{i=1}^{n} \left[-\frac{1}{\alpha} + \frac{1}{\alpha^3} \xi^2(y_i/\beta) \right] = 0,$$

and

$$\frac{\partial L}{\partial \beta} = \sum_{i=1}^{n} \left[\frac{1}{2\beta} - \frac{2}{2\alpha^2\beta} \xi(\beta/y_i)\xi'(\beta/y_i)\frac{\beta}{y_i} \right] = 0.$$

From these two equations in two unknowns we obtain, using the relation $2x\xi(x)\xi'(x) = x - x^{-1}$,

$$\alpha^2 = \frac{1}{n}\sum_{i=1}^{n} \xi^2(y_i/\beta) \quad \text{and} \quad \alpha^2 = \frac{1}{n}\sum_{i=1}^{n} \left[\frac{\beta}{y_i} - \frac{y_i}{\beta} \right],$$

which, after equating and solving, yields the mles, respectively,

$$\hat{\beta} = \bar{y}_+; \quad \text{and} \quad \hat{\alpha}^2 = \bar{y}_+ \left(\frac{1}{\bar{y}_-} - \frac{1}{\bar{y}_+} \right), \quad \text{where} \quad \bar{y}_\pm = \left(\frac{1}{n}\sum_{i=1}^{n} y_i^{\pm 1} \right)^{\pm 1}. \tag{6.1}$$

EXAMPLE 3. Let X_i be iid $\mathcal{LN}(\mu, \sigma^2)$ for $i = 1, \ldots, n$. Then the log-likelihood is

$$L(\mu, \sigma | \mathbf{x}) \cong \sum_{i=1}^{n} \left[-\ln\sigma - \frac{1}{2\sigma^2}(\ln x_i - \mu)^2 \right],$$

from which, by setting $\partial L/\partial\mu = 0$ and $\partial L/\partial\sigma = 0$, we find the mles to be

$$\hat{\mu} = \frac{1}{n}\sum_{i=1}^{n} \ln x_i \quad \text{and} \quad \hat{\sigma}^2 = \frac{1}{n}\sum_{i=1}^{n} (\ln x_i - \hat{\mu})^2. \tag{6.2}$$

EXAMPLE 4. We cannot count the number of bacteria in contaminated water; we can only test whether or not any are present. Let X denote the number of bacteria present in a given volume of water v and assume $X \sim \mathcal{P}(\lambda v)$, i.e.,

$$f_X(x; \lambda) = \frac{e^{-\lambda v}(\lambda v)^x}{x!} \quad \text{for } x = 0, 1, 2, \ldots; v > 0.$$

Here λ is the unknown density of bacteria per unit of water. Since we can count only the number of pure samples, we observe a set of Bernoulli variates, $Y_i = I(X_i = 0)$ for $i = 1, \ldots, n$. Using only the data $\mathbf{y} = \{y_1, \ldots, y_n\}$ we now must estimate λ. Note that $p = \Pr[Y_i = 1] = e^{-\lambda v}$ and $f(y) = p^y(1-p)^{1-y}$ for $y = 0, 1$. Hence the likelihood of λ given $\mathbf{y}$ is

$$L(\lambda | \mathbf{y}) = \sum_{i=1}^{n} [y_i \ln p + (1 - y_i)\ln(1 - p)];$$

whence by using the chain rule

$$\frac{\partial L}{\partial \lambda} = \frac{\partial L}{\partial p} \cdot \frac{\partial p}{\partial \lambda} = \sum_{i=1}^{n} \left[\frac{y_i}{p} - \frac{1 - y_i}{1 - p} \right] e^{-\lambda v}(-v) = 0,$$

we have the maximum likelihood estimators

$$\hat{p} = \bar{y} \quad \text{and therefore} \quad \hat{\lambda} = \frac{-\ln(\hat{p})}{v}. \tag{6.3}$$

In a more general case when the volume of water in the sample is not the same from observation to observation we have both volumes v_i and indicators of purity y_i for $i = 1, \ldots, n$ from which we must estimate λ. We find asas that $\hat{\lambda}$ is the solution of the equation

$$g(\lambda) = \sum_{i=1}^{n} v_i, \quad \text{where} \quad g(\lambda) = \sum_{i=1}^{n} \frac{(1 - y_i)v_i}{1 - e^{-\lambda v_i}}.$$

The solution of this equation, namely $\hat{\lambda}$, requires further numerical techniques. An even more realistic model would be to make λ a function of both the depth of collected sample volume and its distance from pollution sources.

Exercise Set 6.A

1. Find the maximum likelihood estimators of the specified parameter or parameters in the following distributions:
   (a) $f(x) = \theta x^{\theta-1}$ for $0 < x < 1$; $\theta > 0$.
   (b) $f(x) = \lambda e^{-\lambda(x-c)}$ for $x > c$; $\lambda > 0$, for both λ and c.
   (c) $f(x) = (\lambda/2)e^{-\lambda|x-\mu|}$ for $x \in \Re$; $\lambda > 0$, $\mu \in \Re$ for both λ and μ.
   (d) $f(n) = pq^{n-1}$ for $n \in \aleph; 0 < p < 1, q = 1 - p$.
   (e) $f(k) = \binom{n}{k} p^k q^{n-k}$ for $k = 0, 1, \ldots, n; n \in \aleph, 0 < p < 1$.
   (f) $f(x) = \dfrac{1}{\sqrt{2\pi}\sigma t} \exp\{-\frac{1}{2}[\ln(t/\beta)/\alpha]^2\}$ for $t > 0; \alpha > 0, \beta > 0$.

6.3. Properties of the Likelihood

For an observed vector in the sample space, $\boldsymbol{x} \in \mathcal{X}$, assume there is an unknown parameter θ in the known parameter space, Θ for which the log-likelihood is defined by

$$L(\theta|\boldsymbol{x}) = \ln f_{X_1,\ldots,X_n}(x_1, \ldots, x_n|\theta) \quad \text{for } \theta \in \Theta,$$

where the sample point $\boldsymbol{x}$ is an observation from the joint density f_{X}. The mle, label it $\hat{\theta}$, is usually the solution of the system of equations $\frac{\partial L}{\partial \theta} = 0$. Only in special cases is this solution an explicit one. Often one must rely on computational methods such as Newton–Raphson iteration, repeated bi-section, or Aitken acceleration to perform the maximization.

6.3.1. The Likelihood Depends upon the Model

The probability model, that is the choice of sample space, the distribution, and parameter space, denote them by $(\mathcal{X}, F_{\mathrm{X}}, \Theta)$ must not only be in consonance with the data, i.e., cannot be contradicted by it, but also should be in accord with the state of engineering or scientific knowledge of the phenomenon, or process, which is generating the observations. In particular, the parameter space should be constrained by empirically verifiable information.

6.3.2. Relative Likelihoods Are Not Probabilities on Θ

The relative likelihood, say $\mathcal{R}(\theta|\boldsymbol{x})$, defined by

$$\mathcal{R}(\theta|\boldsymbol{x}) = \frac{\mathcal{L}(\theta|\boldsymbol{x})}{\mathcal{L}(\hat{\theta}(\boldsymbol{x})|\boldsymbol{x})} = \exp\{L(\theta|\boldsymbol{x}) - L(\hat{\theta}(\boldsymbol{x})|\boldsymbol{x})\} \quad \text{for } \theta \in \Theta \tag{6.4}$$

has the property that $0 \le \mathcal{R}(\theta|\boldsymbol{x}) \le 1$ but *it is not a probability.* If

$$\Pr_\theta[E] = \int_E f_X(\boldsymbol{x}|\theta)\, d\boldsymbol{x}$$

when $E_1, E_2 \subset \mathcal{X}$ are mutually exclusive events, then we must have for each $\theta \in \Theta$

$$\Pr_\theta[E_1 \cup E_2] = \Pr_\theta[E_1] + \Pr_\theta[E_2]$$

but if $\theta_1, \theta_2 \in \Theta$ with $\theta_1 \ne \theta_2$ then the relative likelihood of "θ_1 or θ_2" is *not defined* because it makes no sense! Subsets of θ-values are not measurable. What is the likelihood that blue is more honest than green? The honesty of colors is not quantifiable! For the similar reasons one cannot use the likelihood

$$\mathcal{L}(\mu|\boldsymbol{x}) = \left(\frac{1}{\sqrt{2\pi}}\right)^n \exp\{-\sum_{i=1}^n (x_i - \mu)^2/2\}$$

to evaluate the probability of say "$\mu < 24$."

There is no single relationship between likelihood on the parameter space and probability on the sample space! The relationship must be determined in each experimental situation by the physics/chemistry of the phenomenon being measured. A simple connection, judged by some to be omphalopsychic, can be made by assigning

$$\Pr[\theta] = \text{the personal belief, hope, or suspicion that } \theta \text{ is true.}$$

The unproven hypothesis, that "personal beliefs" or even "engineering judgment" of the appropriate values of model parameters can be manipulated using the calculus of probabilities, is what fuels *Bayesian analysis.* This method is far different from constructing mathematical models that incorporate engineering knowledge into the model of the physical relation between the test conditions and the service environment. Such a relationship is designed to help isolate the cause and determine the nature of the stochastic variation.

Those statisticians who favor Bayesian analysis often say that the choice of a parametric family to model a phenomenon is already a "semi-Bayesian choice." If so why not then go all the way and use Bayesian methods. In fact, assuming a loss function proportional to squared-error makes most subsequent statistical calculations trivial? But if a statistician has doubts about the proper choice of a distribution then (s)he does not understand well enough the physical conditions generating the distribution and any conclusion reached must be degraded by such lack of knowledge. Any inference can be no better than the knowledge of the validity of the data itself as expressed in its summary as a distribution.

Why do physicists not use the classical Maxwell–Boltzmann occupancy distribution instead of the Bose–Einstein distribution? The answer is that sufficient data have been collected to show that the latter is correct and the former is not!

6.3.3. Likelihoods Invariant under Transformations

Given the relative likelihood $\mathcal{R}(\theta|\boldsymbol{x})$ for $\theta \in \Theta$ as specified in eqn (6.4), with $\hat{\theta}(\boldsymbol{x})$ specified, then for any one-to-one transformation $\omega = g(\theta)$ from Θ to Ω, the mle of ω will be $\hat{\omega} = g(\hat{\theta})$.

There is one instance in which there is symmetry between the mle $\hat{\mu}$ and the parameter μ. Consider a normal distribution and the corresponding relative likelihood

$$\mathcal{R}(\mu|\boldsymbol{x}) = \frac{\left(\frac{1}{\sqrt{2\pi}}\right)^n exp\left[\sum_{i=1}^n (x_i - \mu)^2/2\right]}{\left(\frac{1}{\sqrt{2\pi}}\right)^n exp\left[\sum_{i=1}^n (x_i - \bar{x})^2/2\right]} = e^{-n(\bar{x}-\mu)^2/2}.$$

This accident for normality was the basis for R.A Fisher's "fiducial reasoning" ("fiducia" is the Latin word for "reliable"), but it has not proved, for other distributions, to be usable.

There is another valid use for the relative likelihood and that is the construction of *likelihood regions.* If one defines the set $\{\theta \in \Theta | \mathcal{R}(\theta|\boldsymbol{x}) \geq \gamma\}$ for γ near one, it determines a $100\gamma\%$ likelihood region for θ. Usually it is inverted by plotting $\ln \mathcal{R}(\theta|\boldsymbol{x}) = L(\theta|\boldsymbol{x}) - L(\hat{\theta}|\boldsymbol{x})$; see Figure 6.2.

6.3.4. Likelihoods on Simple Parameter Spaces

Maximum likelihood estimates may not be useful for complicated distributional models when the parameter spaces are of high dimension with many parameters to be estimated. In such circumstances the datasets would have to be very large for sufficient discrimination, and such sample sizes are not usually available in engineering testing. We may have $f(\boldsymbol{x};\theta_1) \doteq f(\boldsymbol{x};\theta_2)$ for all $\boldsymbol{x} \in \mathcal{X}$, when θ, is for different from θ_2 and the mles may not converge for any practical sample size. Essentially, a high-dimensional parameter space may provide more than one explanation for the data. This implies that statistical estimation must often be augmented by physical knowledge of what is modeled, in order to succeed!

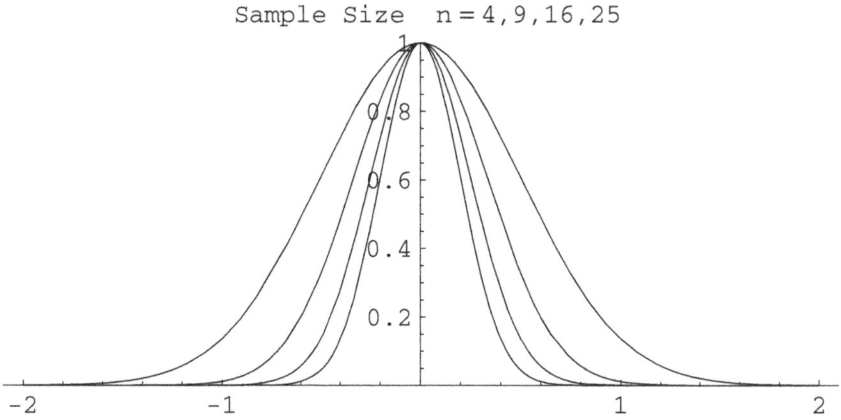

Figure 6.2. Likelihood regions for differing sample sizes.

EXAMPLE 5. Consider the gamma distribution:

$$f(x;\alpha,\beta) = \frac{x^{\alpha-1}e^{-x/\beta}}{\Gamma(\alpha)\beta^{\alpha}} \quad \text{for } x > 0; \alpha, \beta > 0.$$

Case I: If α is known, then the likelihood is

$$L(\beta|\mathbf{x}) \cong \sum_{i=1}^{n} \Big[-\frac{x_i}{\beta} - \alpha \ln \beta \Big].$$

Hence $\partial L/\partial\beta = 0$ gives $\hat{\beta} = \bar{x}/\alpha$.
Case II: If both α and β are unknown, then we find asas

$$\frac{\partial L}{\partial \alpha} = \sum_{i=1}^{n} [\ln x_i - \ln \beta - \zeta(\alpha)]$$

where the "digamma" function $\zeta(\alpha) = \Gamma'(\alpha)/\Gamma(\alpha)$ is known from classical analysis; see Abramowitz and Stegun, p. 258 [2], to be representable, for any complex variable z, as

$$\zeta(1+z) = -\gamma + \sum_{n=1}^{\infty} \frac{z}{n(n+z)} \quad \text{for } z \neq -1, -2, -3, \ldots,$$

and γ is the Euler–Mascheroni constant. Thus, to obtain $\hat{\alpha}$ we must use numerical methods to solve the single equation in α:

$$\zeta(\alpha) - \ln \alpha = \frac{1}{n} \sum_{i=1}^{n} \ln x_i - \ln(\bar{x}); \tag{6.5}$$

then we have $\hat{\beta} = \bar{x}/\hat{\alpha}$.

There are instances where the stationary equations do not apply directly to the determination of the mles.

EXAMPLE 6. Let us consider an iid sample from the uniform distribution on (a, b). Then the likelihood is

$$\mathcal{L}(a, b|\boldsymbol{x}) = \frac{1}{(b-a)^n} \prod_{i=1}^{n} 1(a < x_i < b) = \frac{n! 1(a < x_{1:n} < \cdots < x_{n:n} < b)}{(b-a)^n},$$

where $x_{1:n} < \cdots < x_{n:n}$ are the order observations. Hence we see

$$\frac{\partial \mathcal{L}}{\partial a} = \frac{n \cdot n! 1(a < x_{1:n} < \cdots < x_{n:n} < b)}{(b-a)^{n+1}} > 0 \text{ for } a < x_{1:n},$$

$$\frac{\partial \mathcal{L}}{\partial b} = \frac{-n \cdot n! 1(a < x_{1:n} < \cdots < x_{n:n} < b)}{(b-a)^{n+1}} < 0 \text{ for } b > x_{1:n}.$$

Thus, we reason that $\hat{a} = x_{1:n}$ and $\hat{b} = x_{n:n}$.

Exercise Set 6.B

1. We have acted thoughtlessly in eqn (6.5). We must check whether it is possible that $(\hat{\alpha}, \hat{\beta})$ is a saddle point when $\alpha < 1$ (in which case we would have a minimum likelihood estimator). Determine what must be done if this is the case.
2. Consider an iid sample from an $\mathcal{N}(\mu, \sigma^2)$ population where both parameters are unknown. We want the mles of the mean and the coefficient of variation, i.e., find the mles of α, β, where $\alpha = \sigma/\mu$ and $\beta = \mu$.
3. Verify that the solutions of the stationary equations with $\alpha < 1$ for the Weibull distribution are still maximum likelihood estimates.
4. Consider the problem of deciding which of the following parameterizations is the better. We have two survival distributions, which are given by

$$\bar{F}_1(t) = e^{-\alpha H(t\beta)}, \qquad \bar{F}_2(t) = e^{-\alpha H(t\beta)/\beta} \qquad \text{for } t > 0; \alpha, \beta > 0.$$

   This change of parameters may be regarded as $\alpha \mapsto \alpha/\beta$ and $\beta \mapsto \beta$. Is this transformation $1-1$? By what basis may we choose one parameterization over the other?

6.3.5. Bayes' Theorem and Its Application

All theories (in science) should refer only to quantities that can be measured.
Ernest Mach, a Founder of the Positivist School of Philosophy

There are more things in heaven and earth, Horatio, than are dreamt of in your philosophy.
Hamlet, I.v.174-5, a play by Edward de Vere, 17th Earl of Oxford

A mathematical version of the theorem the Reverend Thomas Bayes first derived in the 18th century to "prove" the existence of a God is

Theorem 1. *Let* $(\Omega, \mathcal{A}, \mathrm{P})$ *be a complete probability space. If* $\{A_i\}_{i=0}^n$ *is a* partition *of* Ω*, i.e.,*

$$A_i \in \mathcal{A} \text{ for } i = 0, \ldots, n, \quad A_i \cap A_j = \emptyset \text{ for } i \neq j, \quad \bigcup_{i=0}^n A_i = \Omega, \tag{6.6}$$

then for any given $B \in \mathcal{A}$ *the probability of* A_0 *becomes*

$$P(A_0|B) = \frac{P(A_0)P(B|A_0)}{\sum_{i=0}^n P(A_i)P(B|A_i)}.$$

This theorem has many applications, all of which are incontestable when the hypothesis is satisfied. One of the foremost is in disease diagnosis. Suppose we have a communicable disease, say D, and a diagnostic test, say T, for it. The frequency of this disease within the given population and the efficiency of diagnostic discrimination for the test, whether positive or negative, writing D' as the complement of D, viz.,

$$\mathrm{P}(D), \quad \mathrm{P}(T^+|D) \quad \text{and} \quad \mathrm{P}(T^-|D'), \tag{6.7}$$

are all known. These are the probabilities, respectively, of a person selected at random within the population having the disease, the probability of the test being positive if administered to a diseased person, and the probability of the test being negative if adminstered to an undiseased person.

It is desirable to determine the probability of a false positive. For example, if a test screening for a serious disease is given to 500 schoolchildren, what percentage of mothers can be expected to endure the needless worry and heartache when their child is wrongfully subjected to a subsequent medical procedure? Using the data in eqn (6.7) Bayes' theorem gives

$$\mathrm{P}(D'|T^+) = \frac{\mathrm{P}(D')[1 - \mathrm{P}(T^-|D)]}{\mathrm{P}(D)\mathrm{P}(T^+|D) + \mathrm{P}(D')[1 - \mathrm{P}(T^-|D')]}.$$

The problematic cases arise when Bayes' theorem is used in situations where the hypothesis of Bayes' theorem is not satisfied, e.g., there is no uncontestable algebra of sets to which conditional probability applies. This often arises when feelings about situations are treated as evidence (events). Can one calculate the probability that aliens visited Roswell, NM, from known evidence (which most investigators interpret as burned air-force test dummies); or that the "crop circles" seen in England are a sure sign of extraterrestrial attempts at communication with earth?

Another type of error arises when one seeks to evaluate the probability of a statement (event) for which no probabilty can be assigned to the evidence. Consider the argument from "Intelligent Design," namely that the interaction necessary among reflected light, the eye, and the brain, which is found in most animals is of such complexity that "Darwinian Evolution," as crude, as wasteful, and as slow as it is, *could never* have developed sight from primitive cells! Might not a "value judgment" be confounded here with a lack of comprehension of genetic heritability. Only opinion, and hence disputation, results unless agreement is reached on the probability measure and the sample space.

"Probability" and "likelihood" are precise mathematical concepts that bear about as much resemblance in common usage as the mathematical concept of "group" does to its meaning in ordinary conversation. These explicit analytic terms are often used, unsuitably, to provide a degree of versimilitude to an otherwise bald and unconvincing argument. Such usage merely denotes belief or hope, which occurs during wagers in college football rivalries or at the race track, and are often independent of evidence. Thus, they are merely "fidelious" (*fides* is Latin for faith) pronouncements. Clear thinkers and writers avoid such usage.

Exercise Set 6.C

1. Identify the measurable space $(\Omega, \mathcal{A})$ in the application of Bayes' theorem to the calculation of the percentage of false-positive tests in medicine.
   (a) Show that if the disease is rare, i.e., $\alpha = \mathrm{P}(D)$ is near zero, then the probability of a false positive is nearly certain, viz., $\mathrm{P}(D'|T^+) \to 1$ as $\alpha \to 0$.
   (b) Consider the situation when even a rare disease is communicable and one memeber of a household tests positive. Does Bayes' theorem predict the probability of a false positive for siblings? To whom would it apply?
   (c) Suppose the disease has a long period of latency after infection before the diagnostic test works. Modify Bayes' theorem to include this condition.
   (d) For some disease an inexpensive screening test T is first applied and if positive an expensive but more accurate test A is administered. Show that for a false positive, using both tests can be written

$$\frac{1}{\mathrm{P}[T^+ \cap A^+|D']} = 1 + \frac{\mathrm{P}(D')}{\mathrm{P}(D)} \cdot \frac{\mathrm{P}(T^+|D')}{\mathrm{P}(T^+|D)} \cdot \frac{\mathrm{P}(A^+|T^+ \cap D')}{\mathrm{P}(A^+|T^+ \cap D)},$$

   and discuss the practical difficulties of obtaining the terms in the final ratio unless one assumes $A \perp T$. What error is made if independence is falsely assumed?

2. Try to formulate a sample space, σ-algbra, and a probability measure for some small part of the intelligent-design hypothesis.

3. Discuss the scientific merit of the following procedure (which was actually employed) to ascertain the safety of nuclear power plants that after construction were found to have been inadvertantly located near or across earthquake faults. (It is called the "Delphi Method" after the Greek oracle.)

   Retain as consultants a group of noted and knowledgeable geologists and ask all of them to write down their estimates of the probabilities of an earthquake of magnitude k for $k = 3, 4, 5, 6, 7$ during the next n years for $n = 10, 20, \ldots, 50$, at each of location of the power plants in question.

   Average these "informed estimates" of the probabilities of quake magnitude into a grand set of prior probability distributions for each site and use them to calculate the probability of rupture of the containment vessel or steam pipes.

4. Would more credibility have been obtained by using historical data to estimate the return time for quakes of a given magnitude at that location, using an extreme value distribution?

5. What other procedure could the Nuclear Regulatory Commission have followed since funds had not been provided by Congress to perform survey excavations at each site of a

nuclear generating plant in order to ascertain whether fault lines were present and their depth so as to estimate the time of the most recent quakes?

6. Some people claim that it was, principally, the 1986 Chernobyl nuclear disaster (causing the evacuation of c. 100,000 people and leaving a core area unsafe for animal habitation for the next 30,000 years) that destroyed the Russian people's confidence in the Soviet Socialist Government and brought and end to the CCCP. What do you think?

6.4. Types of Censoring of Data

There are many similarities between (service) life testing and clinical trials in medicine, namely for each specimen (patient) we observe either the time of failure (death) or, alternatively, the time at which the service-life is terminated for some reason other than failure (the patient is lost to the study) and we know which of these events has occurred. When the latter occurs in engineering testing it is called a *run-out* depending upon circumstances. It is called a *censored* or *truncated* observation by the statistician.

Censoring may take place under a variety of situations. Let X_i for $i = 1, \dots, n$ be iid rvs but due to circumstances the statistician can only observe $\{X_i | a < X_i < b\}$ plus the two counts $\sum_{i=1}^{n} \mathrm{I}(X_i > b)$ and $\sum_{i=1}^{n} \mathrm{I}(X_i < a)$, where $0 < a < b$ are known values. This is called *Type I Censoring*. Of course, one may have one-sided censoring when either $a = 0$ or $b = \infty$.

If we denote the ordered variates (or order-statistics) from an iid sample as

$$X_{1:n} < X_{2:n} < \cdots < X_{n:n}$$

but due to circumstance can observe only the set $\{X_{i:n} : i \in K\}$ for some $K \subset \{1, 2, \cdots, n\}$, this is called *Type II censoring*.

Consider a random sample of identical components with life-lengths $\{X_i : i = 1, \dots, n\}$ for which there is associated by nature a set of circumstances dictated by usage, represented by the rv $(T_1, \dots, T_n)$, such that the statistician can observe for the ith component only the two variates,

$$Y_i = X_i \wedge T_i := \min(X_i, T_i) \quad \text{and} \quad Z_i = \mathrm{I}(X_i < T_i) \quad \text{for } i = 1, \dots, n.$$

This is called *random censoring (on the right)* and $T_1, \dots, T_n$ is the censoring (truncation) vector. We shall call this random censoring, on the left or right, as *Type III censoring*.

This third type of censoring is of practical use in engineering since termination of testing can occur for many different reasons; because of failure of the testing machine, error on the part of laboratory personnel, inadequate time or funds to complete the testing and run-outs. This, of course, has its counterpart in medical trials with the loss-of-contact with the treated person. NB we need neither assume that the T_i are iid nor, as indicated, that $\boldsymbol{X} \perp \boldsymbol{T}$.

6.4.1. Estimation for Type I Censoring

Suppose that a component is to be screened for early failures and the total test time for each component is set at a fixed time t. We would then observe some random number failure times and the unfailed remainder as run-outs. Thus, for some $r = 1, \ldots, n$ the failure data are given by

$$x_1 < x_2 < \cdots < x_r < t, [X_{r+1:n} > t], \ldots, [X_{n:n} > t].$$

In this case, the likelihood becomes

$$\prod_{i=1}^{r} f(x_i) \prod_{i=r+1}^{n} \bar{F}(t) = \prod_{i=1}^{r} h(x_i) \prod_{i=1}^{r} \bar{F}(x_i) \times \bar{F}(t)^{n-r}.$$

We then have the log-likelihood for *Type I censoring* given by

$$L(\theta|x_1, \ldots, x_r) = \sum_{i=1}^{r} [\ln h(x_i : \theta) - H(x_i : \theta)] - (n - r)H(t; \theta). \qquad (6.8)$$

Suppose for purposes of illustration that the population sample is Ep(λ), with a mean of $\mu = 1/\lambda$, then from eqn (6.8) we have

$$L(\lambda|x_1, \ldots, x_r) = r \ln \lambda - \lambda \sum_{i=1}^{r} x_i - (n - r)\lambda t.$$

From $\partial L/\partial \lambda = 0$ we find that

$$\hat{\mu} = \frac{1}{\hat{\lambda}} = \frac{\sum_{i=1}^{r} x_i + (n - r)t}{r} = \frac{\text{total-observed-life statistic}}{\text{number of observed failures}}.$$

Thus, the estimate of mean-life, $\hat{\mu}$, is merely the total life observed divided by the number of failures observed.

6.4.2. Estimation for Type II Censoring

Assume that we observe only $X_{1:n}, \ldots, X_{r:n}$, i.e., the first r-ordered observations in a life test. Here testing is stopped at the rth ordered failure time where r is fixed in advance. Suppose we observe the failure-time data; to simplify our notation write, $x_1 < x_2 < x_3 < \cdots < x_r$, then the actual information obtained during testing is

$$x_1 < x_2 < x_3 < \cdots < x_r, [X_{r+1:n} > x_{r:n}], \ldots, [X_{n:n} > x_{r:n}].$$

The joint density of $X_{1:n}, \ldots, X_{r:n}$ is proportional to

$$\prod_{i=1}^{r} f(x_i; \theta) \prod_{i=r+1}^{n} \bar{F}(x_r; \theta) = \prod_{i=1}^{r} f(x_i; \theta) e^{-(n-r)H(x_r; \theta)}. \qquad (6.9)$$

We recall from previous results that $\bar{F}(x) = e^{-H(x)}$ and $h(x) = f(x)/\bar{F}(x)$ so that we have the log-likelihood for *Type II censoring* given by

$$L(\theta|x_1, \ldots, x_r) = \sum_{i=1}^{r} [\ln h(x_i; \theta) - H(x_i; \theta)] - (n - r)H(x_r; \theta). \qquad (6.10)$$

Again suppose for purposes of illustration that the population sample is $\mathrm{Ep}(\lambda)$ with a mean of $\mu = 1/\lambda$, then from eqn (6.10) we have

$$L(\lambda|x_1, \ldots, x_r) = r \ln \lambda - \lambda \sum_{i=1}^{r} x_i - (n-r)\lambda x_r.$$

From $\partial L/\partial \lambda = 0$ we find again that

$$\hat{\mu} = \frac{1}{\hat{\lambda}} = \frac{\sum_{i=1}^{r} x_i + (n-r)x_r}{r} = \frac{\text{total life statistic}}{\text{number of failures}}.$$

But the remarkable result is that in this case we can write the identity

$$\hat{\mu} = \frac{\sum_{i=1}^{r}(n+1-i)[x_i - x_{i-1}]}{r};$$

and by mle results for $Z_1, \ldots, Z_r$ an iid sample from $\mathrm{Ep}(\lambda)$, we have that

$$\hat{\mu} \sim \frac{\sum_{i=1}^{r} Z_i}{r}. \tag{6.11}$$

6.4.3. Estimation for Type III Random Censoring

Let us consider random censoring of the iid sample of life-lengths $\boldsymbol{X} = (X_1, \ldots, X_n)$ by the censoring vector $\boldsymbol{T} = (T_1, \ldots, T_n)$. We assume $\boldsymbol{T}$ is dependent on $\boldsymbol{X}$ but in the following way:

A° : For any $K \subset \{1, \ldots, n\}$ the rv $\boldsymbol{T}$ is independent of $\{X_i : i \in K\}$ given the event $\bigcap_{i \in K}[X_i > T_i]$.

This means that if K is the index set of censored observations, then the censoring times are independent of the actual times of failure (which in fact were known only to nature). Thus, we are assuming that if a component life-length were censored (for whatever reason), then the component having lived even longer would not have altered that censoring value.

Theorem 2. *If an iid sample, with $X_i \sim F$, is randomly censored of type III and assumption A° holds, then the joint density of $(\boldsymbol{Y}, \boldsymbol{Z})$ is proportional to*

$$\prod_{i \in C_1(\boldsymbol{z})} f_{X_i}(y_i) \times \prod_{i \in C_0(\boldsymbol{z})} \bar{F}_{X_i}(y_i),$$

where $Z_i = I[X_i < T_i]$ counts failures and

$$C_j(z) = \{i : z_i = j\} \quad \textit{for} \quad j = 0, 1,$$

and the constant of proportionality depends upon data $(\boldsymbol{y}, \boldsymbol{z})$ and the distribution of $\boldsymbol{T}$ but not upon the distribution of $\boldsymbol{X}$.

Proof. The proof is straightforward and is left as an exercise. □

6.4.4. Transformation to the Standard Weibull

Let us introduce another parametric model for life length $T \sim G$, where

$$\bar{G}(t) = e^{-H[(t/\beta)^\alpha]} \quad \text{for } t > 0; \alpha, \beta > 0, \tag{6.12}$$

with H a known hazard function. This is the *transformed Weibull model.*

Let W be a standard variate with survival distribution given by

$$\Pr[W > w] = e^{-H(w)} \quad \text{for } w > 0,$$

where H is known. Assume that

$$\bar{F}_Y(y) = e^{-H[(y/\beta)^\alpha]} \quad \text{for } y > 0; \alpha, \beta > 0.$$

and we have shape, α, and scale, β, parameters unknown. Assume that the iid sample $(y_1, \dots, y_n)$ is observed where $Y_i \sim F$. Then we make the transformations

$$W_i = (Y_i/\beta)^\alpha, \qquad U = \hat{\alpha}/\alpha, \qquad V = (\hat{\beta}/\beta)^\alpha.$$

Hence it follows that for mles $\hat{\alpha}$ and $\hat{\beta}$ that

$$(Y_i/\hat{\beta})^{\hat{\alpha}} \sim (W_i/V)^U, \qquad (Y_i/\hat{\beta})^\alpha \sim (W_i/V).$$

If we set $z_i = (w_i/v)^u$ for $i = 1, \dots, n$, then the mle equations can be written, as we show in section 6.6 in eqns (6.22), (6.23), and (6.24):

$$\frac{1}{k}\sum_{i=1}^{n} z_i \psi_{i,k}(z_i) = 1, \qquad \frac{1}{k}\sum_{i=1}^{n} z_i(\ln z_i)\psi_{i,k}(z_i) - \frac{1}{k}\sum_{i=1}^{n} \ln z_i = 1.$$

The standardized failure-times $W_1, \dots, W_k$ are always distributed independently of α and β. So we have

Theorem 3. *In the notation above, under any conditions where $W_{k+1}, \dots, W_n$ are also independent of α and β, it follows that*

$$U = \hat{\alpha}/\alpha \quad \text{and} \quad V = (\hat{\beta}/\beta)^\alpha \tag{6.13}$$

have a joint distribution independent of α and β.

NB the property holds under all types of independent random censoring and type II censoring when the distributional assumptions are met. It thus includes all known results for confidence bounds for log-normal and Weibull distributions and contains Student's seminal result, except that here U and V are not independent.

We may know more. Consider a life-length $T \sim G$ where $\bar{G}(t) = e^{-\lambda H(t)}$ for all $t > 0$ and H is a known increasing nonnegative function with $H(t) \to \infty$ as $t \to \infty$, say $H(t) = \sinh^4(t)$, but the parameter λ is unknown. It follows that $X = H(T)$ has an Epstein distribution with hazard rate λ. Thus, for any set of failure times $t_1, \ldots, t_k$ and survival times $t_{k+1}, \ldots, t_n$ from a set of n randomly censored iid observations one merely makes the transformation $x_i = H(t_i)$ of failure times and the corresponding transformation of survival times and estimates λ using the total-life-statistic.

6.5. Generation of Ordered Observations

Let $Y_1, \ldots, Y_n$ be an iid sample of life-lengths from a population with joint density given by

$$f_{Y_1,\ldots,Y_n}(y_1, \ldots, y_n) = \prod_{i=1}^{n} f(y_i) \quad \text{for } 0 < y_1, \ldots, y_n < \infty.$$

Consider the *ordered observations* (or order statistics)

$$0 < Y_{1:n} < Y_{2:n} < \cdots < Y_{n:n} < \infty, \quad \text{with } Y_{0:n} := 0,\ Y_{n+1:n} := \infty,$$

where in particular we have $Y_{1:n} = \min_{i=1}^{n} Y_i$ and $Y_{n:n} = \max_{i=1}^{n} Y_i$.

Then their joint density has support on the simplex and is given by

$$f_{Y_{1:n},\ldots,Y_{n:n}}(y_1, \ldots, y_n) = n! \prod_{i=1}^{n} f(y_i) \quad \text{for } 0 < y_1 < \cdots < y_n < \infty. \tag{6.14}$$

It is straightforward to show that

$$(i) \qquad \int_0^{\infty} \cdots \int_0^{\infty} f_{Y_{1:n},\ldots,Y_{n:n}}(y_1, \ldots, y_n) \prod_{i=1}^{n} dy_i = 1.$$

(ii) The density of $Y_{1:n}$ is $nf(y)[\bar{F}(y)]^{n-1}$ for $0 < y < \infty$.

(iii) The density of $Y_{n:n}$ is $nf(y)[F(y)]^{n-1}$ for $0 < y < \infty$.

By setting $U_k = F(Y_{k:n}) - F(Y_{k-1:n})$ for $k = 1, \ldots, n+1$, then the vector $\boldsymbol{U} = (U_1, \ldots, U_{n+1})$ has a *Dirichlet distribution* with density $đ$, setting $\boldsymbol{u} = (u_1, \ldots, u_{n+1}), \boldsymbol{a} = (a_1, \ldots, a_{n+1})$, defined by

$$(iv) \quad đ(\boldsymbol{u}; \boldsymbol{a}) = \Gamma(\nu + 1) \prod_{i=1}^{n+1} \frac{u_i^{a_i - 1}}{\Gamma(a_i)} \quad \text{for } 0 < u_i < 1; a_i > 0,$$

where

$$\sum_{i=1}^{n+1} a_i = \nu \quad \text{and} \quad \sum_{i=1}^{n+1} u_i = 1.$$

The kth spacing between order observations is the difference

$$D_k = Y_{k:n} - Y_{k-1:n} \quad \text{for } k = 1, 2, \ldots, n,$$

where for a sample of size n we set $Y_{0:n} = 0$. Recall the *Epstein distribution* is denoted by

$$Y \sim \text{Ep}(\lambda) \text{ iff } f_Y(y) = \lambda e^{-\lambda y} \quad \text{for } y > 0. \tag{6.15}$$

Remark 1. *The variate* $U \sim \text{Unif}(0, 1)$ *iff* $-\ln U \sim -\ln(1-U) \sim \text{Ep}(1)$*. If* $\lambda Y \sim \text{Ep}(1)$*, it follows that*

$$\text{E}[Y] = \frac{1}{\lambda}, \quad \textit{and} \quad \text{Var}[Y] = \frac{1}{\lambda^2}. \tag{6.16}$$

Theorem 4. *If* $Y_i \sim \text{Ep}(\lambda)$ *for* $i = 1, \ldots, n$ *are iid, then*

$$(a)\ D_k \sim \text{EP}[(n-k+1)\lambda] \ \textit{for}\ k = 1, \ldots, n \tag{6.17}$$

and

$$(b)\ D_1, \ldots, D_n \quad \textit{are mutually independent.} \tag{6.18}$$

PROOF. From eqn (6.14) we have

$$f_{Y_{1:n},\ldots,Y_{n:n}}(y_1, \ldots, y_n) = n!\lambda^n e^{-\lambda \sum_1^n y_i} \quad \text{for } 0 < y_1 < \cdots < y_n < \infty.$$

Make the change of variable $d_k = y_k - y_{k-1}$ for $k = 1, \cdots, n$ with $y_0 = 0$, noting that $\sum_{i=1}^n y_i = \sum_{k=1}^n (n-k+1)d_k$, then by the theorem on transformation of rv's we have

$$f_{D_1,\ldots,D_n}(d_1, \ldots, d_n) = n!\lambda^n e^{-\lambda \sum_{k=1}^n (n-k+1)d_k} \cdot \left| J\left(\frac{y_1, \ldots, y_n}{d_1, \ldots, d_n}\right)\right|.$$

From results on Jacobians we know

$$\left| J\left(\frac{y_1, \ldots, y_n}{d_1, \ldots, d_n}\right)\right| = \frac{1}{\left|\left(\frac{\partial d_i}{\partial y_j}\right)\right|_+} = \frac{1}{\begin{vmatrix} 1 & 0 & 0 & \cdots & 0 & 0 \\ -1 & 1 & 0 & \cdots & 0 & 0 \\ 0 & -1 & 1 & \cdots & 0 & 0 \\ \cdot & \cdot & \cdot & \cdots & \cdot & \cdot \\ 0 & 0 & 0 & \cdots & -1 & 1 \end{vmatrix}} = 1.$$

Thus, we have the result, for $0 < d_1, d_2, \ldots, d_n < \infty$, that

$$f_{D_1,\ldots,D_n}(d_1, \ldots, d_n) = \prod_{k=1}^n \left\{\lambda(n-k+1)e^{-\lambda(n-k+1)d_k}\right\}.$$

which proves (a) and (b). □

There follows

Corollary 1. *In Theorem 1, use note the normalized spacings* $(n-k+1)D_k$ *for* $k = 1, \ldots, n$ *are themselves all iid with* $D_k \sim \text{Ep}(\lambda)$*.*

By using the identity $\sum_{i=1}^k D_i = Y_{k:n}$, we have from eqn (6.16),

Corollary 2. *The mean and variance of the ordered observations, from an* Ep(λ) *population, are*

$$\mathrm{E}[Y_{k:n}] = \mathrm{E}\sum_{i=1}^{k} \frac{(n-i+1)D_i}{n-i+1} = \sum_{i=1}^{k} \frac{1}{\lambda(n-i+1)},$$

$$\mathrm{Var}[Y_{k:n}] = \mathrm{Var}\sum_{i=1}^{k} \frac{(n-i+1)D_i}{n-i+1} = \sum_{i=1}^{k} \frac{1}{[\lambda(n-i+1)]^2}.$$

Let us presume that a computer program is available to generate a sequence of pseudo-random numbers uniform on (0, 1). We can use these, by Remark 1, to obtain pseudo-random Epstein variates. Now consider the problem of using a computer to generate pseudo-random variates of the first r ordered observations out of $n > r$, all of which come from a specified life distribution with a given hazard H. This question is answered in

Theorem 5. *Let* $Y_i \sim \mathrm{Ep}(1)$ *for* $i = 1, \ldots, r$ *be iid and set*

$$S_{k,n} = \sum_{i=1}^{k} \frac{Y_i}{n-i+1} \quad \textit{for } k = 1 \ldots, r.$$

Then $X_{1:n} < X_{2:n} < \cdots < X_{r:n}$ *are the first r ordered observations out of n from a population with sdf* $\bar{F} = e^{-H}$ *iff* $X_{k:n} = H^{-1}(S_{k,n})$.

Proof. We proceed by induction. Now $\Pr[X_{1:n} > x] = [\bar{F}(x)]^n = e^{-nH(x)}$ for $x > 0$. And thus, $S_{1,n} = Y_1/n \sim EP(n)$. Therefore, $H^{-1}(Y_1/n) \sim X_{1:n}$. Assume that the assertion is true for all $1, 2, \ldots, k$ with $H(X_{k:n}) \sim \sum_{i=1}^{k} Y_i/(n-i+1)$. Thus, it follows that for $0 < x_k < x_{k+1} < \infty$

$$\Pr[X_{k+1:n} > x_{k+1} | X_{k:n} = x_k] = \exp\{-(n-k)[H(x_{k+1}) - H(x_k)]\},$$

and consequently for this conditional rv, we have

$$Y_{k+1}\,|x_k = (n-k)\,(H(X_{k+1:n}) - H(x_k))\,,$$

and hence

$$H(X_{k+1:n})|x_k = \sum_{i=1}^{k} \frac{Y_i}{n-i+1} + \frac{Y_{k+1}}{n-k}.$$

Thus, $Y_{k+1} \perp Y_i$ for $i \leq k$ and $Y_{k+1} \sim \mathrm{Ep}(1)$. So unconditionally the result is true for $k+1$ and the induction holds. □

Consequently, an arbitrary subset of ordered life observations from any population with specified hazard can be generated by computer as pseudo-random variates from considering an appropriate computation of partial sums. Given an index set $J \subset \{1, 2, \cdots, n\}$, a corresponding subset of ordered observations $\{X_{j:n} | j \in J\}$

can be generated. This is accomplished in the same manner but now

$$X_{j:n} = H^{-1}(S_{j,n}) \quad \text{for } j \in J, \quad \text{where} \quad S_{j,n} = \sum_{i=1}^{j} Y_i/(n-i+1)$$

is computed from $m = \max\{j : j \in J\}$ iid exponential variates $Y_1, \ldots, Y_m$ with unit mean. Clearly, the efficiency of this procedure is best when $m \ll n$. Because the machine generation of the y_i is rapid from the uniform u_i and $H^{-1}(s_{j,n})$ need be computed only for the $j \in J$; this procedure works well in practice unless a large fraction of variates from the whole sample are needed.

Exercise Set 6.D

1. A set of 10 highly reliable, nominally identical devices are tested and all found to work. What would your estimate of their reliability be? How confident would you be that they were perfect?
2. In a particular geographic region the mean life of wooden power poles was desired, i.e., the average time until the pole would have to be reinforced due to deterioration. In the thousands of records of pole replacement only 17 were found not to have been replaced because of automobile collision, street widening, windstorm damage from falling trees, etc. How could the average life be computed when the data were incomplete?
3. For an iid Bernoulli sample of size n, let the mle of p be $\hat{p}$. What is the distribution of $n\hat{p}$? Let $x_1, \ldots, x_k$ be given where $x_i = \sum_{j=1}^{n} x_{ij}$ with x_{ij} iid Bernoulli but both n and p are unknown. Show the mle of n is $\max_{i=1}^{k} x_i$.
4. For an iid sample of size n from Ep(λ), let the mle of λ be $\hat{\lambda}$. What is the distribution of $n\lambda/2\hat{\lambda}$? At a given time GE alternator drives had accummulated 22,324 operating hours in service with 181 failures, while Thompson alternator drives had accumulated 14,036 hours with 60 failures. Is it possible that the two drives are equally reliable? Which drive would you recommend be used? Are there any questions you would like to ask before you make a recommendation?
5. Can we now transform a set of censored data on T with distribution as given in eqn (6.12) to a corresponding one for the Weibull distribution? If so, do it. If not, show why not.
6. Determine the mean and variance of T with distribution as given by the transformed Weibull as in eqn (6.12) in terms of α, β, f, $\bar{F}$, and/or H.
7. Is a variate with survival distribution given in eqn (6.12) much different from one given by $\bar{G}(t) = \exp\{-[H(t)/\beta]^\alpha\}$?

6.6. A Parametric Model of Censoring

Let X, T denote, respectively, the (random) life-length of some given component and the (random) time of censoring of the life-test of that component for any reason other than failure, such as exhaustion of time, money, or operator patience during testing or the failure of an ancillary component including the test machine or any supporting device independent of the failure of the given component. What is observed at each trial is the random couple $Y = X \wedge T$ and $Z = \mathrm{I}(X < T)$. What we record is the time when testing stops, presuming we started at time zero, and whether it was a *failure* or a *run-out* (censored observation). The joint density of this couple is

$$f(y, z) = [f_X(y)\bar{F}_T(y)]^z[f_T(y)\bar{F}_X(y)]^{1-z} \quad \text{for } 0 < y < \infty; z = 0, 1, \tag{6.19}$$

with the probability of an observation rather than a run-out of

$$p = \mathrm{E}Z = \Pr[X < T] = \int_0^\infty F_X(t)\, d\, F_T(t).$$

The likelihood, based on the sample $(\boldsymbol{y}, \boldsymbol{z})$, is

$$\mathcal{L}(\boldsymbol{y}, \boldsymbol{z}) = \prod_{i=1}^{n}[f_X(y_i)\bar{F}_T(y_i)]^{z_i}[f_T(y_i)\bar{F}_X(y_i)]^{1-z_i}.$$

Let $\theta \in \Theta$ be the unknown parameter in the parameter space of X. Then the log-likelihood is, omitting terms independent of θ,

$$L(\theta|\boldsymbol{y}, \boldsymbol{z}) \cong \sum_{i=1}^{n}\left[z_i \ln f_X(y_i|\theta) + (1 - z_i)\ln \bar{F}_X(y_i|\theta)\right].$$

Wlog let us assume, for some $1 \le k \le n$, that

$$z_i = 1 \quad \text{for } i = 1, \ldots, k \quad \text{and } z_i = 0 \quad \text{for } i = k + 1, \ldots, n,$$

and let us specialize to the following transformed Weibull model utilizing shape and scale parameters:

$$\bar{F}_X(y|\theta) = \exp\{-H[(y/\beta)^\alpha]\} \quad \text{for } y > 0; \alpha, \beta > 0, \tag{6.20}$$

where H is a known hazard function but parameters α, β are unknown. Thus asas we obtain the log-likelihood

$$L(\alpha, \beta) = \sum_{i=1}^{k}\{\ln h[(y_i/\beta)^\alpha] + \ln(\alpha/\beta) + (\alpha - 1)\ln(y_i/\beta)\} - \sum_{i=1}^{n} H[(y_i/\beta)^\alpha]. \tag{6.21}$$

We find asas that

$$\frac{\partial L}{\partial \alpha} = \frac{k}{\alpha} + \sum_{i=1}^{k}\ln(y_i/\beta) - \sum_{i=1}^{n}(y_i/\beta)^\alpha \times \ln(y_i/\beta) \times \Psi_{i,k}[(y_i/\beta)^\alpha],$$

and

$$\frac{\partial L}{\partial \beta} = \frac{-k\alpha}{\beta} + \frac{\alpha}{\beta}\sum_{i=1}^{n}(y_i/\beta)^{\alpha} \times \Psi_{i,k}[(y_i/\beta)^{\alpha}],$$

where for $i, k = 1, \ldots, n$, we have introduced the function

$$\Psi_{i,k}(y) = h(y) - I(i \leq k) \times \frac{h'(y)}{h(y)} \quad \text{for all} \quad y > 0. \tag{6.22}$$

Thus, for the model of eqn (6.20), with randomly censored samples, the mles $\hat{\alpha}$, $\hat{\beta}$ are the simultaneous solutions of the two equations in α, β, where for notational simplicity we write here $t_i = (y_i/\beta)^{\alpha}$ for $i = 1, \ldots, n$, (not to be confused with censoring times)

$$\frac{1}{k}\sum_{i=1}^{n} t_i \Psi_{i,k}(t_i) = 1 \quad \text{iff} \quad \frac{\partial L}{\partial \beta} = 0, \tag{6.23}$$

$$\frac{1}{k}\sum_{i=1}^{n} t_i \ln t_i \times \Psi_{i,k}(t_i) - \frac{1}{k}\sum_{i=1}^{k} \ln t_i = 1 \quad \text{iff} \quad \frac{\partial L}{\partial \alpha} = 0. \tag{6.24}$$

We now present three examples where this transformed Weibull model applies.

Example 7. If we let $X \sim \text{Wei}(\alpha, \beta)$, then $h(x) = 1$ for all $x > 0$ and so we see $\Psi_{i,k}(x) = 1$. This results in considerable simplification and if we define

$$A(\alpha) = \frac{\sum_{i=1}^{n} y_i^{\alpha} \ln y_i}{\sum_{i=1}^{n} y_i^{\alpha}} - \frac{1}{k}\sum_{i=1}^{k} \ln y_i, \tag{6.25}$$

then we need only solve the equation

$$A(\alpha) = \frac{1}{\alpha} \quad \text{to obtain } \hat{\alpha}; \text{ then} \qquad \hat{\beta} = \left(\frac{1}{k}\sum_{i=1}^{n} y_i^{\hat{\alpha}}\right)^{1/\hat{\alpha}}. \tag{6.26}$$

Note that A is an increasing function, with $A(0)$ and $A(\infty)$ easily determined so that it is simple to check if a solution for $\hat{\alpha}$ exists. See Figure 6.3.

Example 8. If we let $X \sim \mathcal{LN}(\ln \beta, 1/\alpha)$, then

$$\bar{F}_X(x) = \Phi[-\alpha(\ln x - \ln \beta)] = \exp\{\ln \Phi[-\ln(x/\beta)^{\alpha}]\} \quad \text{for} \quad x > 0$$

and we find the hazard rate to be

$$h(x) = \frac{\varphi(\ln x)}{x\Phi(-\ln x)} \quad \text{for} \quad x > 0.$$

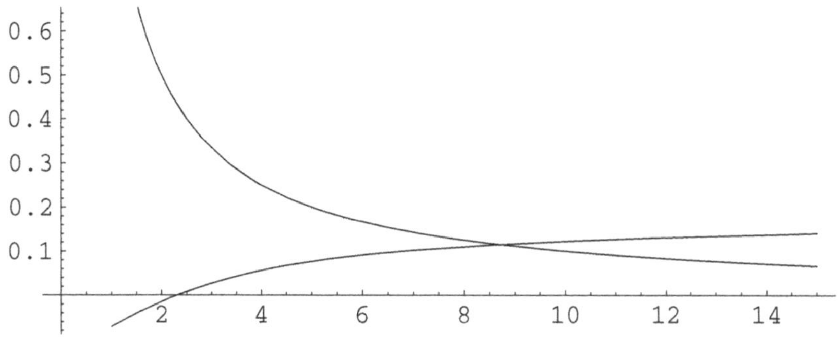

Figure 6.3. Typical behavior determining MLE of the shape parameter α.

We find that the model of eqn (6.20) includes the solution of the stationary equations in this case. We have

$$\Psi_{i,k}(x) = e^x - \mathrm{I}(i \le k) = \begin{cases} (1 + \ln x)/x & \text{if } i \le k, \\ h(x) & \text{if } i > k. \end{cases} \tag{6.27}$$

EXAMPLE 9. The same derivation applies to the Gumbel extreme-value distribution; it has the corresponding definition of $h(x) = e^x$ for $x > 0$. In this case asas we obtain

$$\Psi_{i,k}(x) = e^x - \mathrm{I}(i \le k) \quad \text{for } x > 0; \;\; i, k = 1, \ldots, k. \tag{6.28}$$

6.7. The Empirical Cumulative Distribution

For a complete iid sample, say $X_1, \ldots, X_n$, from a population with unknown cdf F, except that it is continuous, the *empirical cumulative distribution function* (ecdf) is defined by

$$\hat{F}_n(x) = \frac{1}{n} \sum_{i=1}^{n} \mathrm{I}(x \ge X_i) \quad \text{for any } x \in \Re. \tag{6.29}$$

The maximum discrepancy between the ecdf and the true distribution, namely

$$D_n = \sup_{x \in \Re} |\hat{F}_n(x) - F(x)| = \sup_{0<u<1} |\hat{F}_n F^{-1}(u) - u|$$

is called the *Kolmogorov-Smirnov* statistic. It is the “distance ” between $\hat{F}_n$ and F and it is distribution-free (nonparametric) since

$$\hat{F}_n F^{-1}(u) \sim \frac{1}{n} \sum_{i=1}^{n} \mathrm{I}[u \ge U_i],$$

where by the probability integral transform $F(X_i) := U_i \sim \text{Unif}(0, 1)$ are iid.

Exercise Set 6.D

1. Verify eqn (6.21) for the log-likelihood.
2. Prove that the function $A(\alpha)$, defined in eqn (6.25), is monotone increasing, and evaluate both $A(0)$ and $A(\infty)$.
3. The log-normal distribution has been variously applied to analyze statistically the behavior of both the growth of trees and the duration-of-load (fatigue-life) of wood. What do you think of this selection of a distribution?
4. Verify eqn (6.27) and use it to obtain the stationary equations for the mles and show how they can be solved.
5. Evaluate the stationary equations to verify eqn (6.27) and show how they can be solved when one parameter is known: (i) Find $\hat{\alpha}$ when β is known. (ii) Find $\hat{\beta}$ when α is known.
6. Without examining any data present in an argument either for or against using the Galton (log-normal) family to describe: (a) the length of a first marriage. (b) survival time after chemotherapy in cancer treatment.
7. An rv X with density

$$f(x) = \frac{e^x}{(1+e^x)^2} \quad \text{for} \quad -\infty < x < \infty$$

is said to have the (standard) logistic distribution. Find the hazard of the log-logistic and show it is a transformed Weibull.

8. What are the mean and variance of the ecdf, $\hat{F}_n(x)$? Is $\hat{F}_n$ consistent for F? Is it strongly consistent?
9. * Let $X_1, \ldots, X_N$ be an iid sample from a "fat-tailed" distribution like the Cauchy density,

$$f(x) = \frac{1}{\pi[1+(x-\theta)^2]} \qquad \text{for } x \in \Re.$$

Since method-of-moment estimators would likely not be of much utility (why?), find the mle of θ. Hint: try to use the ecdf to express simply the solution of the stationary equation.

10. The forty observations of the diameter of Saturn's rings made by Bessel in the years 1829–1831 have a sample mean of 39.31 and a sample standard deviation of 0.1960. The true diameter of Saturn's rings was then stated, by astronomers, as being 39.31 ± 0.32. What was meant by this? What is random here? Does the diameter of Saturn's ring change? Should the astronomer have used different words?
11. The current value for the speed of light in a vacuum, as stated by NIST SP 966 (March 2001) was 299 792 458 ms^{-1}. It is presented as "exact." What can this mean? Do you think there was no error in a measurement of nine significant figures?

Nature provides every animal with a way to make a living and for the astronomer it is astrology.

Johannes Kepler

Baysian Analysis is to Statistical Science what Astrology is to Astronomy.

Oscar Kempthorne

We are all entitled to our own opinion but nobody is entitled to his own set of facts.

Patrick Daniel Moynihan

Visiting Physicist: "Professor Hilbert, I wanted to inform you that many of the unsolved problems in physics have now been reduced to mathematical problems in Hilbert Space."

David Hilbert: "That is too bad because that means physics has now become too difficult for you physicists."

Einstein's theory of gravitation, first proposed in special form in 1905, has survived every test that it has been possible to make of it; that, with the illumination and understanding it has provided of the natural Universe, is the basis of the contention that it is the outstanding achievement of human intellect and accomplishment during the 20th century.

So say the Physicists

In April 1953, at the University of Cambridge, Francis H.C. Crick and Thomas D. Watson built a model of a DNA molecule, the details of which were sufficient to illustrate how these molecules function as a repository of genetic information. This discovery ranks with that of Copernicus in enlightening mankind as to his true place in Nature and its full potential to science has not yet been revealed.

So say the Biologists

The whole of physics and chemistry have now been reduced to applied mathematics.

P.A.M. Dirac in his *The Principles of Quantum Mechanics,* 1928

I shall adopt the view that there does exist an ultimate set of physical laws (which we do not as yet fully know but which might include quantum gravity) and that those laws do *govern* the Universe. They *force* the Universe to behave the way it does.

Kip S. Thorne

CHAPTER 7

Nonparametric Life Estimators

7.1. The Empiric Survival Distribution

If one wishes an estimate of the survival function S_X of a life variate X, why not use $1 - \hat{F}(x)$, the complement of the ecdf? The answer is that in life studies the data are nearly always censored and how one utilizes censored observations in modifying $\hat{F}(x)$ is not immediately clear. The task in engineering is often different from that of multiple-risk survival studies in medicine, where the hazard rates of all related modes of deterioration (disease) may be sought. Censoring during engineering testing usually comes from accidents during testing, either to the equipment or to the test materials, or the test is terminated by the exhaustion of test funds or of available test time; all of these causes are independent of failure. Moreover the distribution of life, F_X, may be known to be within some parametric family or only within some nonparametric class such as IHR.

7.1.1. Life-Table Methods

One of the oldest methods of estimating survival is called the *life-table method.* Let time be partitioned into $k+1$ intervals with

$$0 = \tau_0 < \tau_1 < \cdots < \tau_k < \infty.$$

The first k intervals are often taken to be of the same length, and in actuarial studies the length is usually one year.

Let us define

n_i = # alive at the beginning of ith interval, i.e., at time τ_{i-1},
d_i = # died during the ith interval, i.e., between time τ_{i-1} and time τ_i,
ℓ_i = # lost to follow-up during the ith interval, i.e., between time τ_{i-1} and time τ_i,
w_i = # withdrew during the ith interval, i.e., between time τ_{i-1} and time τ_i,
$p_i = \Pr\{\text{surviving through time } \tau_i \text{ if alive at time } \tau_{i-1}\}$,
$q_i = 1 - p_i$ for $i = 1, 2, \ldots k+1$.

The Reduced Sample Method

The estimate $\hat{S}$ defined at τ_k is given by

$$\hat{S}(\tau_k) = 1 - \frac{\sum_{i=1}^{k} d_i}{n_1 - \sum_{i=1}^{k}(\ell_i + w_i)},$$

The drawback is that this estimate is badly biased since it ignores the information concerning survival, which is contained in ℓ_i and w_i.

The Actuarial Method

Let X represent the life of a person selected at random from the population under investigation. Then by the law of multiplication of probabilities

$$S(\tau_k) = \Pr[X > \tau_k] = \Pr[X > \tau_1]\Pr[X > \tau_2|X > \tau_1] \cdots \Pr[X > \tau_k|X > \tau_{k-1}],$$

and correspondingly with proper identification of p_i,

$$\hat{S}(\tau_k) = \prod_{i=1}^{k} p_i.$$

But again the question is what do we about losses or withdrawals? The first practical suggestion was to define the effective sample size, during each interval, as $n_i^{'} = n_i - \frac{1}{2}(\ell_i + w_i)$ and then set $q_i^{'} = d_i/n_i^{'}$ with $p_i^{'} = 1 - q_i^{'}$; ultimately the improved estimate was defined as

$$\hat{S}(\tau_k) = \prod_{i=1}^{k} p_i^{'}.$$

An approximation for the variance of this estimator is called *Greenwood's formula*:

$$\widehat{\text{Var}}[\hat{S}(\tau_k)] = [\hat{S}(\tau_k)]^2 \sum_{i=1}^{k} \frac{d_i}{n_i^{'}(n_i^{'} - d_i)}. \tag{7.1}$$

But this is seen to be an approximation with variance and bias unknown.

7.1.2. The Kaplan–Meier Estimator

An improved answer was provided by Kaplan and Meier [53, 1958]. But the exact expectation of the K-M estimator, except asymptotically or in very special cases, was not given until [65] 1993.

Let the observations in the censored sample be given by $(y_i, z_i) : i = 1, \ldots, n$, where $z_i = 1$ indicates y_i, the ith observation, was due to failure (death) while $z_i = 0$ indicates it was censored for some reason and was only known to be alive at that time. Now order these observations so that each sample point $\mathbf{y}$ falls in the simplex $\mathcal{S}$, and the following inequalities are satisfied a.s.,

$$\mathcal{S}_n = \{\mathbf{y} : 0 < y_{(1)} < \cdots < y_{(n)} < \infty\},$$

with $z_{(i)} = 0$ or 1 accordingly as the ith-ordered observation was censored or failed. The product-limit estimate of Kaplan–Meier for the survival distribution S_X when censored from the right, where for notational simplicity and wlog, we omit the round brackets in the subscripts and regard the y_i as ordered, is

$$\tilde{S}_X(t) = \prod_{\{i:y_i \le t\}} \left(\frac{n - i + 1 - z_i}{n - i + 1} \right) \quad \text{for } t > 0. \tag{7.2}$$

This formula was obtained as follows: construct a partition of $\Re_+$ with

$$0 = \tau_0 < \tau_1 < \tau_2 < \cdots < \tau_n < \tau_{n+1} = \infty,$$

then for an continuous unknown distribution F, which we seek to discover, we must have

$$S_X(\tau_k) := \Pr[X > \tau_k] - \Pr[X > \tau_1] \times \prod_{j=1}^{k-1} \Pr[X > \tau_{j+1} | X > \tau_j] = \prod_{i=1}^{k} p_i,$$

where as before

$$p_i = \Pr[X > \tau_i | X > \tau_{i-1}] \quad \text{and } q_i = 1 - p_i \quad \text{for} \quad i = 1, \ldots, n.$$

We then form the estimates

$$\hat{q}_i = \frac{\text{\# of failed observations at time } y_i}{\text{\# of alive observations just before time } y_i} = \frac{\sum_{j=1}^{i} z_i}{n - i + 1},$$

and we set

$$\hat{p}_i = 1 - \hat{q}_i = \begin{cases} 1 - \frac{1}{n_i} & \text{if } z_i = 1, \\ 1 & \text{if } z_i = 0. \end{cases}$$

We then take as the estimate of the survival distribution

$$\hat{S}_X(t) = \prod_{\{i:y_i \le t\}} \hat{p}_i = \prod_{\{i:y_i \le t\}} \left(1 - \frac{1}{n_i}\right)^{z_i} = \prod_{\{i:y_i \le t\}} \left(\frac{n - i}{n - i + 1}\right)^{z_i} \quad \text{for} \quad t > 0, \tag{7.3}$$

which agrees with eqn (7.2).

NB that this estimate is undefined for $t > y_n$ when $z_n = 0$.

Let us now make more explicit the assumptions under which the succeeding derivation follows. Let the observed variate be $Y = \min(X, T)$, where $X \sim F_X$ is the observable life length and $T \sim F_T$ is a censoring variable independent of X. Both variates have densities and support on $(0, \infty)$. Thus, we have the sdf of Y, call it S_Y, and the corresponding hazard rate, h_Y, related by

$$S_Y = S_X \cdot S_T \quad \text{with} \quad h_Y = \frac{f_Y}{S_Y}. \tag{7.4}$$

But note that if it were the sdf of T in which one were interested, with X serving as its censoring variate from the right, then by eqn (7.2) we would merely interchange z_i and $1 - z_i$ to obtain

$$\tilde{S_T}(t) = \prod_{\{i:y_i \le t\}} \left(\frac{n-i}{n-i+1}\right)^{1-z_{(i)}} \quad \text{for} \quad t > 0.$$

The product of these two censored empirical survival distributions would yield, in accord with eqn (7.4), the empirical survival distribution of $Y = \min(X, T)$, since it forms a complete sample, and we would have

$$\tilde{S}_Y(t) = \tilde{S}_X(t) \cdot \tilde{S_T}(t) = \sum_{k=1}^{n} \frac{n-k}{n} \mathrm{I}(y_k < t \le y_{k+1}) \quad \text{for } t > 0.$$

Thus, by division, we have derived a closed-form expression for the K-M estimate, equivalent with the "distributing-to-the-right algorithm" of B. Efron [32, 1967], viz.,

$$\tilde{S}_X(t) = \sum_{k=0}^{n} \prod_{i=1}^{k} \left(\frac{n-i}{n+1-i}\right)^{z_i} \mathrm{I}(y_k \le t < y_{k+1}) \quad \text{for } t > 0.$$

This form proves useful for computation.

Clearly this reasoning can be extended to failure from multiple causes, since $Y = \min(X_1, \ldots, X_m)$ for $m > 2$ as long as the "failing" variate, i.e., the variate that was the minimum, can be correctly identified.

7.2. Expectation and Bias of the K-M Estimator

Under the assumption of random censoring by an independent variate we now derive the exact bias and exact mean-square error of both the Kaplan–Meier estimator of the survival distribution and the Nelson–Aalen estimator of the hazard function. Some explicit small-sample comparisons are made when the hazard of life and the hazard of censoring are proportional. General bounds are then expressed in corresponding terms.

For notational convenience here we write $S = 1 - F$ as the survival distribution and $F^{'} = f$ for the density corresponding to any distribution F with appropriate affix. We observe the couple (Y, Z) where $Y = X \wedge T$, $Z = \mathrm{I}[X < T]$ and $\mathrm{I}[A]$ is the indicator function of any event A. The density of this couple is therefore

$$f(y, z) = [f_X(y)S_T(y)]^z[f_T(y)S_X(y)]^{1-z} \quad \text{for } z = 0, 1; 0 < y < \infty. \tag{7.5}$$

The set of observations $(\mathbf{y},\mathbf{z}) = (y_1,\ldots,y_n,z_1,\ldots,z_n)$ generates a likelihood function

$$L(\mathbf{y},\mathbf{z}) = \prod_{i=1}^{n} f(y_i,z_i) = \prod_{i=1}^{n} [f_X(y_i)S_T(y_i)]^{z_i} \times [f_T(y_i)S_X(y_i)]^{1-z_i}, \quad (7.6)$$

with $\sum_{i=1}^{n} z_i$ counting the number of failures.

We rewrite the K-M (product limit) estimate in terms of observed ordered random variables using a formula equivalent with eqn (7.2), viz.,

$$\hat{S}_X(t) = \tilde{S}_X(t;\mathbf{z},\mathbf{y}) = \prod_{i=1}^{n}\left[\frac{n-i+1-z_i\mathrm{I}(y_i \le t)}{n-i+1}\right]. \quad (7.7)$$

This formulation is defined for all $t > 0$, but $\hat{S}$ remains constant for all $t > y_j$, the jth-order observation, when the corresponding $z_j = 1$ and $z_{j+1} = \cdots = z_n = 0$. Hence $\hat{S}(t)$ does not become zero as $t \to \infty$ unless the last observation was a failure, i.e, $z_n = 1$.

First we find the expected values. We note from eqn (7.5) that $\sum_{z=0}^{1} f(y,z) = f_Y(y)$, and we write using the likelihood of eqn (7.6)

$$\mathrm{E}\hat{S}_X(t) = n!\sum_{\mathbf{z}} \int_{\mathbf{y}\in\mathcal{S}} \tilde{S}_X(t;\mathbf{z},\mathbf{y})L(\mathbf{y},\mathbf{z})\,d\mathbf{y}, \quad (7.8)$$

setting $d\mathbf{y} = dy_1 \cdots dy_n$. We have, asas,

$$\mathrm{E}\hat{S}_X(t) = \int_{0<y_1<\cdots<y_n<\infty} \int \prod_{i=1}^{n} [(n-i+1)f_Y(y_i) - I(y_i \le t)f_1(y_i)]\ dy_1 \cdots dy_n.$$

Now setting

$$g_i(y,t) = (n-i+1)f_Y(y) - I(y \le t)f_1(y), \quad \text{for } i = 1,\ldots,n, \quad (7.9)$$

for short, we have by introducing a distributing-to-the-right form,

$$\mathrm{E}\hat{S}_X(t) = \sum_{k=0}^{n} \int_{0<y_1<\cdots<y_k<t<y_{k+1}<\cdots<y_n<\infty} \int \prod_{i=1}^{k} [g_i(y_i,t)]\,(n-k)! \prod_{j=k+1}^{n} f_Y(y_j)\,d\mathbf{y}.$$

Performing the integration over all variables exceeding t, we find

$$[S_Y(t)]^{n-k} = (n-k)! \int_{0<t<y_{k+1}<\cdots<y_n<\infty} \int \prod_{i=k+1}^{n} f_Y(y_i)\,dy_{k+1}\cdots dy_n,$$

and so we obtain

$$\mathrm{E}\hat{S}_X(t) = \sum_{k=0}^{n} [S_Y(t)]^{n-k} H_k(t),$$

by setting $H_0(t) \equiv 1$ for all $t \ge 0$ and defining, for $k = 1,\ldots,n$,

$$H_k(t) = \int_{0<y_1<\cdots<y_k<t} \int \prod_{i=1}^{k} [(n-i)f_Y(y_i) + f_0(y_i)]\ dy_1 \cdots dy_k. \quad (7.10)$$

Then by denoting

$$G_{n-k}(t) = S_T(t)S_Y(t)^{n-1-k} \quad \text{for } k = 0, \ldots, n, \tag{7.11}$$

we can write, since $S_Y = S_X S_T$, that

$$\mathrm{E}\hat{S}_X(t) = S_X(t)\sum_{k=0}^{n} G_{n-k}(t)H_k(t). \tag{7.12}$$

Before proceeding further, we prove the following:

Lemma 1. *It follows from these definitions that*

$$G_{n-k}(t) \cdot H_k^{'}(t) = -G_{n-k+1}^{'}(t) \cdot H_{k-1}(t) \quad \textit{for } k = 1, \ldots, n. \tag{7.13}$$

PROOF. From the definition in eqn (7.10), with $H_0 \equiv 1$, we see that

$$H_k(t) = \int_0^t H_{k-1}(y)[(n-k)f_Y(y) + f_0(y)]\,dy \quad \text{for } k = 1, \ldots, n. \tag{7.14}$$

From this recursion we have that for $k = 1, \ldots, n$

$$H_k^{'}(t) = H_{k-1}(t)[(n-k)f_Y(t) + f_0(t)]. \tag{7.15}$$

We also find asas from eqn (7.11) that for $k = 0, 1, \ldots, n$

$$G_{n-k}^{'}(t) = -G_{n-k-1}(t)[(n-k-1)f_Y(t) + f_0(t)]. \tag{7.16}$$

Hence from the formulas (7.15) and (7.16), we see eqn (7.13) holds. □

We now show, *mirabile dictu,* that

$$\sum_{k=0}^{n} G_{n-k}(t)H_k(t) - 1 = \int_0^t H_n(x)\,dG_0(x). \tag{7.17}$$

We first check equality at $t = 0$ in eqn (7.16); it follows since $H_0(0) = 1$ and $H_1(0) = \cdots = H_n(0) = 0$ with $G_0(0) = \cdots = G_n(0) = 1$.

Deleting the argument of the integrands in what follows, for notational simplicity, and applying the integration-by-parts formula, we have

$$\begin{aligned}\sum_{k=0}^{n}[G_{n-k}(t) \cdot H_k(t)] - 1 &= \int_0^t \left[\sum_{k=1}^{n} G_{n-k}H_k^{'} + \sum_{k=1}^{n} G_{n-k+1}^{'}H_{k-1} + H_n G_0^{'}\right] \\ &= \int_0^t H_n dG_0.\end{aligned}$$

Using $H_0^{'} = 0$ and the preceding Lemma 1, we find all terms in the summation vanish but one. Thus, we have

Theorem 1. *If the life $X > 0$ is censored by the independent variate $T > 0$, then the Kaplan–Meier estimate $\hat{S}_X$ of survival has expectation (and bias) given by*

$$\mathrm{E}\hat{S}_X(t) = S_X(t)\left[1 + \int_0^t H_n(x)[S_X(x)]^{-2}\, dF_X(x)\right]. \tag{7.18}$$

Proportional Hazards

When it is necessary to perform a duration-of-load test of any kind on a full-scale structural member, several components in the test machine imposing the load are also under heavy stress. Sometimes the specimen fails where it is clamped, not in the intended region. Since "testee" and "testor" both sustain proportional stress repetition throughout, when both are of the same material, it is assumed that the hazard rates of both are proportional. This is also convenient mathematically. We now exhibit a simple, closed-form expression for the expectation in this important special case.

Corollary 1. *If X and T have proportional hazards, i.e., $rH_Y = H_X$ for some $0 < r < 1$, then we have the expectation and the bias of the Kaplan–Meier estimate given by*

$$\mathrm{E}\hat{S}_X(t) = S_X(t)\left[1 - \frac{1}{\mathcal{B}(n+1,-r)}\int_0^p t^n(1-t)^{-r-1}\, dt\right], \tag{7.19}$$

where $p = F_Y(t)$ and $\mathcal{B}(a,b)$ denotes the classical beta function. NB the bias is always positive since for any $n = 1, 2 \ldots,$

$$n\mathcal{B}(n, 1-r) = -r\mathcal{B}(n,-r) = \prod_{j=1}^{n}\left(\frac{j}{j-r}\right). \tag{7.20}$$

Proof. Since in this case we have $f(y,0) = (1-r)f_Y(y)$, we obtain from eqn (7.10) that

$$H_n(t) = \prod_{i=1}^{n}\left[\frac{n-i+1-r}{i}\right] \times [F_Y(t)]^n,$$

which by substituting into eqn (7.19) gives the result. □

We note, from eqn (7.19), the bias is easily computed for even moderate sample sizes n and can be seen to be very small for all values of t for which $p = F_Y(t) \ll 1$.

By using the well-known relationship between the Demoivre and Incomplete Beta distributions, namely,

$$\mathcal{I}_p(n+1,-r) = 1 - \sum_{k=0}^{n}\binom{n-r}{k}(1-p)^{n-k-r}p^k,$$

eqn (7.18) can be converted to the weighted binomial sum, which is the result given by Chen, Hollander, and Langberg (1982) [25].

We now presume the hazard rates of X and T have a bounded ratio.

Lemma 2. *If* $\sup[h_T/h_X] = c_0 < \infty$ *and* $\inf[h_T/h_X] = c_1 > 0$, *then there exists* $0 < r_0 \leq r_1 < 1$ *with* $r_k = c_k/(1+c_k)$ *for* $k = 0, 1$ *such that for* $i = 1, \ldots, n$ *and all* $y > 0$

$$(n-i+1-r_1)f_Y(y) \leq (n-i)f_Y(y) + f(y,0) \leq (n-i+1-r_0)f_Y(y). \tag{7.21}$$

Proof. Simplifying the inequalities in eqn (7.21) we find it equivalent with $r_0 f_Y(y) \leq f(y,1) \leq r_1 f_Y(y)$ for all $y > 0$. But from the definitions of $f_Y(y)$ and $f(y,1)$ in eqn (7.2), we find the lower and upper bounds in eqn (7.21), respectively, are equal to

$$(1-r_0)h_X(y) \geq r_0 h_T(y) \quad \text{and} \quad (1-r_1)h_X(y) \leq r_1 h_T(y) \quad \text{for all } y > 0.$$

By hypothesis we have $h_T(y) \leq c_0 h_X(y)$ and $h_T(y) \geq c_1 h_X(y)$ for all $y > 0$. Hence by setting $c_k = r_k/(1-r_k)$ for $k = 0, 1$ we have the result asas. □

We also obtain

Theorem 2. *In the case the hypothesis of Lemma 2 is satisfied, we have holding, for all* $t > 0$, *the bounds*

$$\frac{[F_Y(t)]^n}{n\mathcal{B}(n, 1-r_1)} \leq H_n(t) \leq \frac{[F_Y(t)]^n}{n\mathcal{B}(n, 1-r_0)} \quad \text{when } 0 < r_0 \leq r_1 < 1, \tag{7.22}$$

as well as the inequalities

$$D_n(t) \leq H_n(t) \leq nD_n(t) \leq [F_Y(t)]^n. \tag{7.23}$$

The lower bound $D_n(t)$ *is given by*

$$D_n(t) = \int_0^t [F_Y(x)]^{n-1} S_Y(x) dH_T(x) \quad \text{for } t > 0. \tag{7.24}$$

Proof. Inequality (7.22) follows from inequality (7.21) by using the definition in eqn (7.10). There remains only the proof of the general bounds. Note inequality (7.23) follows from eqn (7.10) by applying the inequalities $0 \leq f(y,0) \leq f_Y(y)$ for all but the nth term, which contains only $f(y,0) = S_Y(y)h_T(y)$; thus, we obtain the result for D_n. □

Special cases may be required to evaluate H_n in order to determine the exact degree of bias. Remark: If $f_T = 0$ on some interval, then we would have $c_1 = 0$, and correspondingly $r_1 = 0$ and the inequality in expression (7.21) still holds. But if there exists an interval I such that $0 = f_X < f_T$ on I, then expression (7.21) fails on I.

Corollary 2. *If there is an initial period of time during which no censoring can occur, i.e.,* $h_T(t) = 0$ *for* $0 < t < \tau$, *then the K-M estimator is unbiased in that interval and* $E\hat{S}_X(t) = S_X(t)$ *for* $0 < t < \tau$.

PROOF. From $h_T = 0$ on $(0, \tau)$ we see $D_n = 0$ on the same interval. By eqn (7.18) and the inequalities of (7.22), unbiasedness is implied for $0 < t < \tau$. □

NB the unbiasedness is not true for a later interval of time I_1 even if $h_T = 0$ on I_1 whenever there exists a preceding interval $I_0 = (0, \tau)$ on which $h_T > 0$.

7.3. The Variance and Mean-Square Error

To see how closely the K-M estimator, $\hat{S}_X$, approximates S_X as a function of both sample size and time it is advisable to examine the variance and mean-square error; recall

$$\mathrm{Mse}[\hat{S}_X(t)] := \mathrm{E}[\hat{S}_X(t) - S_X(t)]^2. \tag{7.25}$$

By eqn (7.18), we have

$$\begin{aligned}\mathrm{Mse}[\hat{S}_X(t)] :&= \mathrm{Var}[\hat{S}_X(t)] + S_X^2(t)\left[\int_0^t \frac{H_n}{S_X}\,dH_X\right]^2, \\ &= \mathrm{E}[\hat{S}_X(t)]^2 - [S_X(t)]^2\left(1 + 2\int_0^t \frac{H_n(u)}{S_X(u)}\,dH_X(u)\right).\end{aligned} \tag{7.26}$$

The next step must be to evaluate $\mathrm{E}[\hat{S}_X(t)]^2$. From eqn (7.8), we see as before,

$$\begin{aligned}\mathrm{E}[\hat{S}_X(t)]^2 &= n!\sum_{\boldsymbol{z}}\int_{\boldsymbol{y}\in\mathcal{S}}\left[\tilde{S}_X(t;\boldsymbol{z},\boldsymbol{y})\right]^2 L(\boldsymbol{y},\boldsymbol{z})\,d\boldsymbol{y}, \\ &= \sum_{\boldsymbol{z}}\int_{\boldsymbol{y}\in\mathcal{S}}\prod_{i=1}^{n}\left\{\frac{[n-i+1-z_i\mathrm{I}(y_i<t)]^2}{n-i+1}f_{z_i}(y_i)\right\}d\boldsymbol{y}, \\ &= \int_{0<y_1<\cdots<y_n<\infty}\int\prod_{i=1}^{n}\{g_i(y_i,t) - a_i\mathrm{I}(y_i<t)f_1(y_i)\}\,dy_1\cdots dy_n.\end{aligned}$$

where in the third step we set $a_i = (n-i)/(n-i+1)$ and use the definition of g_i in eqn (7.9). Next we partition the sample space a.s. by setting $y_0 = 0$ and $y_{n+1} = \infty$ and introducing $1 = \sum_{k=0}^{n}\mathrm{I}(y_k < t \le y_{k+1})$, then integrate to obtain the expression, omitting functional arguments,

$$\mathrm{E}[\hat{S}_X]^2 = \sum_{k=0}^{n}[S_Y]^{n-k}\Upsilon_k := S_X^2\Omega_n,$$

where we have defined the functions, $\Upsilon_0(t) \equiv 1$ and for $k = 1, \cdots, n$

$$\Upsilon_k(t) := \int_{0<y_1<\cdots<y_k<t} \int \prod_{i=1}^{k} \{(n-i) f_Y(y_i) + f_0(y_i) - a_i f_1(y_i)\} \, dy_1 \cdots dy_k,$$
$$= \int_0^t \Upsilon_{k-1}(x)[(n-k+1) f_Y(x) - (1+a_k) f_1(x)] \, dx. \tag{7.27}$$

Thus

$$\Omega_n = \sum_{k=0}^{n} \Lambda_{n-k} \Upsilon_k, \quad \text{with} \quad \Lambda_{n-k} = [S_T]^2 [S_Y]^{n-2-k}. \tag{7.28}$$

Theorem 3. *The percentage mean-square error of $\hat{S}_X(t)$ is given exactly by*

$$Pmse[\hat{S}_X(t)] := \frac{\mathrm{E}[\hat{S}_X(t) - S_X(t)]^2}{S_X^2(t)} = \Omega_n(t) - 1 - 2 \int_0^t H_n(x) dG_0(x). \tag{7.29}$$

Both asymptotic unbiasedness and mean-square consistency of the K-M estimator have been claimed, since 1958, but sometimes with heuristic arguments. This result makes possible an exact, but tedious, calculation of the Pmse for any prescribed sample size. By noting that $\mathrm{E}\hat{S}_X \geq \mathrm{E}\hat{S}_X^2$, we have an easily calculated upper bound

$$\mathrm{Var}[\hat{S}_X] \leq \mathrm{E}\hat{S}_X[1 - \mathrm{E}\hat{S}_X].$$

Since $\hat{S}_X$ is a biased estimator, we are interested in the mean-square error, namely,

$$\mathrm{E}[\hat{S}_X - S_X]^2 = \mathrm{Var}[\hat{S}_X] + [\mathrm{E}\hat{S}_X - S_X]^2]^2$$
$$\leq S_X(t) \left\{ F_X(t) + [F_X(t) - S_X(t)] \left(\int_0^t \frac{H_n}{S_X} \, dH_X \right) \right\}.$$

But can the exact calculation be reduced as it was in the expectation? Using eqns (7.27), (7.28), and proceeding as before, we obtain asas,

$$\Upsilon_k' = \Upsilon_{k-1}[(n-k+1) f_Y - (1+a_k) f_1], \quad \Lambda_{n-k}' = -\Lambda_{n-k-1}[(n-k) f_Y - 2 f_1]$$

Differentiating Ω_n we find

$$\Omega_n' = \sum_{k=0}^{n-1} \Lambda_{n-1-k} \Upsilon_k [(n-k) f_Y - (1+a_{k+1}) f_1]$$
$$- \sum_{k=0}^{n} \Lambda_{n-1-k} \Upsilon_k [(n-k) f_Y - 2 f_1]$$
$$= 2 f_1 \Lambda_{-1} \Upsilon_n + f_1 \sum_{k=0}^{n-1} \frac{\Lambda_{n-1-k} \Upsilon_k}{n-k}. \tag{7.30}$$

The exact variance has been given previously, in 1982, for the proportional-hazards case, by Chen, Hollander, and Langberg, but even in that reduced case the result is not computationally simple.

If we specialize eqn (7.30) to the proportional-hazards case we have an equivalent expression to that cited above except we use incomplete Beta functions.

Corollary 3. *If the hazards are proportional with* $rH_Y = H_X$, *then*

$$Pmse[\hat{S}_X(t)] = \sum_{k=0}^{n} c_{k,n}(r) \int_0^p u^k(1-u)^{n-1-k-2r}\, du - 2r\mathcal{I}_p(n+1,-r), \quad (7.31)$$

where

$$c_{k,n}(r) = \prod_{i=1}^{k} \left[\frac{(n-i+r)^2 + r(1-r)}{i(n-i+1)} \right].$$

PROOF. We find from eqn (7.28) that

$$\Upsilon_k(t) = c_{k,n}(r) \times F_Y^k(t) \quad \text{for all } t > 0.$$

Then we utilize the results of eqn (7.19). □

7.4. The Nelson–Aalen Estimator

Peterson [91, 1977] and others, have suggested as an estimate of H_X, the hazard function, merely the negative logarithm of the empiric reliability as estimated by the K-M product limit. By eqn (7.3) this is

$$-\ln \hat{S}_X(t) = \sum_{i=1}^{n} z_{(i)} \mathrm{I}(y_{(i)} \le t) \ln \left[\frac{n-i+1}{n-i} \right]. \quad (7.32)$$

This expression is defined for all $t > 0$ and $i < n$, but it remains finite for all $t > 0$ unless the last term has $z_n = 1$, in which case it is infinite. We consider the behavior of estimates of the form

$$\tilde{H}_X(t) = \sum_{i=1}^{n} \omega_{i,n} z_{(i)} \mathrm{I}(y_{(i)} \le t)$$

for some arbitrary weighting factors $\omega_{i,n}$ for $i = 1, \ldots, n$.

Note since

$$\mathrm{E}\tilde{H}_X(t) = \int_0^t \left[\sum_{k=0}^{n-1} (k+1) \binom{n}{k+1} \omega_{k+1,n} [F_Y(y)]^k [S_Y(y)]^{n-k} \right] dH_X(y),$$

we want the term in square brackets to be as close to unity as possible. Thus, we seek estimators that are as close to being unbiased and have as small a mean-square error as possible. Consider the Nelson–Aalen estimate (see [83] and [1]), which is obtained by setting $\omega_{k+1,n} = (n-k)^{-1}$,

$$\hat{H}_X(t) = \sum_{k=1}^{n} \frac{z_{(k)} 1(y_{(k)} \leq t)}{n-k+1}. \tag{7.33}$$

No other simple choice for $\omega_{k,n}$ appears to be as useful. For a discussion of other weights, or "kernels estimates," see [107]. The estimate of S_X formed by using the N-A estimator H_X, namely $e^{-\hat{H}_X}$, was has been studied previously by Altshuler; see [3]. By direct algebraic simplification we can obtain

Theorem 4. *The Nelson–Aalen estimate of the hazard function has expectation*

$$\mathrm{E}\hat{H}_X(t) = H_X(t) - \int_0^t [F_Y(y)]^n \, dH_X(y), \tag{7.34}$$

and mean-square error

$$\begin{aligned}\mathrm{Mse}[\hat{H}_X(t)] = {} & \int_0^t \int_y^\infty \{[F_Y(y) + S_Y(x)]^n - [F_Y(y)]^n\} \, dH_Y(x) \, dH_X(y) \\ & -2 \int_0^t \int_0^y \left[\frac{S_Y(y) F_Y^n(y) - S_Y(x) F_Y^n(x)}{F_Y(y) - F_Y(x)}\right] dH_X(x) \, dH_X(y).\end{aligned} \tag{7.35}$$

The expression for the expectation has been previously given by Aalen in [1, 1976].

Corollary 4. *If for some $0 < r < 1$ we have $rH_Y = H_X$, i.e., proportional hazards between X and T, then the mean-square error is given by the difference of two integrals:*

$$\begin{aligned}\mathrm{Mse}[\hat{H}_X(t)] = {} & r \int_0^p \int_v^1 \frac{t^n - v^n}{(t-v)(1-v)} \, dt \, dv \\ & -2r^2 \int_0^p \int_0^v \frac{v^n - u^n + u^{n+1} - v^{n+1}}{(v-u)(1-v)(1-u)} \, du \, dv,\end{aligned} \tag{7.36}$$

where $p = F_Y(t)$ and the singularities are inessential in both integrands.

7.4.1. Extensions and Generalizations

Of course, we can also consider censoring from the left. This may arise when the initial time of operation is obscured because of circumstance or of insensitivity of the measuring device or when components are initially placed on standby and checked first only when they are put into service. In this case we are concerned with $Y = X \vee T := \max(X, T)$, where $X \sim F_X$ and $T \sim F_T$ where again T is assumed to be independent of X. Hence, we have $F_Y = F_X \cdot F_T$ and as before set $h_X = f_X/S_X$.

Assume that we observe the couple (Y, Z) where we now define $Z = \mathrm{I}[X > T]$. Thus, we obtain the joint density as exactly that of eqn (7.2) with z replaced by $1 - z$. Hence all results given subsequently can be immediately applied in this case, too. Turnbull has studied the K-M estimators under this and more general types of censoring; see [112, 1974] and [113, 1976].

In 1983 Tanner and Wong [110, 1983] obtained an estimate of the hazard rate by using the formal derivative of the N-A estimator smoothed by a class of symmetric nonnegative kernels each of which is $o(t^{-1})$ as $t \to \infty$ and its integral over the sample space is unity. They found the expectation and variance of such estimators and provided asymptotic expressions as well. The flavor of their results is expressed in

Corollary 5. *An estimate of the hazard, say $\tilde{H}_X = (K * \hat{H}_X)$ smoothed by convolution with a distribution K on $\Re$, can be obtained from the N-A estimator as*

$$\tilde{H}_X(t) = \int_0^\infty K(t/x)\, d\hat{H}_X(x) = \sum_{k=1}^{n} \frac{z_{(k)} K(t/Y_{(k)})}{n-k+1} \qquad \text{for } t > 0.$$

Its derivative, namely, $\tilde{h}_X = (K * \hat{H}_X)' = K' * \hat{H}_X$ is an estimate of h_X. It has expectation

$$\mathrm{E}\tilde{h}_X(t) = \int_0^\infty K'(t/x)\, d\,\mathrm{E}\hat{H}_X(x) = \int_0^\infty K'(t/x)[1 - F_Y^n(x)]h_X(x)\, dx,$$

by making use of eqn (7.34).

This agrees with their result for an estimate of the hazard rate by replacing our K' with $K_\varepsilon(y) = K(\ln y/\varepsilon)/\varepsilon$ for any $\varepsilon > 0$ where K is a kernel from their class defined on $\Re$.

Jacobsen has surveyed the state of knowledge in [45, 1988] concerning the K-M and N-A estimators and discussed the use of martingale methods for analyzing failure-time data. in [46, 1989]. A didactic presentation of such methods is given in the text by Shorack and Wellner [104, 1986]. Recent investigations have considered the case when the censoring variates $T_1, \ldots, T_n$ are independent but not identically distributed. Probability bounds have been obtained in 1991 using the $\sup|\hat{H}_X(t) - H_X(t)|$ over the interval $(0, Y_n)$; see [124]. There again the approach was to use the martingale theory of stochastic processes.

Exercise Set 7.A

1. Derive the expectation of Peterson's hazard estimate in eqn (7.32).
2. Verify the result in eqn (7.36) from eqn (7.35).

3. Work out the expectation and mean-square error of the K-M estimator when censoring takes place from the left.

4. Under the same conditions as above, can one find the expectation and variance of the density $\hat{f}_X$, using a convolution-smoothed K-M estimator, from the exact results?

5. *Take the fatigue data in DataSET V and set $K_\varepsilon(x) = \Phi[\xi(x)/\varepsilon]$ for $x > 0; \varepsilon > 0$, then use it to estimate the hazard rate by $\tilde{h}_X(x) = \int_0^\infty K_\varepsilon'(x/y)\, d\hat{H}(y)$, adjusting $\varepsilon > 0$ to obtain the smoothness desired.

6. Sketch the K-M estimate for the sdf and N-A estimate of the hazard function using the data in DataSets II and III in Chapter 7.

7. If both life-length X and left-censor T are Epstein variates, prove the expectation and bias of the K-M estimator in eqn (7.18) are correct.

8. Derive the corresponding expectation and bias for the K-M estimator for a life-length that is independently right censored.

9. (a) If D is the discrete uniform rv on $\{x_1, \ldots, x_n\}$, viz., $\Pr[D = x_i] = 1/n$ for $i = 1, \ldots, n$, and U is uniform on the interval $(-\varepsilon, \varepsilon)$, then show that the rv $D + U$ has the cdf $\hat{F} * K_\varepsilon$, where $\hat{F}$ is the ecdf $\hat{F}(x) = \frac{1}{n}\sum_{i=1}^{n} \mathrm{I}(x \geq x_i)$.
   (b) Why is the smoothed hazard $\tilde{H}_X = K * \hat{H}_X$ said to be a convolution in Corollary 5 when it is true that if X, Y are nonnegative rvs, then the product, not the sum, has the cdf $F_{X \cdot Y} = \int_0^\infty F_Y(t/x) d F_X(x)$ for $t > 0$.

Z.W. Birnbaum began an invited lecture on distribution-free methods before a technical society by saying, "Suppose you are to make observations, for the first time, on a quantity using a metric with unknown properties. Let me assume that you know nothing about the distribution of measurements." After the lecture and questions he remarked, "My first assumption was a VERY APT ASSUMPTION."

CHAPTER 8

Weibull Analysis

8.1. Distribution of Failure Times for Systems

The problem of estimating the hazard rate of a complex system, either of equipment or a biological organism, knowing only that its graph is convex, i.e., initially decreasing and terminally increasing, is both common and difficult. There have been several attempts to introduce multiparameter models in which the hazard rate exhibits such behavior. But these models have often required such large datasets in order to estimate all the parameters that their usefulness has been limited to large samples, as may sometimes be obtained in medical trials but virtually never arise in engineering testing. Because of the paucity of data in engineering applications such multiparameter models are not often favored.

A convex hazard is here regarded as the consequence of failure due to one of a multiple of causes. Thus, we attempt always to utilize information from diagnostic engineering failure analysis and to consider it along with the sample of service-life data. Failures observed during preliminary or qualification testing of any equipment, with subsequent failure diagnosis, are presumed to yield both the service-life (time to failure) plus a determination of the failure mode (cause of failure).

In this study we deal with models for which the existence of changes in the Weibull hazard function are known *ab initio* from knowledge of the engineering utilization with its entailed physical wear. The problem of estimating the hazard under various situations that might arise, with only censored data available, is examined.

8.2. Estimation for the Weibull Distribution

Let the random service-life about which inference is to be made be denoted by X; it has a Weibull distribution when its *hazard function*, say H_X, is given by a power law of the form

$$H_X(y) = (y/\beta)^{\alpha} \quad \text{for all } \ y > 0; \ \text{ for some } \ \alpha > 0, \beta > 0.$$

8.2.1. Right-Censored Estimation

Following the model in Chapter 7 for random censoring on the right during observation of life-length, we define $Y = (X \wedge T)$, where $X \sim F_X$ is the system life-length and $T \sim F_T$ is a censoring variable independent of X.

We observe the couple (Y, Z), where $Z = \mathrm{I}[X < T]$ and $\mathrm{I}[\cdot]$ is the indicator function of the event. The log-likelihood then becomes, by omitting terms independent of α and β, as was discussed in Chapter 6,

$$\ell(\mathbf{y}, \mathbf{z}) \cong \sum_{i=1}^{n} \{z_i[\ln\alpha + (\alpha - 1)\ln y_i - \alpha\ln\beta] - (y_i/\beta)^{\alpha}\}. \tag{8.1}$$

From Chapter 6, the mles, $(\hat{\alpha}, \hat{\beta})$, are the simultaneous solution of the two equations $\frac{\partial\ell}{\partial\alpha} = 0$ and $\frac{\partial\ell}{\partial\beta} = 0$. From eqns (6.17) and (6.18) $\hat{\alpha}$ can be found by solving one equation with a single unknown, viz.,

$$A(\alpha) = \frac{1}{\alpha}, \qquad \text{where} \quad A(\alpha) \equiv \frac{\sum_{i=1}^{n} y_i^{\alpha}\ln y_i}{\sum_{i=1}^{n} y_i^{\alpha}} - \frac{\sum_{i=1}^{n} z_i \ln y_i}{\sum_{i=1}^{n} z_i} \qquad \text{for} \quad \alpha > 0. \tag{8.2}$$

If $\hat{\alpha}$ exists, then the mle $\hat{\beta}$ is given explicitly by

$$\hat{\beta} = \left[\frac{\sum_{i=1}^{n} y_i^{\hat{\alpha}}}{\sum_{i=1}^{n} z_i}\right]^{1/\hat{\alpha}}. \tag{8.3}$$

The solution of eqn (8.2) is the unique intersection of the increasing function A and a decreasing branch of a hyberbola. So the range of $A(\alpha)$ provides a nasc criterion that information in the right-censored dataset ensures the mle $\hat{\alpha}$ exists. It does so iff

$$A(\infty) = \ln[\max_{i=1}^{n} y_i] > \frac{\sum_{i=1}^{n} z_i \ln y_i}{\sum_{i=1}^{n} z_i}.$$

Consider, for example, the case when there is only one failure larger than all the right-censored observations, i.e., all tests but one were terminated for some reason before the single failure was observed. In this circumstance no mles will exist. However, if at least one "run-out" exceeds the failure observation, then both mles would exist.

8.2.2. Left-Censored Estimation

Of course, we may also encounter censoring from the left. This may arise when the initial period of operation is obscured because of circumstance or insensitivity of the device used for failure-detection. Alternatively, it arises when components are inspected for failure modes other than the unknown dominant one, or are initially placed on standby and only tested just before being put into service. In this case we are concerned with $Y = (X \vee T)$, where $X \sim F_X$ and $T \sim F_T$, and again $T \perp X$ is assumed. Hence we have

$$F_Y = F_X \cdot F_T \quad \text{and} \quad h_Y = f_Y/S_Y.$$

Assume that we observe the couple (Y, Z) where we now define $Z = \mathrm{I}[X > T]$. Thus we obtain a joint density which is the same as eqn (8.1) with z_i and $1 - z_i$ interchanged and it leads to the same results as before, properly interpreted.

8.3. Competing Risks

Suppose the service-life of a coherent system is monitored and the cause of each failure is diagnosed; let X_1 denote the time until failure due to improper component installation or substandard material fabrication; let X_2 denote the service-life which is terminated because of an accident, which occurs at random, which defeats the systems capability (such as an accidental overload); let X_3 denote the time until failure due to any damage the nature of which is cumulative such as abrasion, fatigue, or wear. The service-life of a system, call it X, which is subjected only to these three independent competing risks, is

$$X = \min\{X_1, X_2, X_3\} \quad \text{with hazard rate} \quad h_X(t) = h_1(t) + h_2(t) + h_3(t),$$

where $h_i(t)$ is the hazard rate associated with X_i for $i = 1, 2, 3$.

8.3.1. The Bathtub-Shaped Hazard

When the hazard function H_X for life-length X can he written in the form

$$H_X(t) := \sum_{k=1}^{3} \left(\frac{t}{\beta_k}\right)^{\alpha_k} \qquad \text{for } t > 0 \text{ with } \beta_k > 0 \text{ and } 0 < \alpha_1 < \alpha_2 = 1 < \alpha_3,$$

then the hazard rate h_X has a *bathtub shape*, as shown in Figure 2.1. It is called a *competing-risks model* classifying failure as due, respectively, to *imperfect assembly, accident*, or *cumulative damage*. Here $\alpha_1, \alpha_2, \alpha_3$, are the shape parameters, and $\beta_1, \beta_2, \beta_3$ are the characteristic lives, respectively, of each of the failure modes.

Let us assume that α_k, for $k = 1, 2, 3$, are known; then when a complete sample $\{t_1, \cdots, t_n\}$ of failure times is given but *the causes of failure were not noted*, the $\hat{\beta}_k$s, MLEs for the β_k, are the simultaneous solution of equations containing $h_X = H'_X$, where these three parameters are implicit, namely,

$$\sum_{i=1}^{n} t_i^{\alpha_k} = \alpha_k \sum_{i=1}^{n} \frac{t_i^{\alpha_k - 1}}{h_X(t_i)} \qquad \text{for } k = 1, 2, 3. \tag{8.4}$$

Solutions, for any choice of the α_i, can be found using *Mathematica*, or a similar program.

Another specific formulation for the hazard rate of a service-life X, is of the form

$$h_X(t) = at^{-\frac{1}{2}} + b + ct^2 \quad \text{for} \quad t > 0, \tag{8.5}$$

which has application to the service-life of dental-prostheses. A pfm-system (porcelain veneer fused to metal), can be well represented for some unknown positive constants a, b, c. This representation means that failure is due to one of the three modes (categories or causes) with the hazard rate being the sum of the risks contributed by the three terms:

1. b is a constant that determines the chances of failure due to an Act of God. An occurence such as: (i) being accidentally hit in the mouth, or (ii) inadvertently biting down on a hard object, or (iii) accidental staining from food. Such staining may affect any dental prosthesis at any time during its service-life and will occur to new restorations as likely as to old.
2. $at^{-\frac{1}{2}}$ is a term, decreasing in time, that determines the chances of early failure. Its magnitude reflects the likilihood of improper installation or defective materials. (Such failures are similar to infant mortality in biological systems.) The longer a restorative dental system is in service, the more unlikely it is that it will have to be removed because of improper installation.
3. bt^2 is a term, increasing with time, that determines the chances of failure due to *wear-out* from cumulative damage. The longer the prosthesis in service, the more likely that compressive damage will accumulate sufficiently to cause failure; that staining will accumulate until the denture is no longer cosmetic; that crazing will occur on the surface, etc.

Proper analysis depends upon the failure cause being correctly ascribed to the proper failure mode when the time of failure is observed.

8.4. Analysis of Censored Data

8.4.1. Estimation under Independent Competing Risks

Now consider the two random variables (Y, Z), where Y is a life-length variable with failure possible because of three separate causes, i.e., each service-life termination was in one of three independent, distinct, and identifiable modes, or it was censored when still in service, causing termination of testing. Thus, what is observed is a random sample from the couple (Y, Z) where

$$Y = \min\{X_1, X_2, X_3, X_4\} \quad \text{and} \quad [Z = j] \equiv [X_j = Y].$$

Here $X_1, \ldots, X_4$, including the censoring variable, are mutually independent.

Here we have defined the identification sets for each mode and their cardinality as

$$N_k = \{i = 1, \cdots, n : z_i = k\} \quad \text{and} \quad \#(N_k) = n_k \quad \text{for } k = 1, \ldots, 4.$$

Proceeding exactly as in the single failure-mode case we obtain asas

$$\frac{\partial \ell}{\partial \beta_k} = 0 \quad \text{iff} \quad \beta_k^{\alpha_k} = \frac{\sum_{i=1}^n y_i^{\alpha_k}}{n_k}, \tag{8.6}$$

and

$$\frac{\partial \ell}{\partial \alpha_k} = 0 \quad \text{iff} \quad A_k(\alpha_k) = \frac{1}{\alpha_k}, \tag{8.7}$$

where, as before in eqn (8.3), we have

$$A_k(\alpha) \equiv \frac{\sum_{i=1}^n y_i^{\alpha} \ln y_i}{\sum_{i=1}^n y_i^{\alpha}} - \frac{\sum_{i \in N_k} \ln y_i}{n_k} \quad \text{for } k = 1, 2, 3. \tag{8.8}$$

Therefore, the Mle $\hat{\alpha}_k$ is the unique solution of the equation $A_k(x) = 1/x$ for $x > 0$. Correspondingly, we find

$$\hat{\beta}_k = \left[\frac{\sum_{i=1}^n y_i^{\hat{\alpha}_k}}{n_k}\right]^{1/\hat{\alpha}_k} \quad \text{for } k = 1, 2, 3. \tag{8.9}$$

Thus, the estimated hazard of the system life $T = \min(X_1, X_2, X_3)$ is given by

$$\hat{H}(t) = \sum_{k=1}^{3} \left(\frac{t}{\hat{\beta}_k}\right)^{\hat{\alpha}_k} = \sum_{k=1}^{3} \frac{n_k t^{\hat{\alpha}_k}}{\sum_{i=1}^n y_i^{\hat{\alpha}_k}} \quad \text{for } t > 0. \tag{8.10}$$

The estimated values of the shape-parameter $\hat{\alpha}_k$ and the characteristic life $\hat{\beta}_k$ under each failure mode for $k = 1, 2, 3$ collectively determine the shape of the bathtub hazard rate.

8.4.2. Observing Both Time and Cause of Failure

To illustrate the benefits of being able to observe both the *time of failure* and to diagnose the *cause or mode of failure* as well, let us return to the pfm-system disussed previously. Let its random service-life be T with hazard function given by

$$H(t) = \sqrt{\frac{t}{\beta_1}} + \frac{t}{\beta_2} + \left(\frac{t}{\beta_3}\right)^3 \quad \text{for } t > 0, \tag{8.11}$$

with each term corresponding to a distinct failure mode; this pfm-system has a bathtub hazard rate. Assume we have the data-vector $(y_1, z_1, \ldots, y_n, z_n)$ *where we can identify the set of observed service times terminated in the* kth *failure mode, for* $k = 1, 2, 3$ and $k = 4$, which denotes a censored observation. Let n_k denote the number of failure times in the kth mode for $k = 1, 2, 3$. Then we have the three pairs of MLEs for the characteristic lives and shape parameters, one under each failure mode. They are given respectively as the solutions of eqns (8.8) and (8.9).

Notice that *all* the observations, including failures of each kind as well as censored times, are used in the calculation of the estimates of the characteristic life in each failure mode. That is, the failure times in all alternate modes, excepting the one being considered, act as censoring times for that particular mode! It then follows that the estimated hazard and estimated bathtub-shaped hazard rate are, respectively,

$$\hat{H}(t) = \frac{n_1\sqrt{t}}{\sum_{i=1}^{n}\sqrt{y_i}} + \frac{n_2 t}{\sum_{i=1}^{n} y_i} + \frac{n_3 t^3}{\sum_{i=1}^{n} y_i^3},$$

and

$$\hat{h}(t) = \frac{n_1}{2\sqrt{t}\sum_{i=1}^{n}\sqrt{y_i}} + \frac{n_2}{\sum_{i=1}^{n} y_i} + \frac{3n_3 t^2}{\sum_{i=1}^{n} y_i^3}.$$

If we have the more general situation where each of m failure modes can be identified with both shape and scale parameters unknown in each mode, but each mode can be identified by diagnotic analysis of the component at failure, then even when general censoring, from any independent cause, obscures all failures we yet may obtain an estimate of the hazard function, combined from all failure modes, which can be used for any prediction or comparison of that pfm-system; it is given by the expression

$$\hat{H}(t) = \sum_{k=1}^{m}\left(\frac{t}{\hat{\beta}_k}\right)^{\hat{\alpha}_k} = \sum_{k=1}^{m}\frac{n_k t^{\hat{\alpha}_k}}{\sum_{i=1}^{n} y_i^{\hat{\alpha}_k}} \tag{8.12}$$

Just once in a while, let us celebrate the importance of ideas and information.
Edward R. Murrow

Exercise Set 8.A

1. (a) Let $H(t)$ be a hazard function of the form in eqn (8.5) where a, b, and c are unknown. From a censored sample of size n for which $\hat{S}(t)$, the K-M empirical survival function of eqn (7.6), has been obtained determine the minimum-distance estimates for a, b, c by using an appropriate distance between the functions $-\ln[\hat{S}(t)]$ and $H(t)$ for $t > 0$.
   (b) Can you find a distance that gives the least-squares estimates?
   (c) What are the limitations of a censored sample as compared with a complete sample?

2. (a) Consider the demographics of the English village of Stratford-on-Avon during the 16th century. Half the children born there then perished from childhood disease by age 7. By the marriageable age of 18, in the "vigor and health of youth, when the stress of adulthood (work, war, childbirth and omnipresent disease) was encountered," half the surviving population had perished. By age 40 "old age" with its terminal disabilities from a life of hard work and poor nutrition, sets in. Which

hazard $H(t)$ below do you think is the most appropriate

$$(i)\ \ at, \qquad (ii)\ \ a\sqrt{t}+bt, \qquad a\sqrt{t}+bt+ct^3 \tag{8.13}$$

and can be made consistent with the facts above?

(b) What do you think of "the rule of thumb" that sets $H(t) = \frac{1}{2}\left\{\sqrt{\frac{t}{7}} + \frac{t}{18} + (\frac{t}{40})^3\right\}$?

(c) What was the expected life of a child born in Stratford in 1520?

(d) What was the expected life remaining for a person of age 20?

(e) William Cecil, Queen Elizabeth's Secretary of State, died in 1598 at the remarkable age of 78. How can one account for this near miraculous survival?

(f) Suppose that by examination of the gravestones and church registry in the Stratford township one can determine the fraction of those born that have have died at age τ_i (which gives the fraction of ξ_i which have survived) for $i = 1, 2, 3$; we assume both τ_i and ξ_i both decrease with i. Can this information be beneficially used to determine the parameters of eqn (8.13)? What are the cautions of adopting such a procedure?

3. When the first calculations were made of life contingency by Gompertz in the 19th century, there was only one term in the hazard, namely, αt!
   (a) Why, if the hazard was only αt, was its use considered such a big deal?
   (b) What do you think accounts for the difference between simple calculations used then and more complex ones required now?

8.4.3. Estimation with Dependent Failure Modes

What happens if we cannot assume that the failure modes are independent? Even in this case we can still obtain a lower bound on the reliability of the system whenever the times to failure under the separate failure modes are *associated* random variables, in this regard see [7, 1981] or Chapter 4.

Let $T_1, \ldots, T_m$ be the random times until failure in each of the m possible modes. Then if they are associated, it follows that

$$\Pr[\min_{i=1}^{m} T_i > t] \geq \prod_{i=1}^{m} \Pr[T_i > t] \quad \text{for all } t > 0.$$

If we let $Y = \min(T_1, \ldots, T_m)$, then the result above says that

$$\exp\{-H_Y(t)\} \geq \exp\{-\sum_{i=1}^{m} H_{T_i}(t)\},$$

or equivalently that the true hazard function is really less than that estimated from the assumption of independence, namely, $H_Y(t) \leq \sum_{i=1}^{m} H_{T_i}(t)$. From Chapter 4 we recall one can determine if the times of failure in each mode are associated if they are all increasing functions of associated random variables. This can be argued since in each dental restoration the time-to-failure in each mode is extended, or at least not affected, if brushing is done more carefully and regularly, if more care is exercised when biting hard foods, consuming foods that stain dentures is avoided, if less tobacco is chewed, and so on. All types of failure are posited to be extended if better oral hygene is practiced.

8.4.4. Estimation under Random Censoring on Both Sides

Consider the data available from a time-to-failure study of type IV, which we now define to be of three distinct kinds. Each observation is one of three classifications of events, viz., $[X < y]$, $[X = y]$, $[X > y]$. These three types of observations are collected into subsets. These subsets are comprised of all: (1) *failure times*, (2) *run-outs (nonfailures)*, and (3) *the times of discovery of failures which had occurred previously*. Let us, wlog, assume each subset of observations is ordered.

$$IV = \begin{cases} y_1 < \cdots < y_k & \text{are service-times terminating in failure,} \\ y_{k+1} < \cdots < y_n & \text{are right-censored observations,} \\ y_{n+1} < \cdots < y_m & \text{are left-censored observations.} \end{cases}$$

NB we assume $1 \le k \le n \le m$ but we assume no order relation among the maximum observation of each set, here to minimize notation, denoted simply by y_k, y_n, and y_m.

Dropping the subscript X on all life-functions the log-likelihood from data IV can be written, by omitting terms independent of life-functions,

$$\ell \cong \sum_{i=1}^{k} \ln h(y_i) - \sum_{i=1}^{n} H(y_i) + \sum_{i=n+1}^{m} \ln F(y_i),$$

from which we obtain, asas,

$$\frac{\partial \ell}{\partial \alpha} = \frac{k}{\alpha} + \sum_{i=1}^{k} \ln(y_i/\beta) - \sum_{i=1}^{n} H(y_i)\ln(y_i/\beta) + \sum_{i=n+1}^{m} \varepsilon(y_i/\beta;\alpha)\ln(y_i/\beta), \tag{8.14}$$

and

$$\frac{-\beta\partial\ell}{\alpha\partial\beta} = k - \sum_{i=1}^{n} H(y_i) + \sum_{i=n+1}^{m} \varepsilon(y_i/\beta;\alpha), \tag{8.15}$$

where we have introduced, for notational convenience, the function

$$\varepsilon(t;\alpha) = \frac{t^\alpha}{e^{t^\alpha} - 1} = \frac{1}{\sum_{j=0}^{\infty}[t^{j\alpha}/(j+1)!]} \quad \text{for any } t > 0. \tag{8.16}$$

It is easy to see that $\varepsilon(t;\alpha)$ is a slowly varying function of α. If $t < 1$, then $\varepsilon(t;\cdot)$ is an increasing map of $\Re_+$ onto the real interval $(.5819, 1]$, while if $t > 1$ it is a decreasing map of $\Re_+$ onto $[0, .5819)$. Clearly $\varepsilon(\cdot;\alpha)$ is a decreasing function mapping $\Re_+$ onto $(0, 1)$ for any $\alpha > 0$.

The equation $\frac{\partial \ell}{\partial \beta} = 0$ can be written as

$$\beta^\alpha = \frac{\sum_{i=1}^n y_i^\alpha}{k + \sum_{i=n+1}^m \varepsilon(y_i/\beta;\alpha)}. \tag{8.17}$$

The problem in this case is that ε is a function of both α and β so that no explicit solution to eqn (8.17) can be given. Instead we solve the equivalent expression for the variable $x = \beta^\alpha$ in the equation

$$\psi_\alpha(x) = \sum_{i=1}^n y_i^\alpha, \qquad \text{where} \qquad \psi_\alpha(x) := kx + \sum_{i=n+1}^m \frac{y_i^\alpha}{\exp[y_i^\alpha/x] - 1}.$$

This can be done numerically since one notes that $\psi_\alpha(\cdot)$ is an increasing function mapping, for each fixed $\alpha > 0$, the domain $\Re_+$ onto $\Re_+$. So there always exists a unique solution for each given $\alpha > 0$ of eqn (8.17) that can be written as

$$\beta = C(\alpha) := \left[\psi_\alpha^{-1}(\textstyle\sum_{i=1}^n y_i^\alpha)\right]^{1/\alpha}. \tag{8.18}$$

From eqn (8.14), one finds asas, that $\frac{\partial \ell}{\partial \alpha} = 0$ iff

$$(\ln \beta)\left[k + \sum_{i=n+1}^m \varepsilon(\tfrac{y_i}{\beta};\alpha) - \sum_{i=1}^n (\frac{y_i}{\beta})^\alpha\right]$$
$$= \frac{k}{\alpha} + \sum_{i=1}^k \ln y_i + \sum_{i=n+1}^m (\ln y_i)\varepsilon(\tfrac{y_i}{\beta};\alpha) - \sum_{i=1}^n \left(\frac{y_i}{\beta}\right)^\alpha (\ln y_i).$$

But from eqn (8.17) we see the term above in square brackets on the lhs is zero; thus by substituting for β^α from eqn (8.17) we find at the stationary point $\frac{\partial \ell}{k\partial \alpha} = 0$, that

$$\frac{1}{\alpha} + \frac{1}{k}\sum_{i=1}^k \ln y_i = B(\alpha) + \frac{1}{k}\sum_{i=n+1}^m \varepsilon(\tfrac{y_i}{\beta};\alpha)[B(\alpha) - \ln y_i], \tag{8.19}$$

where $B(\alpha)$ is the increasing function of α defined by

$$B(\alpha) := \frac{\sum_{i=1}^n y_i^\alpha \ln y_i}{\sum_{i=1}^n y_i^\alpha} = A(\alpha) + \frac{1}{k}\sum_{i=1}^k (\ln y_i), \tag{8.20}$$

with $A(\alpha)$ having been defined previously in eqn (8.3) by setting $z_i = 0$ for $i > k$.

We now have

Theorem 1. *If any sample data, such as given in display IV, satisfy the hypothesis that for each given $\beta > 0$ the function $D(\cdot|\beta)$ is concave increasing, where $D(\cdot|\beta)$ is defined, for $\alpha > 0$, by*

$$\frac{k}{D(\alpha|\beta)} := \sum_{i=1}^m [B(\alpha) - \ln y_i]\left[I(i \le k) + I(i \ge n+1)\varepsilon(\tfrac{y_i}{\beta};\alpha)\right], \tag{8.21}$$

then the mles $(\hat\alpha, \hat\beta)$ both exist uniquely with $\hat\alpha = \lim_{j\to\infty} \alpha_j$ and $\hat\beta = \lim_{j\to\infty} \beta_j$.

The iterates can be easily computed numerically from the initial values (α_0, β_0), *where* $\beta_0 = C(\alpha_0)$ *and* α_0 *is the unique solution in* α *of the equation*

$$\frac{1}{\alpha} + \frac{1}{k}\left[\sum_{i=1}^{k} \ln y_i + \frac{1}{e-1}\sum_{i=n+1}^{m} \ln y_i\right] = A(\alpha)\left[1 + \frac{m-n}{k(e-1)}\right]. \tag{8.22}$$

For a given pair of iterates $(\alpha_{j-1}, \beta_{j-1})$, *define* (α_j, β_j) *as*

$$\alpha_j = D(\alpha_{j-1}|\beta_{j-1}) \quad \textit{and} \quad \beta_j = C(\alpha_j) \quad \textit{for} \quad j = 1, 2, \ldots.$$

PROOF. The unique existence of α_0 follows from eqn (8.22), since as a function of α the lhs is decreasing while the rhs is increasing. Since $D(\cdot|\beta)$ is concave, increasing the solution of the equation $D(\alpha|\beta) = \alpha$ in α can be found by iteration since $D(\cdot|\beta)$ will be a contractive map. □

We comment that if a type IV sample is obtained in which the right censored data does not come early but constitutes a significant portion of the information so that monotonieity holds but not concavity of $D(\cdot|\beta)$ then computation of the (possibly) multiple roots to determine the mles must be accepted. It is not true that the mles will exist or be unique for all possible points in our censored sample space. However, intractable samples may indicate the assumption of the sample being from a Weibull population may itself be at fault.

We now consider a simple example, namely, the situation for all six permutations of order with three observations: one failure at y_1, one run-out at y_2, and one previous failure detected at y_3. So with k = 1, n = 2, m = 3, we have the inequality $k + m \leq 2n$ satified and the following contingencies:

	Existence $\sqrt{y_1 y_3} < \max(y_1, y_2)$	Uniqueness $y_3 < \sqrt{y_1 y_2}$
(1): $y_3 < y_2 < y_1$	Yes	Yes
(2): $y_3 < y_1 < y_2$	Yes	Yes
(3): $y_2 < y_3 < y_1$	Yes	If $y_1 >> y_3$
(4): $y_2 < y_1 < y_3$	No	
(5): $y_1 < y_3 < y_2$	Yes	If $y_2 >> y_3$
(6): $y_1 < y_2 < y_3$	Yes if $y_2 \approx y_3$	No

8.4.5. Censoring for the Reciprocal Weibull

One might also encounter a reciprocal-Weibull distribution, i.e., an rv that is the reciprocal of a Weibull variate. It is, of course, one of the asymptotic distribution of maxima, viz., one of the extreme-value distributions studied by Gumbel, related by logarithm to eqn (5.17). In this case if $X \sim F_X$ we introduce the *safety function* defined for some $\alpha < 0 < \beta$

$$Q_X(x) := -\ln F_X(x) = (x/\beta)^{\alpha} \qquad \text{for all } x > 0,$$

and the corresponding *safety rate*

$$q_X(x) = \lim_{\epsilon \downarrow 0} \Pr[X > x - \epsilon | X < x] = Q'_X(x).$$

NB when the value of the safety is high the probability of failure is low. This corresponds to having the shape parameter of a Weibull hazard which is negative. Then the distribution and density are defined respectively for all $x > 0$ and some $\alpha < 0 < \beta$ by

$$F_X(x) = \exp[-(x/\beta)^\alpha], \qquad f_X(x) = \frac{-\alpha}{\beta}\left[\frac{x}{\beta}\right]^{\alpha-1} \cdot \exp[-(x/\beta)^\alpha]. \quad (8.23)$$

If we consider a random sample from the pair (Y, Z) where $Y = \max(X, T)$ and $Z = \mathrm{I}(X < T)$ where T is an independent censoring left-variate and Z counts failure, then we find the joint density of (Y, Z) to be exactly the same as in Example 7, Chapter 6, remembering that F_X and $\bar{F}_X$ have been interchanged in the two models.

Thus, given a left-censored sample, the two stationary equations in α and β remain the same. To obtain the mles as given in the solutions of eqn (8.3) and eqn (8.4), we must solve

$$A(\alpha) = \frac{1}{\alpha} \quad \text{for } \alpha < 0, \tag{8.24}$$

where A is the monotone increasing function

$$A(\alpha) = \frac{\sum_{i=1}^n x_i^\alpha \ln x_i}{\sum_{i=1}^n x_i^\alpha} - \frac{1}{k}\sum_{i=1}^k \ln x_i. \tag{8.25}$$

The solution $\hat{\alpha} < 0$ will exist iff

$$A(-\infty) = \ln[\min_{i=1}^n y_i] < \frac{1}{k}\sum_{i=1}^k \ln y_i. \tag{8.26}$$

Then compute

$$\hat{\beta} = \left[\frac{1}{k}\sum_{i=1}^n x_i^{\hat{\alpha}}\right]^{1/\hat{\alpha}} \tag{8.27}$$

Thus, we see the same equations hold for the mles of the parameters of the reciprocal Weibull as given in eqns (8.3) and (8.4) in the case of left censoring. But if $\sum_{i=1}^k (\ln x_i)/k$ is larger than $\sum_{i=1}^n (\ln x_i)/n$ the solution for $\hat{\alpha}$ will be negative.

This same identification holds for both right censoring and double censoring models in this case, and so the same equations hold for all estimators.

Exercise Set 8.B

1. (a) The economic viability of a communications satellite, e.g., for cell phones, depends upon its length of service. Its operation can be permanently interrupted by the

in-service failure of one of a multitude of its components as well as being destroyed by the impact (to use the word correctly) of dust particles or by intense bursts of solar or cosmic radiation. If the service-life, T, is the waiting time until the minimum of these possible modes of failure and the anticipated duration of service until obsolescence is D years, calculate the probability of profit when a revenue stream of \$R/year of operation is anticipated and the cost of construction-launch is \$C; assume all variables are stochastic.

(b) Justify your assumptions, perhaps $C \perp D \perp T$.

(c) Argue that your stochastic formulation provides more insight than just the expected profit. What assurance can you provide for the 'vagueness' of your answer?

(d) What other information would you like to know?

2. Suppose that observations of service-lives of previous satellites, that have been subjected to the same environment, are given in years with plus sign denoting run-out, as

$$t_1 = 0.3,\ t_2 = 2.1+,\ t_3 = 2.7,\ t_4 = 3.3+,\ t_5 = 4.2,\ t_6 = 5.7.$$

Then if D is specified as being between 4 and 5 years, make a more specific calculation for the distribution of profit and upgrade your statement of confidence.

8.5. Change Points and Multiple Failure Mechanisms

The problem of estimating the hazard function of equipment (or biological organism) when its behavior is bilinear is here regarded as a change-point problem for which we propose the utilization of engineering information determinable usually from the physical model or the circumstances of the sample. Also failures observed during preliminary/qualification testing and analysis are presumed to yield both the time of failure (service-life) and a diagnosis of the failure-mode. In other instances the physics of the material may dictate the change point. A few varieties of circumstances from various fields under which this occurs are illustrated with examples in the exercises.

The problem of detecting change points in the hazard function is not new. In this section we deal only with models for which the existence of changes in the Weibull hazard function are known *ab initio* from engineering or physical knowledge. The problem of estimating the hazard under various situations that have arisen, with only censored data available, is examined.

Let the random service life about which inference is to be made be denoted by $X \sim F_X$; it has a Weibull distribution with *hazard function* given by a power law of the form

$$-\ln[1 - F_X(y)] := H_X(y) = (y/\beta)^\alpha \quad \text{for all } y > 0; \text{ for some } \alpha > 0, \beta > 0.$$

The *survival distribution, density*, and *hazard rate* are defined, respectively, by

$$S_X = 1 - F_X = \exp[-H_X], \quad f_X = F_X', \quad \text{and } h_X = H_X' = f_X/S_X.$$

This notation will be used for any random variable by altering the affixes.

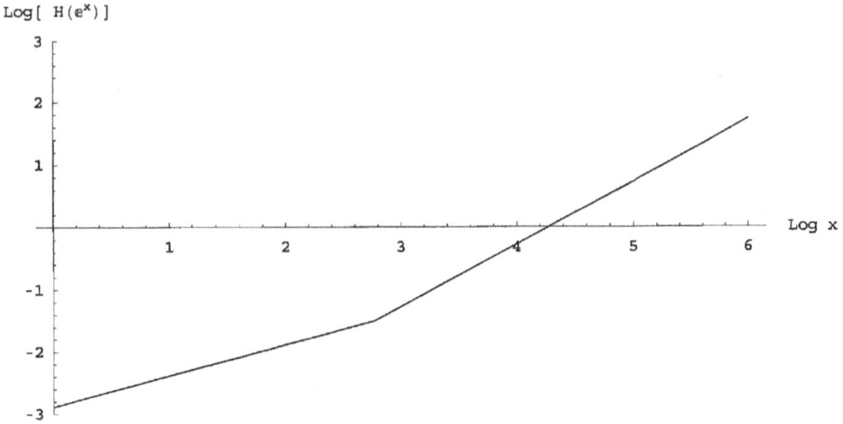

Figure 8.1. Log-hazard plot of flexed-Weibull with slopes $\alpha_1 > \alpha_0$ and with intercepts at $(\log_e \tau, y)$, where $y = \frac{\alpha_0}{\alpha_0 - \alpha_1}[\log_e \beta_0^{\alpha_1} - \log_e \beta_1^{\alpha_0}]$, $\log_e \tau = \frac{1}{\alpha_0 - \alpha_1}[\alpha_0 \log_e \beta_0 - \alpha_1 \log_e \beta_1]$.

8.5.1. A Known Change Point

In 1968, after seeing a few large uncensored datasets of fatigue lives on metallic wire, which exhibited nonlinear behavior in the plot of the log-hazard of the empiric cumulative, V. Weibull proposed an adaptation of the Weibull distribution. He proffered physical reasons why two failure mechanisms can be postulated with the stronger one in the early part of life and a second of less severity during the latter. It can often be inferred from the log-hazard plot of a large dataset when a bilinear hazard is appropriate. In fatigue the second line has the smaller slope.

The situation in Figure 8.1, from de Rijk et al. [29, 1990], represents compressive strength for a dental composite when the second slope increases. It is a resin-based restorative and if the theoretical proportions are exact should exhibit a change point that can be theoretically determined. If a Weibull distribution is appropriate, then a bilinear log-hazard with a single change of slope in the data would apply. A model can be constructed by proposing a single flex-point in the log-hazard as follows: let

$$H_k(x) = (x/\beta_k)^{\alpha_k} \quad \text{for } x > 0; \; \alpha_k, \beta_k > 0; \; \text{with } k = 0, 1$$

denote the hazard for two Weibull distributions. Then the hazard for the flexed-Weibull was defined, for known τ, by

$$H(x) = \max[H_0(x), H_1(x)] \quad \text{for } x > 0 \text{ with } H_0(\tau) = H_1(\tau). \tag{8.28}$$

Of course, we could have data for which the hazard function in eqn (8.28) should be defined using the minimum. Note the hazard function in these cases is different than in mixtures given by $pS_0(t) + (1 - p)S_1(t)$ for some $0 < p < 1$.

So making no presupposition about the relative magnitudes of $\ln H_0$ and $\ln H_1$ on $(0, \tau)$, we set

$$H(x) = \mathrm{I}[x \geq \tau]H_0(x) + \mathrm{I}[x < \tau]H_1(x) \;\; \text{with } H_0(\tau) = H_1(\tau), \tag{8.29}$$

then we have the log-hazard rate for a double-mechanism failure with a known change point at τ given by

$$\ln h(x) = \mathrm{I}[x \geq \tau] \ln h_0(x) + \mathrm{I}[x < \tau] \ln h_1(x) \quad \text{for } x > 0. \tag{8.30}$$

We then obtain the density

$$f_X(x) = \mathrm{I}(x \geq \tau) \cdot f_0(x) + \mathrm{I}(x < \tau) \cdot f_1(x) \quad \text{for } x > 0. \tag{8.31}$$

NB that $f_X(\tau+)$ and $f_X(\tau-)$ may not be equal.

If independent censoring by T occurs and as before we observe $Y = \min(X, T)$ along with $Z = \mathrm{I}[X < T]$, then by following the development in eqns (8.3) and (8.4) we have for the log-likelihood, given the sample vector and the change point τ asas and omitting terms containing no unknown parameters, the expression

$$\ell(\alpha_0, \beta_0, \alpha_1, \beta_1 | \mathbf{y}, \mathbf{z}, \tau) \cong \sum_{k=0}^{1} \sum_{i \in N_k} [z_i \ln h_k(y_i) - H_k(y_i)]. \tag{8.32}$$

We utilize the notation for fixed $\tau > 0$ for both sets of indices

$$N_k = \{i = 1, \ldots, n : I[y_i < \tau] = k\}, \quad N_k^* = \{i \in N_k : z_i = 1\} \quad \text{for } k = 0, 1, \tag{8.33}$$

with

$$\#\{N_k\} = n_k, \quad \#\{N_k^*\} = n_k^* \quad \text{for} \quad k = 0, 1. \tag{8.34}$$

The likelihood must be maximized subject to the constraint $\ln H_0(\tau) = \ln H_1(\tau)$. We utilize a Lagrange multiplier λ to obtain the augmented log-likelihood, say

$$\ell^* = \sum_{k=0}^{1} \left\{ \lambda_k \ln H_k(\tau) + \sum_{i \in N_k} [z_i \ln h_k(y_i) - H_k(y_i)] \right\}, \tag{8.35}$$

where for notational convenience we set $\lambda_k = (-1)^k \lambda$ for $k = 0, 1$.

Thus, we must solve the four equations:

$$\frac{\partial \ell^*}{\partial \alpha_k} = 0, \qquad \frac{\partial \ell^*}{\partial \beta_k} = 0 \quad \text{for} \quad k = 0, 1,$$

along with the constraint, which we rewrite as

$$(\tau/\beta_0)^{\alpha_0} = (\tau/\beta_1)^{\alpha_1}. \tag{8.36}$$

We find from eqn (8.35) that for $k = 0, 1$

$$\frac{\partial \ell^*}{\partial \alpha_k} = \sum_{i \in N_k} \left\{ z_i \left[\frac{1}{\alpha_k} + \ln(y_i/\beta_k) \right] - (y_i/\beta_k)^{\alpha_k} \ln(y_i/\beta_k)] \right\} + \lambda_k \ln(\tau/\beta_k),$$

and

$$\frac{\partial \ell^*}{\partial \beta_k} = \frac{\alpha_k}{\beta_k} \left[\sum_{i \in N_k} (y_i/\beta_k)^{\alpha_k} - (n_k^* + \lambda_k) \right].$$

We note since τ is known that we can wlog replace each y_i by y_i/τ and each β_k by β_k/τ within each ratio y_i/β_k; thus, τ can be assumed to be unity and so disappears in both the above partials.

We see that $\frac{\partial \ell^*}{\partial \beta_k} = 0$ implies

$$\beta_k^{\alpha_k} = \frac{\sum_{i \in N_k} y_i^{\alpha_k}}{n_k^* + \lambda_k}; \tag{8.37}$$

and that $\frac{\partial \ell^*}{\partial \alpha_k} = 0$ implies, asas following substitution from eqn (8.37),

$$\sum_{i \in N_k} z_i \ln y_i = B_k(\alpha_k, \lambda), \tag{8.38}$$

where we define for $\alpha > 0$ the right-hand side of eqn (8.38) by

$$B_k(\alpha, \lambda) := [n_k + \lambda_k] \psi_k'(\alpha) - \frac{d_k \psi_k(\alpha) + c_k(\lambda)}{\alpha}. \tag{8.39}$$

Here we have, for notational simplicity, introduced the sample functions

$$\psi_k(\alpha) = \ln \left[\sum_{i \in N_k} y_i^{\alpha}/n_k \right] \quad \text{and} \quad \psi_k'(\alpha) = \frac{\sum_{i \in N_k} y_i^{\alpha} \ln y_i}{\sum_{i \in N_k} y_i^{\alpha}}, \tag{8.40}$$

using the constants from eqn (8.34)

$$d_k = n_k - n_k^*, \quad \text{and} \quad c_k(\lambda) = n_k^* - d_k \ln[1 + (\lambda_k/n_k)]. \tag{8.41}$$

Now one checks that ψ_k is convex with $\psi_k(0) = 0$; thus, ψ_k is star-shaped. It follows that the difference

$$\Delta_k(\alpha) = \psi_k'(\alpha) - \frac{\psi_k(\alpha)}{\alpha} = \frac{1}{\alpha} \int_0^{\alpha} x \psi_k''(x)\, dx \geq 0. \tag{8.42}$$

Now we will show, except when the percentage of censored observations is very high, that in a right-censored Weibull sample $(\boldsymbol{y}, \boldsymbol{z})$ we have both the existence and uniqueness of the mles of β_k and α_k.

Lemma 1. *If we define the two constants*

$$u_k = \exp\{n_k/d_k\} - n_k \quad \textit{for } k = 0, 1 \textit{ and restrict } \lambda \in (-u_1, u_0) \tag{8.43}$$

then $B_k(\alpha, \lambda)$ is a monotone increasing of α for $k = 0, 1$.

Proof. Since clearly the first term of eqn (8.39) is increasing, we see $B_k(\alpha, \lambda)$ will be increasing if we show that the second term, viz., $[d_k\psi_k(\alpha) + c_k(\lambda)]/\alpha$, is decreasing in α. The derivative of this term is nonpositive iff $\alpha\Delta_k(\alpha) \leq c_k(\lambda)/d_k$. (In the case when $d_k = 0$ the claim is trivially true.) By eqn (8.42) to see that $\alpha\Delta_k(\alpha)$ is an increasing function of α, it is sufficient to check that

$$\frac{c_k(\lambda)}{d_k} \leq \lim_{\alpha\uparrow\infty} \alpha\psi_k(\alpha) = \lim_{\alpha\uparrow\infty}[\alpha\psi_k'(\alpha) - \psi_k(\alpha)]$$
$$= \lim_{\alpha\uparrow\infty}\left\{\frac{\sum_k t_i^\alpha \ln t_i^\alpha}{\sum_k t_i^\alpha} - \ln\left[\frac{\sum_k t_i^\alpha}{n_k}\right]\right\} = \ln n_k,$$

with the obvious notation for summation where we defined

$$m_k = \max_{i\in N_k}(y_i) \quad \text{and set} \quad t_j = y_j/m_k \;\; \text{for } j \in N_k. \tag{8.44}$$

Hence a sufficient condition for the monotoneity of $B_k(\cdot, \lambda)$ is that $c_k(\lambda) \geq d_k \ln m_k$, which is equivalent with the restriction on λ given in eqn (8.43). □

Lemma 2. *If $\lambda \in (-u_1, u_0)$, as specified in eqn (8.43), and we define the constants*

$$v_k = n_k^* - \frac{\sum_{i\in N_k^*}}{\ln m_k} \quad \textit{for } k = 0, 1 \;\; \textit{and restrict} \;\; \lambda \in (-v_0, v_1), \tag{8.45}$$

then we find

$$\lim_{\alpha\uparrow\infty} B_k(\alpha, \lambda) > \sum_{i\in N_k^*} \ln y_i. \tag{8.46}$$

Thus $\hat{\alpha}_k$ exists uniquely since eqn (8.38) has exactly one solution.

Proof. The restriction on λ given in eqn (8.45) is equivalent with

$$\ln m_k > \frac{\sum_{i\in N_k} \ln y_i}{n_k^* + \lambda_k},$$

which one checks to be the same as eqn (8.46). □

The two mles $\hat{\alpha}_k$ for $k = 0, 1$ and the Lagrange multiplier $\hat{\lambda}$ can be determined uniquely by an iterative process. They are the simultaneous solution of three equations, namely, eqn (8.38) for $k = 0, 1$ and $\lambda = \Lambda(\alpha_0, \alpha_1)$ where, from the constraint given in eqn (8.36) after substitution from eqn (8.37) and replacing each y_i by y_i/τ, we have introduced the function

$$\Lambda(\alpha_0, \alpha_1) = \frac{n_1 \sum_{i\in N_0}(y_i/\tau)^{\alpha_0} - n_0 \sum_{i\in N_1}(y_i/\tau)^{\alpha_1}}{\sum_{k=0}^{1}\sum_{i\in N_k}(y_i/\tau)^{\alpha_k}}. \tag{8.47}$$

To see this note that eqn (8.43) ensures $B_k(\cdot, \lambda)$ is monotone increasing (which is always true in the complete sample case since $n_k = n_k^*$) so that eqn (8.38) has at most one solution. For a solution to exist we must have $\lim_{\alpha\to\infty} B_k(\alpha, \lambda) > \sum_{i\in N_k^*} \ln y_i$, which is quaranteed by eqn (8.45).

Algorithm. Let the right-censored sample (y, z) be divided at a given τ into the two subsets indexed by N_k for $k = 0, 1$ as defined by eqn (8.33). Here wlog the scale measurement is chosen so that

$$m_k = \max_{i\in N_k} y_i > 1 \quad \text{for both } k = 0, 1, \tag{8.48}$$

with n_k, n_k^*, d_k as defined in eqn (8.34).

Let the slopes $\hat{\alpha}_{k,0}$ for $k = 0, 1$ be estimated (probably by least-squares) from a log-plot of the hazard. Then given $\hat{\alpha}_{k,j}$ for both $k = 0, 1$ and any $j = 0, 1, 2, \ldots$ one computes the jth iterate, namely,

$$\hat{\lambda}_j = \Lambda(\hat{\alpha}_{0,j}, \hat{\alpha}_{1,j})$$

from eqn (8.47). If $\hat{\lambda}_j \in (-u_1, u_0) \cap (-v_0, v_1)$, then $\hat{\alpha}_{k,j+1}$ can be computed as the unique solution to eqn (8.38).

To solve for the three variables $\hat{\alpha}_0$, $\hat{\alpha}_1$, and $\hat{\lambda}$ in the three expressions given in eqns (8.38) and (8.47), we must have $\hat{\lambda}_j \in (-u_1, u_0) \cap (-v_0, v_1)$ at each stage of the iteration in order to compute the next iterate, after the initial quess at alpha-values. The induction step is: given $\alpha_{0,j}$ and $\alpha_{1,j}$, we then use eqn (8.47) to compute $\hat{\lambda}_j$. Then, from eqn (8.38), we solve for $\alpha_{0,j+1}$ and $\alpha_{1,j+1}$. The solution for the mles is obtained, to within a preset degree of accuracy, in the limit as

$$\hat{\alpha}_0 = \lim_{j\to\infty} \alpha_{0,j}, \qquad \hat{\alpha}_1 = \lim_{j\to\infty} \alpha_{1,j}, \qquad \hat{\lambda} = \lim_{j\to\infty} \lambda_j.$$

Then from eqn (8.37) the mles of β_k are computed explicitly as

$$\hat{\beta}_k = \left[\frac{\sum_{i\in N_k} y_i^{\hat{\alpha}_k}}{n_k + (-1)^k \hat{\lambda}}\right]^{1/\hat{\alpha}_k} \quad \text{for} \quad k = 0, 1. \tag{8.49}$$

8.5.2. A Change Point at an Unknown Location

In case the change point τ must itself be estimated, we might resort to "brute computational force" and calculate, from eqn (8.9), the log-likelihood maximized with respect to all parameters except the assumed value of $\tau > 0$, namely, we have by setting

$$g(\tau) = \ell^*(\hat{\alpha}_0, \hat{\beta}_0, \hat{\alpha}_1, \hat{\beta}_1 | y, z, \tau) = \sum_{k=0}^{1} \sum_{i\in N_k} [z_i \ln \hat{h}_k(y_i) - \hat{H}_k(y_i)],$$

$$\cong \sum_{k=0}^{1} n_k^* \left\{ \ln \hat{\alpha}_k + \hat{L}_k[\sum_{i\in N_k^*} \ln y_i / n_k^*] \right\}, \tag{8.50}$$

where we obtain eqn (8.50) by omitting constants independent of τ and we recall $L_k(x) = \ln H_k(x)$. We note that $\sum_{i\in N_k} \hat{H}_k(y_i) = n_k$ and $\hat{h}_k(y) = \hat{\alpha}_k \hat{H}_k(y)/y$. We also see that $\hat{\alpha}_k$, $\hat{\beta}_k$, n_k^*, and N_k^* *are all implicit functions of* τ for $k = 0, 1$.

This calculation could be repeated for a suitable range of values of τ to obtain $\hat{\tau}$, the mle of the flex-point. This has been done explicitly for the exponential distribution by Loader (see [61, 1991]), but it is computationally slow and inelegant.

However, one can see that if all the data, viz., $\ln y_i$ for $i = 1, \ldots, n$, were to follow exactly a "broken line," then $g(\tau)$ would attain its supremum at the change point since from eqn (8.50), omitting all terms independent of τ,

$$g(\tau) \cong \sum_{k=0}^{1} \left\{ n_k^*[\ln \hat{\alpha}_k - \hat{\alpha}_k \ln \hat{\beta}_k] + [\hat{\alpha}_k - 1] \sum_{i\in N_k^*} \ln y_i \right\}. \tag{8.51}$$

We note $g(\tau)$ is a step function with saltuses at the points of the right-censored data set $\{y_1, \ldots, y_n\}$ which wlog we can assume to be ordered. To fix ideas assume there is a unique change point τ, as might be appropriate for the data as given in Figure 8.1, such that

$$\operatorname{sgn}(t - \tau) = \operatorname{sgn}[H_0(t) - H_1(\tau)] \quad \text{and} \quad H_0(\tau) = H_1(\tau). \tag{8.52}$$

Let us consider the two subsets of observations divided at τ, which has been chosen arbitrarily, viz., $N_k(\tau)$ and $N_k^\star(\tau)$ are as defined in eqn (8.33) for $k = 0, 1$. Our problem is to develop an easy procedure to obtain a good estimate of the true change point. We utilize the (nonparametric) Nelson–Aalen estimator, c.f. Chapter 7.5.2, for each section of the hazard, namely,

$$\tilde{H}_0(t) = \sum_{i\in N_0(\tau)} \frac{z_i \mathrm{I}(y_i \le t)}{n - i + 1} \quad \text{for } 0 < t < \tau,$$

$$\tilde{H}_1(t) = \tilde{H}_0(\tau) + \sum_{i\in N_1(\tau)} \frac{z_i \mathrm{I}(y_i \le t)}{n - i + 1} \quad \text{for } \tau < t < \infty.$$

This is known to be a slightly biased, but consistent, estimator of the hazard for any reasonable censoring distribution and its mean-square error is known, *loc. cit.*

In theory if τ is the true change point, then each section of the log-hazard is (nearly) linear. Hence we propose to measure the degree of linearity for various choices of τ. The first method depends on the linear correlation coefficient.

I: Set

$$u_k(i) = \ln \tilde{H}_k(y_i) \quad \text{for } i \in N_k(\tau) \text{ and } k = 0, 1 \tag{8.53}$$

and compute the linear correlation coefficient for each set of pairs $\{(u_k(i), y_i) : i \in N_k(\tau)\}$ for $k = 0, 1$. This is

$$\rho_k(\tau) = \frac{\sum_{i\in N_k^\star(\tau)} (y_i - \bar{y}_k)(u_k(i) - \bar{u}_k)}{\sqrt{\sum_{i\in N_k^\star(\tau)} (y_i - \bar{y}_k)^2 \sum_{i\in N_k^\star(\tau)} (u_k(i) - \bar{u}_k)^2}}. \tag{8.54}$$

We then seek to maximize the sum $\rho_0(\tau) + \rho_1(\tau)$ as a function of τ. Fortunately this can be easily done with extant software. Let the maximizing value be labeled $\tilde{\tau}$.
II: Another method is merely to minimize the sum of squared deviations. A moments reflection on the data seen in Figure 8.1 shows the mle of the flex-point, as computed from eqn (8.50), being a function of slopes and intercepts, must be near the intersection of the two least-square lines. These are easily obtained by computer using the two divided data sets.

Thus, having determined from calculations by either method that $g(\tau)$ is maximized on $y_j^* < \tau < y_{j+1}^*$ we then compute the final mle's α_k^*, β_k^* for $k = 0, 1$. Then based on this bisection of the data we take as our estimate

$$\hat{\tau} = \exp\left\{\frac{\alpha_0^*(\tau)\ln\beta_0^* - \alpha_1^*\ln\beta^*(\tau)}{\alpha_0^* - \hat{\alpha}_1^*}\right\}. \tag{8.55}$$

Of course, this process could theoretically be repeated for as many change-points $\tau_1 < \cdots < \tau_m$ as could be inferred from the data. But here the equations have been derived for data, including right-censored observations, only from a distribution with a known number of change points.

8.5.3. Conclusions

In a number of situations in the estimation of the hazard rate of a complex system, not only is the time of each failure recorded but a careful diagnosis is made as to the mechanism of failure and the causal component identified. Some observed failures, caused by mechanisms of which the designer was ignorant, require that certain components be redesigned. But the tested life of the system, counting as censored observations those failure times which have been observed occuring in modes that are no longer possible on the redesigned system, can sometimes be utilized to construct an estimate of the new reliability.

Exercise Set 8.C

1. There are three methods that might be utilized to estimate the change point in bilinear log-hazard from a Weibull distribution.
   (a) Use mles to obtain the $\hat{\alpha}_k$, $\hat{\beta}_k$ on each segment $k = 0, 1$ using data truncated at known τ. Then we have $\hat{\delta}_k = \hat{\alpha}_k \ln[\hat{\beta}_k]$ for $k = 0, 1$.
   (b) Use the empiric sdf, viz., $\ln\ln[\hat{S}_n(x)]^{-1}$, and least-squares on each linear segment from the "guess-timated" change point.
   (c) Use the N-A estimate, $\hat{H}_k$ on each segement from the "guess-timated" change point and apply least-squares to the saltus points of $\ln[\hat{H}_k]$.
   (d) Discuss the statistical properties, such as bias and consistency, of each of the methods presented to estimate the change point when contrasted with the ease of getting the answer. Compare these methods with Bayesian procedures and discuss their relative merits.

2. If it is supposed that data in the K-M estimator were from a Weibull population, then it is anticipated that the log-log-plot , namely, $\ln[-\ln \tilde{S}(t)]$, should be approximately linear (at least in early life).
   (a) What do you think of this method for supporting this supposition?
   (b) Some investigators recommend that to avoid difficulties outside the interval betwen the minimum and maximum failure that $\tilde{S}(t)$ be replaced by $[1 + 2n\tilde{S}(t)] \cdot [2(n+1)]^{-1}$. Explain why and when you favor, or disfavor, this practice.

Therefore, I have now proved by means of numbers gotten on one side from astronomy and on the other side from harmonies that, taken in every which way, harmonic ratios hold between the twelve termini or movement of the six planets revolving around the sun or they approximate such ratios within an imperceptible part of least accord.

Harmonies of the World, Johannes Kepler—1618

These volumes of Kepler's contained his third law of planetary motion, which despite its inelegance, showed how the planets orbited. His "harmonies" are what we would call his "theory of elliptical orbits." This astonishing agreement between theory and observation for the movement of all six planets is what later inspired Newton to seek to discover why, and more succinctly how, they did so.

CHAPTER 9

Examine Data, Diagnose and Consult

9.1. Scientific Idealism

Michael Faraday was a boy of the London Slums, whose education consisted only of the rudiments of reading, writing, and arithmetic. He was, at age 14, apprenticed to a bookbinder. Being both curious and intelligent, he read the books he bound and soon became fascinated by the scientific problem of "the electric fluid."' He then believed, he later recalled, that all men of science were admirable, "philosophers" diligent in their pursuit of truth. Later, as a man, his discovery of electromagnetic induction afforded him introductions to the renown European scientists of his age. He was disappointed to learn, instead of pursuing truth with humility, they were, in the main, noticeably jealous of one another and driven, like many other men, by a desire for fame and wealth.

Ultimately Faraday's discoveries made him so well known that he was offered large fees to study technical problems confronting industrial and military leaders who were seeking to exploit "magnetism" for wealth or aggrandizement. By 1832 he was earning more than £1000 per year. Remaining true to his altruistic ideals for scientists he determined not to accept paid assignments when it interfered with his concentration on pure science; thus in 1833 his income dropped to £152. Moreover, he neglected to take out proprietary patents on his discoveries, which allowed others to profit undeservedly from his work!

This experience pales to insignificance when compared with the explotation of Nicola Tesla by George Westinghouse, during the expansion of his industrial empire. This tragedy is trumped by Tesla's own refusal of a Nobel prize because he would have had to share it with, in his view, 'that undeserving tinkerer,' Thomas Edison.

Persons of such idealistic dedication, or naivete, are regarded as foolish today. Most students who study science cannot expect to be employed in the mythical 'Ivory Tower' with time to investigate whatever suits their fancy. But students can hope to be employed in businesses that profit from research using scientific advancement or technological development.

Organized scientific investigation is not expected to uncover basic knowledge for its own sake so much as to work on project developments that are, perhaps narrow, but of foreseeable economic return. In the current view, allowing scientists to work without direction only affords them an opportunity for engaging in forms of expensive and profitless "research" yielding only personal gratification.

To some persons such efforts are like the failed expedition of Hernando De Soto to find the Seven Cities of Cibola. If he had just mapped the rivers, the harbors, and other geographic features, his expedition would have been of great value, instead of being known only for its failure to find gold. In our current litigious society scientists are often expected to behave like lawyers,[1] i.e, not to do unbiased research but to amass only data that supports the position which their management has already determined. Usually this means following the employer's belief that 'the greatest good to the greatest number' comes from maximizing the return to their stockholders (indirectly to themselves) and too often without regard for consequences to society or posterity.

9.2. Consultation and Diagnosis

Many students of mathematical reliability will eventually be, occasionally, engaged, at least temporarily, as technical consultants. In such cases it is advisable for the consultant to have clearly in mind guidelines for evincing, most rapidly, the nub of the problem with which he is presented. These include

1. Know the client's business interest in obtaining a solution, along with the entailed business bias.
2. Learn the expectations of the client both as to what can be accomplished and when it is to be completed.
3. Ascertain what relevant data are currently extant to support the investigation or confirm the solution.
4. Keep your own counsel until you are sure you understand the real problem.

There are some additional homilies that might well be kept in mind.

- Consultants are like the moon, they can only shine in reflection.
- Clients need help more often with problem formulation than in data analysis.
- The *Petitio Principii* is: do not assume anything, however small, that has yet to be established, e.g., how do you know which factors make no difference? What is the sample space?
- Giving the right answer to the wrong question is a Type III error which occurs too frequently. For example, what is the correct responce to the question "How do you apply the paired t-test to this data?"
- Statistical significance is not practical significance.
- Every consultant should be knowledgeable about the phenomenon being tested, not just on the methods of analysis advocated.
- Statistics is not a subset of Mathematics. If 'A implies B ' has been proved, then it always makes sense to ascertain what evidence supports A.

[1] "I would not wish to traduce any man, but wasn't that gentleman a lawyer?": Samuel Johnson.

More specifically one ought to begin with these cautions.

1. Clarify Vocabularly:
   Make certain that the jargon, which always arises to facilitate communication between members of common interest groups, is clearly defined and understood. You must be able to understand the problem as the client views it and ultimately to communicate an understanding of your findings, and their consequences, to him. Otherwise your efforts, however definitive, are in vain.
2. Clarify the Problem:
   Let the client state the problem and then you restate it to him, in general terms as you understand it, elucidating all the enthymemes. Repeat this iteration until you reach agreement.
3. Qualify the data:
   If data are to be collected, it is very important that you review the research design, the questionnaire or measurement-instrument, and the mathematical/statistical analysis that is to be performed *prior to the data collection*. If the data are already in hand, make sure they satisfy the assumptions required in the analysis that is to be used. (Grab-sets are not good datasets.) If the hypothesis is false, any conclusion can correctly be deemed true.
4. Understand the Client and his interests:
   Clients often think they know the answer to the problem they pose and come to you only to have you bolster (prove?) their case. They can then avoid responsibility, in case an error is discovered later, by saying it was approved by your authority.

Besides having the requisite knowledge, a consultant must possess suitable interpersonal skills. (Just as with physicians, where bedside manner counts as well as medical knowledge.) A consultant wants to be judged a wise person, not a wise guy. Don't act like John von Neumann unless you are as knowledgable as he. A successful consultant must acquire background and experience in areas related to his mathematical speciality. This enables a consultant to provide viable alternative interpretations or to suggest other approaches to the problem's solution.

Paul Halmos, although an exceptionally lucid expositor and textbook author, in his *Automathography* [43], said: Anybody can do consulting in applied mathematics. "I never needed high-power mathematics: a little calculus, a rare bit of linear algebra, and a lot of trignometry." This really means that Halmos knew insufficient Physics or Aeronautical Engineering or Economics, *et al.* for consultation in any of those fields so that his supervisor could direct only the most trivial consulting problems to him. You must know some science; mathematics is not a science.

A consultant must learn to develop cooperative relationships with the clients, and to respect practical experience. Most significant scientific problems today are so complex they require a team of specialists from several fields. An apparent measure of professional achievement is successful interaction between oneself and persons who were known to be involved in meritorious scientific endeavor. For consultants in mathematical modeling the overweaning requirement is that the assumptions be correct for the application intended. In statistics this often means the sample must be properly related to the inference sought. Mere "collection of data" is frought with difficulty. The redoubtable Sir Ronald Fisher once said in regard to a "grab-set" "That Sir is not an experiment you've run, that is an experience you encountered."

After clarifying the assumptions and deriving the implications from the entailed model, one must insure it is justified. If there is an error the consultant will be blamed but if the model is right and successful the consultant will likely get little or no credit. Remember Alan Turing's remark: "This model is a simplification of an idealization and hence a falsification, yet it provides theoretical insight where none was before and allows predictions of practical importance."

A consultant can have the business or the credit, not always both. A consultant can increase his "business" by markedly improving the scientific content of each project (or publication) through his contributions but not insisting that he be credited as a co-author. (Giving away something of value is always popular.) Consequently, by increasing one's business one's scientific recognition is penalized. If you always insist upon fee with credit, your desirability is slightly reduced. One's clients must come to regard one not as an employee but as a deserving conferee. To the experimentalists it seems basically unfair for them to have worked for months or years in the field or laboratory obtaining data and then have to share publication with someone who has worked only a few days or weeks.

Moreover, if consultants get paid why should they get credit too? This difficulty is at least as old as l'Hôpital's Rule.[2] The crux of the problem is that justice is a thin line. Remember the comment of the German Professor, "As a graduate student and instructor I received no credit for much work I did, but as a Full Professor I received much credit for work my graduate students did. Thus, on the whole justice was done."

9.3. Datasets in Service-Life Prediction

In many situations in engineering, which are to confirm structural reliability, the longest observed life (latest failures) to occur may require 10 times as much test time as the earliest failures. Since the cost of completing such testing is so expensive there is always a tendency to stop the test before the data are complete. But how can one utilize only the early data?

There are some duration-of-load tests for structural materials that take years to complete. Must one wait until all test units have failed before one can estimate the average time to failure? The proper method of analyzing incomplete data sets depends upon both the sampling situation and upon the proper mathematical model

[2] The Bernoullis were paid for their mathematical results, one of which was the aforesaid rule. They were included in l'Hôpital's calculus textbook, one of the first produced, which remained in publication for a century.

from which is derived the distribution of life. The following datasets are all from actual engineering tests.

DATASET I	DATASET II	DATASET III
$t_1 = 21,388$	$y_{(1)} = 4,183$	$x_{1:25} = 4,513$
$t_2 = 21,252$	$y_{(2)} = 5,578$	$x_{2:25} = 5,562$
$t_3 = 23,747$	$y_{(3)} = 6,568$	$x_{3:25} = 6,242$
$t_4 = 24,503$	$y_{(4)} = 7,105$	$x_{4:25} = 6,590$
$t_5 = 22,970$	$y_{(5)} = 7,339$	$x_{5:25} = 6,737$
$t_6 = 27,340$	$y_{(6)} = 7,840$	$x_{6:25} = 7,043$
$t_7 = 20,379$	$y_{(7)} = 8,323$	$x_{7:25} = 7,325$
$t_8 = 27,681$	$y_{(8)} = 8,493$	$x_{8:25} = 7,422$
$t_9 = 24,239$	$y_{(9)} = 8,540$	$x_{9:25} = 7,448$
$t_{10} = 16,415$	$y_{(10)} = 9,020$	$x_{10:25} = 7,713$

DATASET IV		DATASET V	
13061	46274+	$n = 10$	$n = 24$
14801	24396	4232	14839
15088+	37242	6988	16044
	44508+	10021	18515
3520	44751	14212	19323
3802	13860	15850	25873
2830	51766+	16156	28363
5705+		16393	29795
	57234	17387	
37909	29032		
34074			

Exercise Set 9.A

1. A complete sample of size $n = 10$, of lifetimes, recorded in hours, is given in Dataset I. Assuming a log-normal distribution, estimate the scale and shape parameters. Use these mle estimates to calculate an estimate of a safe-life, i.e. a life time t_0 in hours such that if T is the lifetime of a component selected at random from the same population, then $\Pr[T > t_0] = .999$.

2. The first 10 out of 25 observations from a Weibull distribution are recorded in cycles during a life-test are given in Dataset II. Estimate both shape and scale parameters.

3. The smallest 10 out of 25 observations from a Gumbel extreme value distribution are recorded in Dataset III. Estimate both shape and scale parameters.

4. In #4 in EXERCISE SET 6.D there were unstated facts that might have been pertinent. The 'better' alternator drive had (i) an inspection every 100 hours while the other drive had an inspection every 200 hours. (ii) the 'better' drive had 4 catastrophic in-service failures, causing economic loss of far greater value than just the alternator drive. Did you think to ask any of these questions? If not your job would be on the line.

Exercise Set 9.B

1. The five groups of observations in Dataset IV are from different Weibull populations with each having the same shape parameter but different scale parameters. Compute the 2-out-of-n unbiased estimate of α for each group of size n using the estimate $Z_n = n\ln[n/(n-1)]\ln(X_{1:n}/X_{2:n})$ and then calculate the sharp estimate formed by their appropriate combination as the global estimate of α. Note that $a+$ denotes a run-out at time a, not a failure at time a.

2. Assume that the variance of $\hat{\alpha}_n$ from a complete sample of size n is proportional to n^{-1}. Then use an appropriate linear combination of the mles of α from each group to form a sharp estimate of α.

3. Two sets of fatigue data with type-II censoring, i.e., the smallest k failures were observed out of n specimens, are given in cycles to failure in Dataset V. Assuming a Weibull distribution is the correct model estimate for both shape and scale parameters. Plot both empiric distributions on Weibull paper and guestimate the parameters from a straight line fitted by eye.

Exercise Set 9.C

1. Assume that the life-length of a strand tested at 80% of mean-rupture strength has an exponential distribution. Using the data presented in Dataset VI, Table 9.1:
   (a) censor the data at the 25th observation and estimate the mean.
   (b) censor the data at the 50th observation and estimate the mean.
   (c) censor the data at the 75th observation and estimate the mean.
   (d) using all the data estimate the mean.

2. A set of observations has a sample mean of $\bar{x}^{(1)} = 59.1$ and a second moment sample value of $\bar{x}^{(2)} = 3529.7$. If we know that the observations come from a normal population with a known coefficient of variation $\kappa = \sigma/\mu = .2$, what is the mle of the population mean μ? What assumptions did you make? Why? Can you determine an engineering rule-of-thumb about estimation when κ is known?

3. Derive specialized computational procedures for computing the mles, implicitly defined from the likelihood, for censored life data with $x_1 < x_2 < \cdots < x_k$ as failure times and $x_{k+1} < x_{k+2} < \cdots < x_n$ as alive times assuming that life has a:
   (a) Weibull distribution
   (b) Gumbel extreme-value distribution
   (c) log-normal distribution
   (d) log-logistic distribution.

If you torture data long enough, you can usually make it confess.

The Grand Inquisitor

An education is what remains after what you've memorized has been forgotten.

B. F. Skinner

After all, there are really only 10 kinds of people in the world. Those who understand binary and those who don't.

Anon.

Table 9.1. Dataset VI: Times-to-failure in hours of strands subjected to stress at 80% of mean-rupture strength listed in increasing order

1.8	3.1	4.2	6.0	7.5	8.2	8.5	10.3	10.6	24.2
29.6	31.7	41.9	44.1	49.5	50.1	59.7	61.9	64.4	69.7
70.0	77.8	80.5	82.3	83.5	84.2	87.1	87.3	93.2	103.4
104.6	105.5	108.8	112.6	116.8	118.0	122.3	123.5	124.4	125.4
129.5	130.4	131.6	132.8	133.8	137.0	140.2	140.9	148.5	149.2
152.2	152.8	157.7	160.0	163.6	166.9	170.5	174.9	177.7	179.2
183.6	183.8	194.3	195.1	195.3	202.6	220.2	221.3	227.2	251.0
266.5	267.9	269.2	270.4	272.5	285.9	292.6	295.1	301.1	304.3
316.8	329.8	334.1	346.2	351.2	353.3	369.3	372.3	381.3	393.5
451.3	461.5	574.2	653.3	663.0	669.8	739.7	759.6	894.7	974.9

Table 9.2. Dataset VII: Fatigue lifetimes under cyclic loading, maximum stress 21,000 psi, 18 cycles per second, 6061-T6 Aluminum coupons cut parralel to the direction of rolling. Observations listed in increasing order.

370	706	716	746	785	797	844	855	858	886
886	930	960	988	990	1000	1010	1016	1018	1020
1055	1085	1102	1102	1108	1115	1120	1134	1140	1199
1200	1200	1203	1222	1235	1238	1252	1258	1262	1269
1270	1290	1293	1300	1310	1313	1315	1330	1355	1390
1416	1419	1420	1420	1450	1452	1475	1478	1481	1485
1502	1505	1503	1522	1522	1530	1540	1560	1567	1578
1594	1602	1604	1608	1630	1642	1674	1730	1750	1750
1763	1768	1781	1782	1792	1820	1868	1881	1890	1893
1895	1910	1923	1940	1945	2023	2100	2130	2215	2268
2440									

Problem Set 9.D

1. Assume that a complex aeronautical system has four principle subsystems with operational reliability goals, respectively, of
   - Structure (.990)
   - Propulsion and pay load operation (.900)
   - Flight Control Avionics (.900)
   - Electric Systems (.950)

   The nominal flight time is five units; the fleet size may be as large as 250.

   A particular diode has failed in a noncritical circuit within the avionics system during an acceptance test.

   (a) thirteen such diodes are used in the subsystem; 10 in critical and 3 in noncritical circuits.

   (b) three hundred avionics packages containing this diode have been delivered.

   The prior service history of this component (diode) is

   (a) zero flight anomalies reported in 20 flights.

(b) six ambient ground failures in the field, 2 critical, 4 non-critical.
(c) a 20% failure rate was observed on a sweep of the 600 diodes in stock.
Assuming the utilization of this diode is essentially the same in all circuits, what is your advice to the fleet authority about this problem? Please consider all the costs and consequences of your recommendation.

2. What would your recommendation be in Problem 1 if, instead of no flight anomalies, one probable-cause failure was attributed to the diode?

3. We have the following service history of the pay load operation.
   (a) Lot #2 of 500 operation items had 3 outliers during a functional test measurement. Subsequently 50 were tested with no functional failures.
   (b) Lot #1 and lot #3 reported no problems.
   (c) One possible-cause flight failure is attributed to item from lot #2; 30 flights have been made to date.
   (d) The critical material had the following yield (passed critical evaluation and testing) during an inspection of each lot.

   lot#1 – 70%, lot#2 – 40%, lot#3 – 70%,

   (e) Flight data showed that 12 systems used lot #1, 7 systems used lot #2, and 11 systems used lot #3.
   (f) This operation item is used in three places, two of which are redundant.
   (g) Lots #1, #2, #3 have been installed in the fleet; 40% from lot #1, 20% from lot #2, 40% from lot #3.

   Assuming this operation item is utilized in a "1-out-of-3 system" in the subsystem, what is your advice to the fleet about its reliability problem? Please consider the consequences and costs of your recommendation.

4. Suppose in Problem 3 that field information from the fleet showed crystalization occurring in the critical material used in the ordnance from lot #1. A material sample was chosen from lot #1 and returned for analysis. A 5% loss of velocity-of-operation was detected but it was still within specification (10% is allowable.)

5. At this time we have the following service history of flight control.
   (a) Three rate-gyros used in the flight control system were tested for 60 minutes in a new production assessment. (There is one gyro per system.) One gyro drifted so as to result in flight failure; 2 were still operating succesfully when the test terminated.
   (b) Ground test history of 250 gyros, each having been tested for 5 minutes, resulted in 26 failures (3 critical, 23 noncritical). The failure modes were different from that occuring in production assessment.
   (c) No flight history subsequent to new flight test program after new ATP (see below). One flight failure attributed to rate-gyro during old flight test program.
   (d) The manufacturer shows 85% pased critical evaluation during his ATP test.
   (e) ATP test prior to operation will be
       - 100% of pkgs will see 5 units of vibration at flight level.
       - 10% of pkgs will see 10 units of vibration at flight level.
       - 2% of pkgs will see 15 units of vibration at flight level.
   (f) Three hundred gyros have been produced and delivered to the fleet.

   Assuming only one gyro is utilized in each flight control system, what is your advice to the fleet authority about this reliability problem?

6. Suppose we assume that one probable failure during flight nine was attributable to gyro drift and 100 gyros were tested to new ATP with one more failure, what changes would you make in your recommendation?

7. At this time we have the following service history of the structure.
   (a) A particular metal casting shows hairline cracks prior to usage (due to processing methods).
   (b) A special high pressure test shows no deterioration of structural integrity.
   (c) Temperature cycle tests reveal the additional stress to reach the point of failure.
   (d) The usual operating cycle would not see these extremes but the normal logistics cycle could see these temperatures.
   (e) 20% of the fleet (one per system) is supplied with this type of casting.

   Assuming only one casting is utilized in each structure, what is your advice about this reliability problem? Please consider all the consequences and the costs of your recommendation.

Dataset VIII Time in minutes;
$k = 3, n = 14$

Failed	Censored			
1	59	113	145	182
8	72	117	149	320
10	76	124	153	

Dataset IX Time in minutes;
$k = 2, n = 9$

Failed	Censored			
37	60	66	72	123
53	64	70	96	

Dataset X Time in hours;
$k = 10, n = 42$

Failed	Censored			
0.10	2.35	2.63	3.18	4.50
0.11	2.37	2.87	3.23	5.12
0.20	2.43	2.88	3.42	5.42
0.27	2.48	2.92	3.45	6.82
0.37	2.50	3.00	3.75	8.15
0.43	2.52	3.07	3.87	11.15
1.17	2.57	3.08	3.93	11.30
1.50	2.58	3.10	3.97	
1.98	2.63			
2.32				

Dataset XI Time in hours; $k = 7, n = 42$

Failed	Censored				
0.18	1.48	2.90	4.30	4.55	5.18
0.50	1.48	3.53	4.33	4.60	5.30
0.70	2.18	3.63	4.46	4.95	5.45
0.90	2.25	3.75	4.52	5.00	5.47
0.95	2.63	4.30	4.55	5.00	5.85
1.25					
6.83					

Table 9.3. Dataset XII: cycles-to-failure of cylindrical steel specimens of .41% C. annealed for 30 min. at 870°C; after surface finish with coarse emery, tested by Ono's Rotary Uniform Bending Testor at 2000rpm with ± 35.6 kg/mm^2.

44.9	50.5	53.0	58.3	67.7	71.5	82.1	91.3	104.6	119.9
122.8	123.7	138.9	141.0	142.8	143.1	145.3	147.8	151.8	157.2
163.2	172.8	180.5	185.3	188.1	192.1	202.0	213.2	223.6	240.2
243.8	244.4	284.4	287.2	298.8	324.7	330.3	350.1	354.5	380.5
385.8	392.0	394.1	406.6	423.0	423.3	443.6	481.2	548.2	671.1
740.7	748.7	815.6	819.5	834.7	967.9	1190.8	1286.6	1315.7	1379.3
1691.7	2187.8	2272.2	2505.9	2580.7	3643.5	3694.0	4212.7		

9.4. Data, Consulting, and Modeling

There have been serious environmental problems in the burning of low-quality, high-sulfur-content coal to produce electricity in the large (billion-dollar) power plants due to the enormous quantities of sulfur dioxide, which is released into the air. The EPA wants such a power company to install scrubbers (costing a few hundred million) to wash the sulfur from the coal before it is burned. But this would also force the power company to spend millions to dispose of the pools of weak suluric acid so generated.

In one instance the preliminary calculations seemed to predict the highest concentration of sulfur dioxide that would occur within the smoke plume over a 3-year period would, by law, necessitate the use of scrubbers. Of primary importance in this prediction is the concentration of the sulfur in the coal from the various strata within the coal fields which will be mined and subsequently burned.

Deep core samples were made by the power company of each coal field with subsequent chemical analysis which gave the sulfur concentration by weight for each strata within each field. The plan was to dilute the sulfur to an acceptible level by grinding and combining the higher sulfur coal with the lower. Suitable averages were obtained for different areas for each specific field. At the power plant the coal that was brought in by train would be mixed and ground and then be sampled daily for sulfur content analysis. These testing procedures were in accord with those specified by *The American Society for Testing and Materials.* Finally the concentration of sulfur in the power plant's smoke plume was also measured.

Data were available from all strata in three different mines within the coal fields that had provided coal in the recent past. Thus, there was a large database of both core-sample averages and power plant daily averages that showed that the averaged sulfur content from the core sample was consistently significantly larger than the corresponding average from the power-plant coal! However, these differences were just ignored because there seemed to be no adequate explanation. In one of the data sets there was 1700 observations, which gave an accurate empirical frequency distribution of the sulfur concentration. In N.R. Mann's Lecture Notes on Reliability given at NIST it said:

> The (density from the power plant) was narrow and looked almost like a bell-shaped curve, but was not quite symmetric and had thin trailing tails on either end. The (densities for the core-samples from the three mines) looked similar but were considerably more spread out.

Ray Schafer, who had discussed the distribution of the reciprocal-gamma in [69], noted that the empirical sulfur densities were similar to it in form; this recommendation was utilized. (For this density the harmonic mean is used as an estimator of the population mean.) The fit to the data was very good, and the discrepancy was explained because the harmonic mean is always less than the arithmetic mean with the discrepancy increasing when the variability of the sample increases.

But is this change in distribution just "tergiversation by number-crunching"? Does the harmonic mean really impart more information than the arithmetic mean? Yes, because the tacit assumption, in using the arithmetic mean, is that the densities of coal and sulfur are constant! The density of coal (weight per unit volume) varies considerably while weight of sulfur per unit volume of coal was almost constant. Thus, the variation in the weight of sulfur per unit weight of coal is due to the randomness in the denominator.

This more accurate analysis by statisticians was considered a successful application of science since it purportedly showed that: (1) the amount of sulfur in the coal was less than formerly calculated, (2) the scrubbers were unnecessary, (3) the disposition of the sulfuric acid was obviated, (4) the power subscribers were spared a rate increase.

Problem Set 9.E

1. The coal-mixing and grinding from the various mines was designed to keep the sulfur content per unit of mixed coal nearly constant at a prescribed level. Should not this variation in cost stemming from the purity of the raw material also be taken into account in the analysis?

2. The density of the ratio of two Gaussian variates is difficult to calculate and is always a bimodal density, but how difficult is it to calculate the density of the ratio of two independent CD-distributions? What does the density look like?

3. Could the utilization CD-distributions make a significant difference in the conclusion?

4. Did any mixing, grinding of coal, or data "massaging" make a difference in the total amount of sulfur dioxide that would be deposited by the smoke plume downwind of the power plant? What difference did the analysis make to cleaning the environment?
5. Is it the business of statisticians to interfere in the conduct of business the way W. Edwards Deming recommends in his book *Quality, Productivity, and Competitive Position* about statistical process control?
6. Would you have made a different recommendation?

Remember what Polonius said in his advice to his son Laertes:

> "These few precepts (keep) in thy memory .... Give every man thine ear but few thy voice; Take each man's censure, but reserve thy judgment.
>
> *Hamlet* Act I, Sc.iii

> Experimentation, without mathematical modelling for confrontation and predictive verification, is no more science than is butterfly collecting.
>
> Anon.

> If you set out to be a mathematician, you must learn the profession, every part of it, then work at it, profess it, live it as best you can. If you keep asking "What's in it for me?" you're in the wrong business. If you're looking for comfort, money, fame and glory you probably won't get them, but if you keep trying to be a mathematician you might.
>
> Paul R. Halmos

CHAPTER 10

Cumulative Damage Distributions

10.1. The Past as Prologue

The density of the first-passage time for a particle, with drift, undergoing Brownian motion was first obtained by E. Schrödinger [101, 1915]. He considered N particles in Brownian motion, all initially at zero and white. When one reaches a distance ℓ it becomes green. Let the probability of first passage beyond ℓ in the interval $(t, t+\Delta t)$ be $\int_t^{t+\Delta t} p(x)\,dx$ with $p(\cdot)$ unknown. If $N\mathrm{P}_w(t)$ is the expected number of white particles at $t > 0$, then

$$N\int_t^{t+\Delta t} p(x)\,dx = N[N\mathrm{P}_w(t) - N\mathrm{P}_w(t+\Delta t)] \quad \text{implies} \quad p(t) = -\mathrm{P}_w'(t).$$

From assumed Brownian motion of velocity ν the density of particles, at position x at time $t > 0$, is

$$\rho(x,t) = \frac{N}{\sqrt{2\pi\sigma^2 t}} e^{-\frac{(x-\nu t)^2}{2\sigma^2 t}}.$$

Make the transformation $y = x - \nu t$ to obtain particles without drift, namely,

$$\rho(y,t) = \frac{N}{\sqrt{2\pi\sigma^2 t}} e^{-\frac{y^2}{2\sigma^2 t}}.$$

Schrödinger recognizes this as the solution of the heat-diffusion equation, viz.,

$$\frac{\sigma^2}{2}\frac{\partial^2\rho}{\partial y^2} = \frac{\partial\rho}{\partial t} \qquad \text{with b.c. } \rho_w(\ell - \nu t, t) = 0.$$

He then obtained the solution for the density using the *reflection principle*, viz.,

$$\rho_w(y,t) = \frac{N}{\sqrt{2\pi\sigma^2 t}}\left[e^{-\frac{y^2}{2\sigma^2 t}} - e^{2\ell\nu/\sigma^2} e^{-\frac{(y-\nu t)^2}{2\sigma^2 t}}\right].$$

His knowledge of classical analysis enabled the use of the *principle of superposition* with a change of coordinates to obtain

$$N\mathrm{P}_w(t) = \frac{N}{\sqrt{2\pi}}\left[\int_{-\infty}^{\frac{\ell-\nu t}{\sigma\sqrt{t}}} e^{-y^2/2}\,dy - \int_{-\infty}^{\frac{\ell+\nu t}{-\sigma\sqrt{t}}} e^{-y^2/2}\,dy\right].$$

Differentiation wrt t yields

$$p(t) = \frac{\ell}{\sigma\sqrt{2\pi}} t^{-3/2} e^{-\frac{(\ell-\nu t)^2}{2\sigma^2 t}}.$$

This same density with different parameterization was later obtained, independently, by A. Wald [117, 1944] in his study of sums of independent and identically distributed (iid) random variables.

During Wald's investigation of the distribution of the sample size for his sequential probability-ratio tests, he sought the distribution of the random number, say N, of iid random summands X_i each having mean $\nu > 0$ and variance $\sigma^2 > 0$, necessary to exceed for the first time a given bound, say $\omega > 0$. In this case the stopping event becomes

$$[N = n] = \left\{X_1 \le \omega, X_1 + X_2 \le \omega, \ldots, \sum_{i=1}^{n-1} X_i \le \omega, \sum_{i=1}^{n} X_i > \omega\right\}.$$

Wald's lemma says that under these assumptions $\mathrm{E}N \cdot \mathrm{E}X = \omega$ and if we set $\beta = \omega/\nu$ and $\alpha^2 = \sigma^2/\omega\nu$, then we have the asymptotic approximation

$$\mathrm{P}[N \le y\beta] \approx \int_0^y \frac{1}{\sqrt{2\pi}\,\alpha} e^{\alpha^{-2}} x^{-3/2} \exp[-\frac{1}{2\alpha^2}(x + x^{-1})]\, dx \quad \text{for } y > 0.$$

By "neglecting the excess over the boundary," Wald's phrase, we take this approximation as exact and so have, for all practical purposes, the distribution of N. (The exact limiting distribution is attained as $\beta \to \infty$.) This density is the standard form of what is now called the *Wald distribution*, see [47, 1970]. It was subsequently, and independently, reformulated by Zigangirov in Russia, [123, 1962] and by Shuster [105, 1968].

We write

$$X \sim \mathcal{W}(\alpha, \beta) \tag{10.1}$$

when $X \sim F$ with the cdf F defined by

$$F(y) = \int_0^{y/\beta} \frac{1}{\sqrt{2\pi}\,\alpha} e^{\alpha^{-2}} x^{-3/2} \exp[-\frac{1}{2\alpha^2}(x + x^{-1})]\, dx \quad \text{for } y > 0. \tag{10.2}$$

Later the name "inverse-Gaussian" was applied to an rv identified by Tweedie [114, 115, 116, 1945, 1957] because of the inverse relationship between its cumulant generating function (cgf) and the cgf of a Gaussian variate. Using the parameterization of Tweedie, the inverse-Gaussian density for $Z \sim \mathrm{IG}(\mu, \lambda)$ is written

$$\sqrt{\frac{\lambda}{2\pi}\frac{1}{z^{3/2}}} \exp\left\{-\frac{\lambda}{2}\frac{(z-\mu)^2}{z\mu^2}\right\}, \quad \text{for } z > 0; \lambda, \mu > 0. \tag{10.3}$$

NB this is only a reparametrization of the density of Wald! To see this set

$$\mu = \beta, \qquad \lambda = \beta/\alpha^2. \tag{10.4}$$

The distribution of a reciprocal inverse-Gaussian variate was also examined by Tweedie. If X has the density given by eqn (10.2) and $Y/\beta \sim \beta/X$, then Y has the density expressed as follows:

$$\frac{1}{\sqrt{2\pi\beta}}\frac{y^{-1/2}}{\alpha}\exp\left\{-\frac{(y-\beta)^2}{2y\alpha^2\beta}\right\} \quad \text{for } y > 0; \alpha, \beta > 0. \tag{10.5}$$

We denote these families, respectively, in the IG-parmeterization by

$$Z \sim \mathrm{IG}(\mu, \lambda) \quad \text{and} \quad Y \sim \mathrm{RIG}(\mu, \lambda).$$

In our parameterization, we call the reciprocal variate a *Tweedie* variate and denote it by $Y \sim \mathcal{T}(\alpha, \beta)$. $X \sim \mathcal{W}(\alpha, \beta)$ iff

$$\frac{X}{\beta} \sim \frac{\beta}{Y}, \quad \text{where } Y \sim T(\alpha, \beta). \tag{10.6}$$

10.2. The Fatigue-Life Distribution

This distribution too has had multiple inseminations by independent investigators. It was suggested by Freudenthal and Shinozuka in an unpublished technical report in [36, 1961]. Its investigation had been advocated even earlier by E. Parzen in an article reprinted in [90]. Moreover T. von Karman discussed its utility in the 1920s. The difference between the presentation in [18, 1969] and the earlier reports was in interpretation and parameterization plus the degree of fit shown to actual fatigue data in [19, 1969]. In the earlier studies a distribution had been presented as the discrete probability of failure given by

$$\Phi\left[\frac{n\mu-\omega}{\sqrt{n}\sigma}\right] \quad \text{for number of cycles } n = 1, 2, 3, \cdots.$$

First we illustrate the range of applicability to cumulative damage by examining its derivation. Let $X_i > 0$ for $i = 1, 2, \ldots$ be an iid sequence of variates each having mean ν and variance κ^2. Each variate represents incremental damage per cycle to a structure having a critical damage level, call it $\omega > 0$. Let the waiting time until cumulative damage first exceeds the critical level be N_ω. In this case the stopping event is

$$[N_\omega = n] = \{S_{n-1} \le \omega, S_n > \omega\}.$$

where $S_n = \sum_{i=1}^n X_i$ is the cumulative damage after n cycles. Then by duality

$$[N_\omega \ge n] = [S_n \le \omega] \equiv \left[Z_n \ge \frac{\sqrt{\nu\omega}}{\kappa}\xi(n\nu/\omega)\right] \tag{10.7}$$

where ξ is the monotone increasing function defined by

$$\xi(z) = \sqrt{z} - \frac{1}{\sqrt{z}} \quad \text{for } z > 0, \tag{10.8}$$

and by the CLT, since $\mathrm{E}S_n = n\nu, \quad \mathrm{Var}[S_n] = n\kappa^2$,

$$Z_n := \frac{n\nu - S_n}{\sqrt{n}\kappa} \rightsquigarrow \mathcal{N}(0,1) \text{ as } n \longrightarrow \infty.$$

Now set $\omega = \omega_0\sqrt{\tau}$ and $\nu = \nu_0/\sqrt{\tau}$ where τ is an arbitrary parameter and define $\beta = \omega_0/\nu_0$ with $\alpha = \kappa/\sqrt{\nu_0\omega_0}$, then making the change of variable $n = x\tau$ in eqn (10.7) and letting $[x]$ denote the largest integer not exceeding x, we have from eqn (10.7),

$$\mathrm{P}[N_{\omega_0\sqrt{\tau}} \geq x\tau] = \mathrm{P}[Z_{[x\tau]} \geq \frac{1}{\alpha}\xi(x/\beta)] \quad \text{for any } \tau > 0.$$

Now let $\tau \to \infty$ to obtain the distribution for T where $T = \lim_{\tau\to\infty} \frac{N_{\omega_0\sqrt{\tau}}}{\tau}$.

Birnbaum-Saunders [18, 1969] obtained this limiting distribution and applied it to the fatigue life of metals. An rv $T > 0$ has the *fatigue-life* distribution and we write $T \sim \mathcal{FL}(\alpha, \beta)$ when for some $\alpha, \beta > 0$

$$\mathrm{P}[T \leq x] = \Phi[\tfrac{1}{\alpha}\xi(x/\beta)] \quad \text{for } x > 0, \tag{10.9}$$

with the corresponding density

$$\frac{1}{\alpha\beta}\xi'(t/\beta)\varphi[\frac{1}{\alpha}\xi(t/\beta)] \quad \text{for } t > 0; \alpha, \beta > 0. \tag{10.10}$$

Here, and subsequently, Φ is the standard normal distribution, with $\Phi' = \varphi$ denoting the standard normal density and $\bar{\Phi} = 1 - \Phi$. Clearly such variates are self-reciprocal after scaling, i.e., $\frac{X}{\beta} \sim \frac{\beta}{X}$.

The reason for distinguishing between these two distributions the Wald and fatigue-life, is in their application to physical or biological processes of cumulative damage it may be important to make the distinction between incremental change, which can be negative or positive, such as in smoking where there may be temporary "healing" when no smoking is occurring, as contrasted with nonnegative damage accumulation such as in metallic fatigue where incremental damage is an irreversible microscopic fatigue-crack extension per cycle. One last point concerning application, if the bound ω, the critical-flaw size in fatigue-life studies, or the critical stress intensity necessary during an earthquake in seismology, is itself a random variable the number of summands for it to be exceeded converges to the $\mathcal{FL}$-distribution if the summands are all nonnegative but not if they have a positive probability of being negative.

The obvious relationship between the $\mathcal{FL}$-distribution and normality is shown in [18, 1969] by

Theorem 1. *If we let $X \sim \mathcal{FL}(\alpha, \beta)$, then*

$$\frac{1}{\alpha}\xi(X/\beta) \sim \mathcal{N}(0,1) \quad \textit{and therefore} \quad \frac{1}{\alpha^2}\xi^2(X/\beta) \sim \chi^2(1).$$

The relationship of any variate to normality is of interest. For example, in the monograph by Chhikara and Folks [26, 1989] is the statement: "Although the authors have been able to develop many analogous results (between the normal and the inverse-Gaussian) they do not at this writing understand the underlying reason. There is no (known) transform by which the inverse-Gaussian variable can be obtained from the normal." (Of course, this is not strictly true since the probability integral transform will provide one.)

The next class of densities that is examined was introduced by Jørgenson, Sehradri and Whitmore [49, 1991], and independently in the thesis of Minogue, [80, 1991]. It is the mixture generated by all convex combinations of an inverse-Gaussian density and the corresponding density of the appropriate reciprocal inverse-Gaussian.

Exercise Set 10.A

1. Retrace Tweedie's derivation as follows:
   (a) Using the density of an inverse-Gaussian variate T as parameterized in eqn (10.3), show its cumulant generating function $C_T := \ln m_T(t)$, where m_T is the mgf, is given by
   $$C_T = \frac{\lambda}{\mu}(1 - \sqrt{1 - \frac{2t\mu^2}{\lambda}}).$$
   (b) Thus, prove the rth cumulant of T is
   $$\kappa_r = (-1)^{r-1}\frac{\lambda}{\mu}\left(\frac{1}{2}\right)^{r\downarrow}\left(\frac{2\mu^2}{\lambda}\right)^r \quad \text{for } r = 1, 2, \cdots.$$
   (c) Using the cumulants verify that $\mathrm{E}T = \mu$ and $\mathrm{Var}(T) = \mu^3/\lambda$.
   (d) Recall if $Y \sim \mathcal{N}(\mu, \sigma^2)$ then $C_Y(s) = -s\mu + \frac{s^2\sigma^2}{2}$. Now set $t = C_Y(s)$ and solve for $s = C_Y^{-1}(t)$ then reverse and invert the parameters by setting $\lambda = \sigma^{-2}$, $\nu = \mu^{-1}$ to demonstrate that $C_Y^{-1}(t) = C_T(t)$ when $\nu \mapsto \mu$.
2. Using the definition in eqn (10.8) show that $\xi(x) = 2\sinh[\ln\sqrt{x}]$ for $x > 0$.

Knowledge, without thinking, is useless. Thinking, without knowledge, is futile.
The twelfth analect of Confucious (c. 551–479 BCE)

One is never ever excused from the moral obligation to act intelligently.
Anon.

10.3. The Mixed Class of Cumulative Damage Distributions

The relationships among these three distributions can be more easily seen if we utilize the same parameterization. We say that the random variable Z has a *cumulative damage distribution* and write $Z \sim CD(\alpha, \beta, p)$ iff its density is given by the mixture

$$f(z; \alpha, \beta, p) = \frac{\sqrt{\beta}}{\alpha}\left(\frac{p}{z^{3/2}} + \frac{q}{\beta z^{1/2}}\right)\varphi[\tfrac{1}{\alpha}\xi(z/\beta)] \quad \text{for } z > 0, \qquad (10.11)$$

where $p + q = 1$ with its distribution given by (see Exercise Set 10.B.1 for definition of ψ)

$$F(z;\alpha,\beta,p) = \Phi[\tfrac{1}{\alpha}\xi(z/\beta)] + (p-q)e^{2/\alpha^2}\Phi[\tfrac{-1}{\alpha}\psi(z/\beta)], \qquad (10.12)$$

and the parameters must satisfy $\alpha, \beta > 0$, and $0 \le p \le 1$.

One also finds that the mgf is given, for $2(\alpha\beta)^2 < t^{-1}$, by

$$m_Z(t) = \exp\{\tfrac{1}{\alpha^2}\left(1-\sqrt{1-2t\alpha^2\beta}\right)\} \times \left[p + \frac{q}{\sqrt{1-2t\alpha^2\beta}}\right]. \qquad (10.13)$$

Our parameters are related to those initially chosen by Tweedie, and continued in the monograph by Chhikara and Folks [26, 1989], through the transformations of eqn (10.4).

In the study by Jøorgensen, Seshradri, and Whitmore [49, 1991] of the properties of this family of distributions they introduced a parameterization that presumed that $\gamma = q\beta/p$ was known. In this study it is p that is known, that is, e.g., it is $p = 0, 1, 1/2$. Moreover we restrict our parameters to those that make physical sense in the engineering applications that are in mind.

One reason for the adoption of our parameterization is that for the $\mathcal{W}$-distribution the shape parameter, α, now becomes exactly the *coefficient of variation* while β, remains the scale parameter and mean. Often in the engineering sciences the coefficient of variation has physical significance (in some applications its reciprocal, is called the signal-to-noise ratio) and so provides not only better understanding of the meaning of parameteric changes, but frequently alternative methods, other than by statistical estimation, become available for its determination.

Exercise Set 10.B

1. Define $\psi(x) := \sqrt{x} + \frac{1}{\sqrt{x}} = 2\cosh[\ln\sqrt{x}]$ for $x > 0$ then show the following identities: label them *the $\psi-\xi$ (hyperbolic trignometric) identities*.
   (a) $\psi^2(x) - \xi^2(x) = 4, \quad \xi'(x) = \frac{\psi(x)}{2x}, \quad \psi'(x) = \frac{\xi(x)}{2x}$ for $x \in \Re$.
   (b) $\xi^{-1}(y) = \left(\frac{y}{2} + \sqrt{\frac{y^2}{4}+1}\right)^2, \quad \psi\xi^{-1}(y) = \sqrt{y^2+4}$ for $y \in \Re$.
   (c) $\psi^{-1}(y) = \left(\frac{y}{2} - \sqrt{\frac{y^2}{4}-1}\right)^2, \quad \xi\psi^{-1}(y) = -\sqrt{y^2-4}$ for $y > 2$.
   (d) $2\xi(xy) = \psi(y)\xi(x) + \psi(x)\xi(y), \quad 2\psi(xy) = \xi(y)\xi(x) + \psi(x)\psi(y)$.
2. Use the $\psi-\xi$ identities to show the density for $Z \sim CD(\alpha,\beta,p)$ distribution can be written as
$$\frac{1}{\alpha z}\left(p\sqrt{\frac{\beta}{z}} + q\sqrt{\frac{z}{\beta}}\right)\varphi[\tfrac{1}{\alpha}\xi(z/\beta)] \quad \text{or} \quad [\psi(z/\beta) - (p-q)\xi(z/\beta)]\,\frac{\varphi[\frac{1}{\alpha}\xi(z/\beta)]}{2\alpha z}.$$
3. In this latter form, since $2z\psi'(z) = \xi(z)$ and $2z\xi'(z) = \psi(z)$ integrate the density to obtain the distribution in eqn (10.12).
4. Discover the distribution of $[\frac{1}{\alpha}\xi(Z/\beta)]^2$ when $Z \sim CD(\alpha,\beta,p)$.

10.4. Elementary Derivation of Means and Variances

Derivations of the mean and variance for inverse-Gaussian (and its reciprocal) variates, obtained by integrating their densities, are never given, in part, because the moments are easily obtained using the cgf (see [26, p.11–12]), and in part because the direct evaluation of such integrals requires a knowledge of the recursion relations among modified Bessel functions of the third kind (realized after the fact) [48, 1982]. The direct integration now presented for a $\mathcal{CD}(\alpha, \beta, p)$ density utilizes the relationship between the fatigue-life and the standard Gaussian distribution, and provides, *a fortiori,* elementary derivations of the mean and variance of the IG and RIG distributions. Let $Z \sim \mathcal{CD}(\alpha, \beta, p)$: α is the parameter governing spread while β is the *characteristic life* and p is the *mixing fraction* with

$$\mathrm{E}Z = \int_0^\infty tf(t;\alpha,\beta,p)dt = \beta \int_0^\infty \left(\frac{p}{t^{1/2}} + qt^{1/2}\right)\frac{1}{\alpha}\varphi[\frac{1}{\alpha}\xi(t)]\,dt. \tag{10.14}$$

Consider the first integral; label it $p\beta A$, where

$$A = \int_0^\infty \frac{t^{-1/2}}{\alpha}\varphi[\frac{1}{\alpha}\xi(t)]\,dt = \int_0^\infty \frac{t^{-1/2}}{2\alpha}\varphi[\frac{1}{\alpha}\xi(t)]\,dt + \int_0^\infty \frac{y^{-3/2}}{2\alpha}\varphi[\frac{1}{\alpha}\xi(y)]\,dy.$$

This identity is obtained by adding two halves of A and making the change of variable $t = 1/y$ in the second. We then find by recombining them,

$$A = \int_0^\infty \frac{\xi'(t)}{\alpha}\varphi[\tfrac{1}{\alpha}\xi(t)]\,dt = \textstyle\int_{-\infty}^\infty \varphi(z)\,dz = 1. \tag{10.15}$$

In like manner we obtain for the second integral in eqn (10.14)

$$B = \int_0^\infty \frac{t^{1/2}}{\alpha}\varphi[\frac{1}{\alpha}\xi(t)]\,dt = \int_0^\infty \frac{[\xi'(t) + \xi^2(t)\xi'(t)]}{\alpha}\varphi[\tfrac{1}{\alpha}\xi(t)]\,dt = 1 + \alpha^2. \tag{10.16}$$

By adding eqns (10.15) and (10.16) we find the mean

$$\mathrm{E}Z = \beta[1 + q\alpha^2] \quad \text{or} \quad \mathrm{E}\left(\frac{Z-\beta}{\beta}\right) = q\alpha^2. \tag{10.17}$$

The same method applied again shows

$$\begin{aligned}
\mathrm{E}\left(\frac{Z-\beta}{\beta}\right)^2 &= \int_0^\infty \frac{\xi^2(t)}{\alpha}\left[pt^{-1/2} + qt^{-3/2}\right]\varphi[\frac{1}{\alpha}\xi(t)]\,dt \\
&= \int_0^\infty p\xi^2(t)\frac{\xi'(t)}{\alpha}\varphi[\frac{1}{\alpha}\xi(t)]\,dt \\
&\quad + \int_0^\infty q\xi^2(t)\left[1 + \xi^2(t)\right]\frac{\xi'(t)}{\alpha}\varphi[\frac{1}{\alpha}\xi(t)]\,dt \\
&= p\alpha^2 + q[\alpha^2 + 3\alpha^4].
\end{aligned}$$

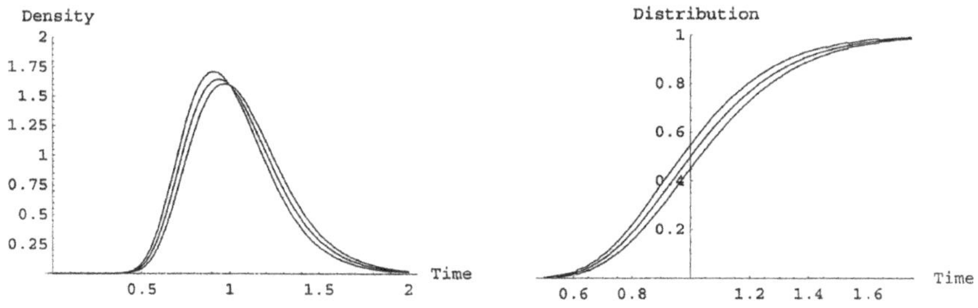

Figure 10.1. *CD*-densities and distributions for $p = 0, 1/2, 1 : \alpha = .25, \beta = 1$.

Thus we find, using the preceding result,

$$\begin{aligned} \mathrm{Var}(Z) &= \beta^2 \mathrm{Var}\left(\frac{Z-\beta}{\beta}\right) = \beta^2\left[\mathrm{E}\left(\frac{Z-\beta}{\beta}\right)^2 - q^2\alpha^4\right] \\ &= (\alpha\beta)^2[1 + q\alpha^2(2+p)]. \end{aligned} \tag{10.18}$$

Note the parameter β is the mean and α is the coefficient of variation iff $p = 1$.

To illustrate differences among related *CD*-distributions, we present some graphs in Figures 10.1 and 10.2.

Remark. *When $\alpha < 0.30$ there is no difference detectable by sample among the CD-distributions so a desirable property of any one might be utilized in a "quick but not dirty" analysis with another.*

We now show this mixed class, CD, is the most general class of continuous distributions on $\Re_+$ for which the exponent of e in the density has a Chi-square distribution with one degree of freedom.

Theorem 2. *Let Z, be an rv having a density on $\Re_+$ which is analytic when extended to the annulus excluding the origin. Let $U = \frac{1}{\alpha}\xi(Z/\beta)$ for some $\alpha, \beta > 0$. Then*

$$U^2 = \frac{1}{\alpha^2}\xi^2(Z/\beta) \sim \chi^2(1),$$

is a nasc that $Z \sim CD(\beta, \alpha, p)$ for some $p \in [0, 1]$.

Proof. Let $Z \sim CD(\alpha, \beta, p)$ with density given by eqn (10.11); then it is easily shown that $U^2 = \frac{1}{\alpha^2}\xi^2(Z/\beta)$ has a Chi-squared distribution with one degree of freedom; see [116, 1957]. Now consider sufficiency. Assume that $\frac{1}{\alpha^2}\xi^2(Z/\beta) \sim \chi^2(1)$. Then it follows, since ξ is one-to-one that

$$\mathrm{P}[\beta\xi^{-1}(-y) < Z < \beta\xi^{-1}(y)] = 2\Phi(y/\alpha) - 1,$$

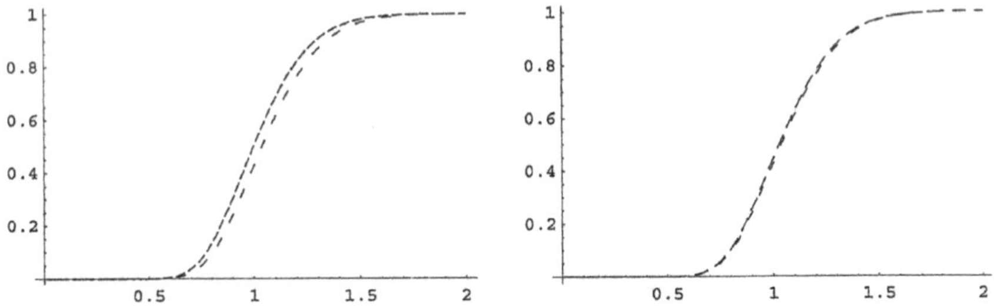

Figure 10.2. Two pairs of $\mathcal{CD}$-distributions compared: (L) For $p = 0, 1$ with same mean $\mu = 1.04$ and variance $\sigma^2 = .0432$. (R) $p = 1/2$ with $\alpha = .2$, $\beta = 1.031$ and $p = 1$ with $\alpha = .199852$, $\beta = 1.04$.

and thus by assuming $Z \sim F$, we must have $F(\beta x) - F(\beta/x) = 2\Phi[\xi(x)/\alpha] - 1$. We differentiate with respect to x to obtain

$$\beta f(\beta x) + \frac{\beta}{x^2} f(\beta/x) = \frac{2\xi'(x)}{\alpha}\varphi[\frac{\xi(x)}{\alpha}].$$

We replace the unknown function $f(x)$ with the unknown function $g(x)$ by substituting $g(x) = \{x\beta f(\beta x)\} \div \varphi[\xi(x)/\alpha]$, and by setting $h(y) = g(y^2)$ to obtain a functional equation in h, viz.,

$$h(y) + h(1/y) = \frac{y}{\alpha} + \frac{1}{\alpha y} \quad \text{for} \quad y > 0. \tag{10.19}$$

Since h is analytic by assumption its Laurent expansion is unique. Solving the resulting equations for the coefficients shows that h must have the form

$$h(y) = \frac{qy}{\alpha} + \frac{p}{\alpha y} \quad \text{for some} \quad 0 \leq p \leq 1.$$

Reversing the steps shows that the solution is indeed $f(x; \beta, \alpha, p)$ as defined by eqn (10.11). □

This relationship will be examined further to provide elementary derivations of some of the basic properties of the $\mathcal{W}$- and $\mathcal{T}$-distributions.

10.5. Behavior of the Hazard Rate

One of the initial objections to the use of the $\mathcal{FL}$-distribution was that it was not IHR. It was shown in [19] that if h is the hazard rate when $\alpha = \beta = 1$ that h reaches a maximum and then decreases slowly. However, it was later discovered that only IHRA behavior, not IHR, was necessary to encompass life-length behavior for coherent systems. But does this IHRA classification also hold when life is terminated by an accumulation of damage?

The behavior of the coefficient of variation is known to be related to the IHRA (DHRA) class, for which this parameter must always be less than or equal to

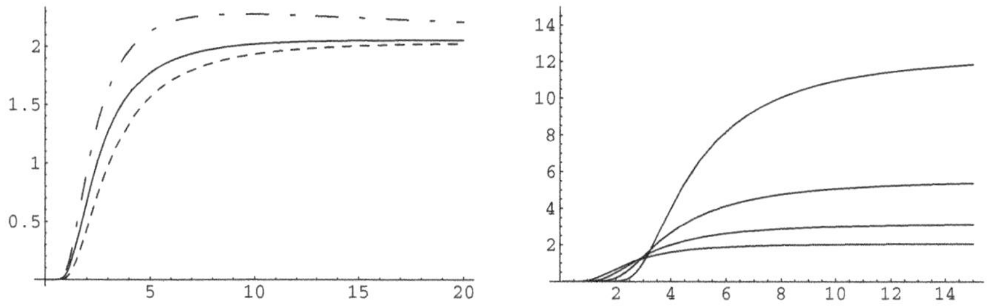

Figure 10.3. Hazard rates for $p = 0, 1/2, 1$ and with $p = 1/2$ for $\alpha = .1(.1).4$.

(greater or equal) to unity; see [6]. For the $\mathcal{CD}$-distributions, recall, β is always a scale parameter so we can wlog assume $\beta = 1$ but the parameter α is only approximately the coefficient of variation, unless $p = 1$. So what is the range of α for which a $\mathcal{CD}$-distribution is IHRA? To answer this we examine the asymptotic behavior of the hazard rates.

The density, survival, and hazard functions in the general case when $\beta = 1$ become, respectively, for $x > 0$

$$\begin{aligned} f_p(x;\alpha) &= \frac{1}{\alpha\sqrt{x}}\left(\frac{p}{x}+q\right)\varphi[\tfrac{1}{\alpha}\xi(x)], \\ \bar{F}_p(x;\alpha) &= \bar{\Phi}[\frac{1}{\alpha}\xi(x)] + (2p-1)e^{2/\alpha^2}\bar{\Phi}[\frac{1}{\alpha}\psi(x)], \end{aligned} \tag{10.20}$$

with

$$h_p(x\,;\,\alpha) = \frac{f_p(x\,;\,\alpha)}{\bar{F}_p(x\,;\,\alpha)}. \tag{10.21}$$

See Figure 10.3 for some graphs of the behavior of the hazard and hazard rate.

Theorem 3. *The asymptotic limit of the hazard rate is*

$$\lim_{x\to\infty} h_p(x;\alpha) = q/2p\alpha^2 \textit{ for } p > 0.$$

Proof. Since we seek the limit of h_p as $x \to \infty$, wlog let $y = \xi(x)/\alpha$ and consider the limit of $\hbar(y) := h_p[\xi^{-1}(\alpha y)]$. Asas, we find

$$\begin{aligned} \hbar(y) &= \frac{1}{\alpha}\left\{\left[q + \frac{p}{\xi^{-1}(\alpha y)}\right]\cdot\left[\frac{\varphi(y)}{\frac{\alpha y}{2}[1+\sqrt{1+4(\alpha y)^{-2}}]}\right]\right. \\ &\quad \left.\times\left[\bar{\Phi}(y) + (p-q)e^{2/\alpha^2}\bar{\Phi}[\sqrt{y^2+4\alpha^{-2}}]\right]^{-1}\right\}, \\ &\asymp \frac{2q}{\alpha^2}\left\{[1+\sqrt{1+4(\alpha y)^{-2}}]^{-1}\cdot\frac{\varphi(y)}{y\bar{\Phi}(y)}\right. \\ &\quad \left.\times\left[1 + (p-q)e^{2\alpha^{-2}}\frac{\bar{\Phi}[\sqrt{y^2+4\alpha^{-2}}]}{\bar{\Phi}(y)}\right]^{-1}\right\}. \end{aligned}$$

Since, by applying l'Hôpital's rule, we find the respective limits of the three factors in braces above are $1/2,\ 1,\ 1/2p$ and obtain $\lim_{y\to\infty} \hbar_p(y) = q/2p\alpha^2$. □

From the definition we see for given p, α the distribution is IHRA on $\Re_+$ iff

$$x f_p(x;\alpha) \geq -\bar{F}_p(x;\alpha)\ln[\bar{F}_p(x;\alpha)] \qquad \text{for all } x > 0. \tag{10.22}$$

If we again let $\alpha y = \xi(x)$ the requisite inequality then becomes, asas,

$$\frac{1}{\alpha}\left\{p[\xi^{-1}(\alpha y)]^{-1/2} + q[\xi^{-1}(\alpha y)]^{1/2}\right\}\phi(y) \geq -\bar{G}_p(y;\alpha)\ln[\bar{G}_p(y;\alpha)], \tag{10.23}$$

where we have written

$$\bar{G}_p(y;\alpha) = \bar{\Phi}(y) - (2p-1)\bar{\Phi}\left(\sqrt{4\alpha^{-2}+y^2}\right) \qquad \text{for } \ y \in \Re.$$

From Exercise 2 in 10.A it follows that $\sqrt{\xi^{-1}(y)} = \left[\frac{y}{2} + \sqrt{(\frac{y}{2})^2+1}\right]$ so asas we find eqn (10.23) is equivalent to

$$\frac{1}{2}\left\{\sqrt{4\alpha^{-2}+y^2} - (p-q)y\right\}\varphi(y) \geq -\bar{G}_p(y;\alpha)\ln[\bar{G}_p(y;\alpha)] \text{ for all } x > 0. \tag{10.24}$$

If we consider the $\mathcal{FL}$-distribution, we see $\bar{G}_{1/2}(y;\alpha) = \bar{\Phi}(y)$ is independent of α; thus so is the rhs of eqn (10.24). Hence a nasc that the $\mathcal{FL}$-distribution is IHRA on $\Re_+$, is,

$$\frac{1}{\alpha^2} \geq \sup_{y\in\Re} A(y) \quad \text{where} \quad A(y) = \left\{\left[\frac{\Phi(y)\ln[\Phi(y)]}{\varphi(y)}\right]^2 - \frac{y^2}{4}\right\}. \tag{10.25}$$

The last equation follows since $\bar{\Phi}(y) = \Phi(-y)$. The behavior of $A(y)$ is shown in Figure 10.4, q.v.

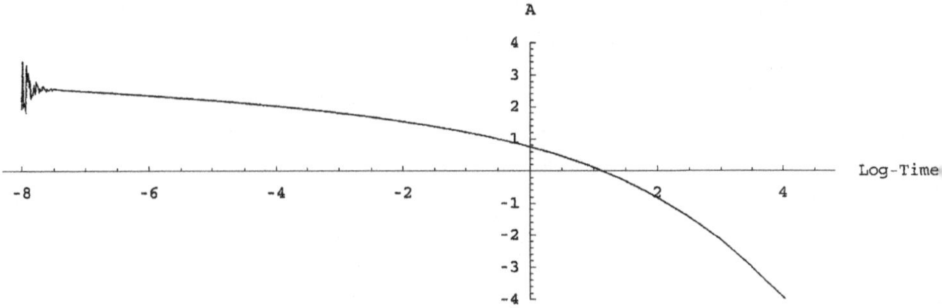

Figure 10.4. *Mathematica* unable to compute $A(y)$ for $y < -7.5$.

The hazard rate for the $\mathcal{FL}$-distribution, label it h, can be simply writtem

$$h(x) = \frac{1}{\alpha}\xi'(x)\,\hbar[\tfrac{1}{\alpha}\xi(x)] \qquad \text{for all } x > 0,$$

where $\hbar := \varphi/\bar{\Phi}$ is the standard Gaussian hazard rate.

Let $y = \frac{1}{\alpha}\xi(x)$ with $y' = \frac{1}{\alpha}\xi'(x)$, $y'' = \frac{1}{\alpha}\xi''(x)$ and note $\varphi'(y) = -y\varphi(y)$; thus since $\hbar' = -y\hbar + \hbar^2$, we find asas

$$h'(x) = \hbar(y)\left[y'' + (y')^2(\hbar - y)\right] = (y')^2\left[\frac{y''}{(y')^2} - y + \hbar\right].$$

Since $(y')^2$ has no zeros on $(0, \infty)$, we see the maximizer of h, namely the solution of $h'(x) = 0$, can be found as a zero of the function $L(y, \delta)$, by setting $\delta = 2/\alpha$ and,

$$\mathrm{L}(y : \delta) = (y^2 + \delta^2)(\hbar(y) - y) + y - 2\sqrt{y^2 + \delta^2} \quad \text{for } y \in \Re.$$

Clearly a unique zero exists; label it $\hat{y}_\alpha$. It can be found numerically quite easily if $\alpha \gg 0$, e.g., if $\alpha = .798$ then $\hat{y}_\alpha = 0$. But exact determination is difficult if α is small. Note in all cases there is a unique maximum for the hazard rate for a $\mathcal{FL}$-distribution.

In the general case, by noting $\bar{G}_p(y;\alpha) = G_q(-y;\alpha)$ we have the condition

$$\frac{1}{\alpha^2} \geq \sup_{y\in\Re}\left\{\left[\frac{G_q(y;\alpha)\ln[G_q(y;\alpha)]}{\varphi(y)} + (2q-1)\frac{y}{2}\right]^2 - \frac{y^2}{4}\right\},$$

is a nasc that $\mathcal{CD}(\alpha, \beta, p)$ is IHRA.

Exercise Set 10.C

1. An rv X is said to be *infinitely decomposable* iff for every positive integer n one can express $X = \sum_1^n Y_j$ as the sum of n iid rvs. Show that the Wald and Tweedie variates are infinitely decomposable. Why are the $\mathcal{FL}$-variates not?
2. Using Mathematica, or a similar program, graph the behavior of the density and hazard rate in eqns (10.20) and (10.21) for $\alpha > \beta > 1$.
3. Show that $\lim_{x\to\infty} h_1(x, \alpha) = 0$; so can the $\mathcal{W}$-distribution be IHRA for any practical purposes? What about the $\mathcal{T}$-distribution?
4. The $\mathcal{FL}$-distribution is IHRA on $\Re_+$ iff $\alpha^{-2} \geq \sup_{y\in\Re} A(y)$. Show, using l'Hôpital's rule, that $A(y) \to \infty$ as $y \to -\infty$, to conclude there is no value of α for which IHRA on $\Re_+$ holds. Is this a practical consideration? For what values is it IHRA on $[0, 10^6]$?
5. If, say $\alpha < 1/2$, check the range of the initial interval on the sample space, say $[0, n\beta]$, on which we can be certain the distribution is IHRA.
6. To determine the rate at which $A(-y)$ goes to ∞, show that $\bar{\Phi}(t) \leq \varphi(t)/t$ and use the asympotic expansion of Mill's ratio, given in 2.23, to prove $\lim_{y\to-\infty} A(y) = \lim_{y\to\infty}\left[\ln y - \ln m(y) + \ln\sqrt{2\pi} + o(y)\right]$.

7. Find the range of parameters of $X \sim CD(\alpha, \beta, p)$ for which the shape parameter α remains virtually the coefficient of variation. Let the ratio of the exact coefficient of variation to α be $g(\alpha) = \sqrt{\mathrm{Var}(X)}/\alpha \mathrm{E}(X)$. Show that
   (a) $g(0+) = g(\alpha_2) = 1$ when $\alpha_2 = \sqrt{p/q}$.
   (b) the maximum of g on the interval $[0, \alpha_2]$ occurs at $\alpha_1 = \sqrt{p/q(2+p)}$ and that $g(\alpha_1) = (2+p)\sqrt{1+p}/(2+2p) > 1$.
   (c) over the interval $[0, \alpha_2]$ the maximum discrepancy, $g(\alpha_1) - 1$, is very small, e.g., for the $\mathcal{FL}$-distribution on the interval $[0, 4/3]$ the discrepancy is less than 2.06%.
   (In engineering applications the coefficient of variation is usually much less than unity.)

10.6. Mixed Variate Relationships

We first state a few lemmas about the mixed CD-variates from [80, 1991], some of which are also given in Jørgensen et al. [49, 1991] in different notation. In what follows let $I \sim \mathcal{B}(1, p)$ denote a Bernoulli variate with probability of success p; not to be confused with the indicator function I.

Lemma 1. *Let $X \sim CD(\alpha, \beta, 1)$, $Y \sim CD(\alpha, \beta, 0)$ with $I \sim \mathcal{B}(1, p)$ independent of both for some $0 \leq p \leq 1$. Then $Z \sim CD(\alpha, \beta, p)$ when and only when $Z \sim IX + (1 - I)Y$.*

Lemma 2. *$Z \sim CD(\alpha, \beta, p)$ iff $1/Z \sim CD(\alpha, 1/\beta, q)$, where $q = 1 - p$.*

If $p = q = 1/2$ in this lemma, it restates the reciprocal property of the $\mathcal{FL}$-variates first given in [19, 1969]. But there also follows:

Lemma 3. *If $Z \sim CD(\alpha, 1, p)$ and $I \sim \mathcal{B}(1, r)$, for some $0 < r < 1$, are independent, then the mixture $[IZ + (1 - I)/Z] \sim CD(\alpha, 1, rp + (1 - r)q)$.*

The proof depends upon the identity $x^2\xi'(x) = \xi'(1/x)$, which follows from $\xi(x) = -\xi(1/x)$.

Corollary 1. *In Lemma 3 above, iff $r = 1/2$ do we have, for $Z \sim CD(\alpha, 1, p)$ and any $0 \leq p \leq 1$,*

$$[IZ + (1 - I)/Z] \sim CD(\alpha, 1, 1/2) \sim \mathcal{FL}(\alpha, 1).$$

Lemma 3 and its corollary show that reciprocal mixing with probability one-half of any CD-variate yields a FL-variate. This implies a stochastic relationship between CD-variates and normal variates.

Theorem 4. *If we let $Z \sim CD(\alpha, \beta, p)$ for any $0 \leq p \leq 1$ and $I \sim \mathcal{B}(1, 1/2)$ with $I \perp Z$, then*

$$\frac{2I - 1}{\alpha}\xi(Z/\beta) \sim \mathcal{N}(0, 1); \tag{10.26}$$

but the corresponding inverse transformation always gives, by Corollary 1, an $\mathcal{FL}$-variate.

PROOF. We merely note that $\xi\left(\frac{IZ}{\beta} + \frac{(1-I)\beta}{Z}\right) = (2I-1)\xi(Z/\beta)$. □

In particular this yields a stochastic relationship between Wald (inverse-Gaussian) and normal variates previously quoted from [26, 1989].

Corollary 2. *If we let $X \sim \mathcal{W}(\alpha, \beta)$ and $I \sim \mathcal{B}(1, 1/2)$ with $I \perp X$, then*

$$\frac{2I-1}{\alpha}\xi(X/\beta) \sim \mathcal{N}(0,1), \tag{10.27}$$

but the corresponding inverse transformation gives an $\mathcal{FL}$-variate.

We must relax the independence between I and X to obtain an inverse transformation.

Theorem 5.

$$Z \sim \mathcal{N}(0,1) \quad \textit{iff} \quad X = \beta\xi^{-1}[(|Z|J_{\alpha|Z|})] \sim \mathcal{W}(\alpha, \beta), \tag{10.28}$$

where the binary variate J_u is defined for $u > 0$ by

$$J_u = \begin{cases} 1 & w.p. \quad \frac{1}{1+\xi^{-1}(u)} \\ -1 & w.p. \quad \frac{1}{1+\xi^{-1}(-u)}. \end{cases} \tag{10.29}$$

PROOF. We make us of the $\psi - \xi$ identities in Exercise Set 10.A without mention. Thus, if we let $Y = \frac{1}{\alpha}\xi(X/\beta)$ with $X \sim \mathcal{W}(\alpha, \beta)$, it is straightforward to show that the distribution is

$$F_Y(y) = \Phi(y) + e^{2/\alpha^2}\bar{\Phi}[\sqrt{y^2 + (2/\alpha)^2}] \quad \text{for} \quad y \in \Re, \tag{10.30}$$

and hence the density is

$$f_Y(y) = \varphi(y)\left[1 - \frac{y}{\sqrt{y^2 + (2/\alpha)^2}}\right]. \tag{10.31}$$

This distribution of eqn (10.30) and density of (10.31) agree with results given in [26] by setting $\alpha = \sqrt{\mu/\lambda}$.

Thus, if we define $U = |Z| \cdot J_{\alpha|Z|}$, it is sufficient to show that $U \sim Y$. Noting from the identities in Exercise Set 10.A that $\xi^{-1}(-y) \cdot \xi^{-1}(y) = 1$ we have from eqn (10.29) that

$$\begin{aligned} U &= |Z| \quad \text{w.p.} \quad \frac{1}{2}\left[1 - \frac{|z|}{\sqrt{z^2 + (2/\alpha)^2}}\right], \\ U &= -|Z| \quad \text{w.p.} \quad \frac{1}{2}\left[1 + \frac{|z|}{\sqrt{z^2 + (2/\alpha)^2}}\right]. \end{aligned}$$

We find that $f_{|Z|}(z) = 2\varphi(z)$ for $z > 0$ is a necessary and sufficient condition, by the usual transformation of variates, that $f_U = f_Y$. □

Remark. *We note that this result provides a method to generate pseudo-random $\mathcal{W}(\alpha, \beta)$ variates from pseudo-random Gaussian variates. This is equivalent to the method proposed by Michael* et al. *[76, 1976] and restated in [26].*

Many of the results that show certain statistics from samples of inverse-Gaussian variables possess distributions related to normality become more transparent by using this relationship. For example, the transformation to a Chi-square distribution, the result mentioned in [116, 1957], is immediate. We have noted that if $Z/\beta \sim \mathcal{CD}(\alpha, 1, p)$, then $\frac{1}{\alpha}\xi(Z/\beta)$ is not standard normal unless $p = 1/2$.

We now state the important characterization theorem for this family.

Theorem 6. *If $X_1, \ldots, X_n$ is an iid sample from the $\mathcal{W}(\alpha, \beta)$ family, with $\bar{X}_+$ and $\bar{X}_-$ denoting the arithmetic and harmonic sample means, respectively, then U and V are independent where*

$$U^2 = \frac{1}{\alpha^2}\xi^2(\bar{X}_+/\beta) \sim \chi(1) \quad and \quad V = \frac{n\beta}{\alpha^2}\left(\frac{1}{\bar{X}_-} - \frac{1}{\bar{X}_+}\right) \sim \chi^2(n-1). \tag{10.32}$$

Comment: It is an important result, see Chhikara and Folks [26, 1989], that the independence of

$$\bar{X}_+ \quad \text{and} \quad \frac{1}{\bar{X}_-} - \frac{1}{\bar{X}_+} \tag{10.33}$$

characterizes inverse-Gaussian variates. An elementary, inductive proof of independence is given in [94, 1991].

Exercise Set 10.D

1. A Putnam Examination problem in 1985 was the evaluation of the integral

$$\int_0^\infty x^{-1/2}\exp\{-a(x + x^{-1})\}\,dx,$$

and several clever derivations were published subsequently using variation of parameters and other techniques; see [14, 1990]. Recognize the integrand as being proportional to the Wald density and so find its value easily.

2. Draw some illustrative graphs showing how the density of a nonnegative $\mathcal{W}$-variate behaves for fixed β and varying α and contrast that with the behavior of eqn (10.3) with fixed μ and varying λ.

3. Define the (complex) cumulant generating function here by $K_X(t) = \ln \mathrm{E}[e^{\imath X t}]$, where $\imath = \sqrt{-1}$.

   (a) Let $X \sim \mathcal{W}(\alpha, \beta)$ and show that

$$-K_X^{-1}(t) = K_Z(t) \quad \text{where} \quad Z \sim \mathcal{N}(1/\beta, \alpha^2/\beta).$$

(b) Let $Z \sim \mathcal{N}(\mu, \sigma^2)$ and show that

$$K_Z^{-1}(t) = \imath K_X(-t) \quad \text{where} \quad X \sim \mathcal{W}(\sigma/\sqrt{\mu}, 1/\mu).$$

Thus, X is "inverse-Gaussian" or nearly so.

4. Let $X_i > 0$ be any sequence of incremental damages for which the CLT holds, viz.,

$$\frac{S_n - m_n}{s_n} \rightsquigarrow \mathcal{N}(0,1) \quad \text{as } n \longrightarrow \infty,$$

where $S_n = \sum_i^n X_i$ is the cumulative damage after n cycles and

$$\mathrm{E}S_n = m_n, \quad \mathrm{Var}S_n = s_n^2.$$

Assume also that the critical damage level decreases with the load cycle, i.e., $\omega_1 \geq \omega_2 \geq \cdots$ with $\omega_n \to \omega$. (This corresponds to relaxation of the yield stress over time or environmental degradation of strength.) What additional assumptions, if any, are necessary to obtain the $\mathcal{FL}$–distribution in the limit?

5. Let $X_i \sim CD(\alpha_i, \beta_i, p_i)$ be independent for $i = 1, 2$. Make the definitions

$$\alpha_\ominus = \frac{\alpha_1\alpha_2}{\sqrt{\alpha_1^2 + \alpha_2^2}}, \quad \beta_+ = \beta_1 + \beta_2, \quad \text{with } F \simeq CD(\alpha_\ominus, \beta_+, 1), \quad H_\nu \simeq \chi_\nu^2 \text{ for } \nu = 1, 2.$$

Show if $\alpha_1^2\beta_1 = \alpha_2^2\beta_2$ then $X_1 + X_2 \simeq p_1p_2F + (p_1q_2 + p_2q_1)F * H_1 + q_1q_2F * H_2$.

6. Show solutions of the functional eqn (10.19) exist. Find a discontinuous one defined in terms of indicator $I_A(y) = 1$ or $= 0$ accordingly as $y \in A$ or $y \in A'$;

$$h(y) = \frac{y}{\alpha} I_A(y) + \frac{1}{\alpha y} I_{A^*}(y) + \left(\frac{y}{\alpha} + \frac{1}{\alpha y}\right) I_{(A \cup A^*)'}(y)$$

for sets A and A^* such that $y \in A$ iff $1/y \in A^*$.

7. The behavior of the coefficient of variation is related to the the IFRA and DFRA classes, for which this parameter must be less than or equal to or greater than or equal to unity, respectively, see [7, 1981]. Moreover this reparameterization clarifies the anomalous behavior of the hazard rate.

8. Prove that if $Z \sim \mathcal{N}(0, 1)$ then

$$X = \beta\xi^{-1}[(2I^* - 1)|Z|] \sim CD(\alpha, \beta, p), \tag{10.34}$$

where I is here a binary variate depending upon Z and α given by

$$I = \begin{cases} 1 \text{ w.p.} & q + \frac{p-q}{1+\alpha|z|\sqrt{\xi^{-1}(\alpha|z|)}} \\ 0 \text{ w.p.} & p - \frac{p-q}{1+\alpha|z|\sqrt{\xi^{-1}(\alpha|z|)}}. \end{cases} \tag{10.35}$$

9. Show directly that $\frac{\lambda(Z-\mu)^2}{\mu^2 Z} \sim \chi_1^2$ when either $Z \sim \mathrm{IG}(\mu, \lambda)$ or $Z \sim \mathrm{RIG}(\mu, \lambda)$.

10. Prove the general result concerning the transformation to Chi-square, namely, if we take $Z \sim CD(\alpha, \beta, p)$ for any $0 \leq p \leq 1$, then $\frac{1}{\alpha^2}\xi^2(Z/\beta) \sim \chi_1^2$.

10.7. Estimation for Wald's Distributions

10.7.1. Estimation for Complete Samples

In this section we summarize results from [26, 1989] in terms of our alternate parameterization. These are presented here for convenience in engineering application. From the density of an $\mathcal{W}(\alpha, \beta)$ variate, the likelihood is

$$L(\alpha, \beta : \boldsymbol{x}) = \prod_{i=1}^{n} f_X(x_i; \alpha, \beta) = \prod_{i=1}^{n} \frac{\sqrt{\beta}}{\alpha} x_i^{-3/2} \varphi[\frac{1}{\alpha}\xi(x_i/\beta)]. \tag{10.36}$$

Let us introduce notation for the *arithmetic and harmonic means* and the *mean-mean* of the sample. They are, respectively,

$$\bar{X}_{\pm} = \left[\frac{1}{n}\sum_{i=1}^{n} X_i^{\pm 1}\right]^{\pm 1} \quad \text{and} \quad \bar{\bar{X}} = \sqrt{\bar{X}_{-}\bar{X}_{+}}.$$

When both parameters are unknown, then straightforwardly we obtain the mles of α^2 and β. They are, respectively, given by

$$\hat{\alpha}^2 = \frac{\bar{X}_{+}}{\bar{X}_{-}} - 1 \quad \text{and} \quad \hat{\beta} = \bar{X}_{+}. \tag{10.37}$$

The distribution of $\hat{\beta}$ was found by Tweedie (see Tweedie [115, 1957] and [116]). It is, if based on a sample of size n,

$$\hat{\beta} = \bar{X}_{+} \sim \mathcal{W}(\alpha/\sqrt{n}, \beta). \tag{10.38}$$

This estimator $\hat{\beta}$ is both unbiased and consistent. To check if these properties hold for $\hat{\alpha}^2$ we utilize the characterization theorem for the Wald family, in Theorem 6.

Let $T = \hat{\beta}/\beta = \bar{X}_{+}/\beta$ and V be as defined in eqn (10.32). Then the expectation of $\hat{\alpha}^2 \sim \alpha^2 TV/n$ is given by

$$\mathrm{E}\hat{\alpha}^2 = \frac{\alpha^2}{n}\mathrm{E}T \cdot \mathrm{E}V = \frac{(n-1)\alpha^2}{n}. \tag{10.39}$$

Thus, an unbiased estimate of α^2 can be defined, namely,

$$\tilde{\alpha}^2 = n\hat{\alpha}^2/(n-1). \tag{10.40}$$

Theorem 7. *The symmetric distribution of* $U = \frac{\sqrt{n}}{\alpha}\xi\left(\bar{X}_{+}/\beta\right)$ *is given by*

$$F_U(u) = \Phi(u) + e^{2n/\alpha^2}\bar{\Phi}\left(2\sqrt{u^2 + \frac{n}{\alpha^2}}\right) \quad \textit{for} \quad -\infty < u < \infty, \tag{10.41}$$

and consequently the t-statistic ratio, below depends an α *but not* β*:*

$$\frac{U}{\sqrt{V/(n-1)}} = \frac{\sqrt{(n-1)}\left(\frac{\bar{X}_{-}}{\beta} - 1\right)}{\sqrt{\frac{\bar{X}_{+}}{\bar{X}_{-}} - 1}}. \tag{10.42}$$

However, exact statistical inference is possible for two-sided confidence intervals on β from existing tables of the $\mathcal{F}$-distribution since $(n-1)U^2/V$ has an $\mathcal{F}_{1,n-1}$ distribution (functionally) independent of α.

Nevertheless, if we use randomization, we obtain

Corollary 3. *Statistical inference is also possible since if $I \sim \mathcal{B}(1, 1/2)$ is independent of the sample, then*

$$T = \frac{(2I-1)U}{\sqrt{V/(n-1)}}$$

has the t-distribution with $(n-1)$ degrees of freedom independent of α.

Nota bene a significant justification for Tweedie's parameterization is that exact one-sided confidence intervals on the parameter $\lambda = \beta/\alpha^2$ can be obtained using the pivotal statistic $\hat{\lambda} = \frac{\bar{X}_+ \cdot \bar{X}_-}{\bar{X}_+ - \bar{X}_-}$.

Estimation of α When β Is Known

First consider estimating the coefficient of variation, α, for Wald's distribution in the case when β is known, then the mle of the shape parameter is

$$\breve{\alpha} = \left[\frac{1}{n}\sum_{i=1}^{n} \xi^2(X_i/\beta)\right]^{1/2} = \left[\frac{\bar{X}_+}{\beta} - 2 + \frac{\beta}{\bar{X}_-}\right]^{1/2}, \tag{10.43}$$

with $\mathrm{E}(\breve{\alpha})^2 = \alpha^2$ and $n(\breve{\alpha}/\alpha)^2 \sim \chi^2(n)$.

Estimation of β When α Is Known

Now we consider estimating the mean when α is known; in this instance the mle of β (call it $\breve{\beta}$), is found, after straightforward maximization of the likelihood from eqn (10.36), to be

$$\breve{\beta} = \bar{X}_- \left[\frac{\alpha^2}{2} + \sqrt{\frac{\alpha^4}{4} + \frac{\bar{X}_+}{\bar{X}_-}}\right], \tag{10.44}$$

$$\asymp \bar{\bar{X}} + \frac{\alpha^2 \bar{X}_-}{2}\left[1 + \frac{\alpha^2}{4}\sqrt{\frac{\bar{X}_-}{\bar{X}_+}} - \frac{\alpha^6}{64}\sqrt{\left(\frac{\bar{X}_-}{\bar{X}_+}\right)^3} + O(\alpha^{10})\right] \quad \text{for} \quad \alpha \ll 1.$$

There are important instances, e.g., in measuring distances with an optical rangefinder, in astronomy while measuring parallax, or in engineering while surveying, when the accuracy, as measured by the coefficient of variation, of the measuring device is known. Thus, if repeated measurements are made, then the arithmetic mean of the sample, which is always used because it is unbiased, robust and because of the universal assumption of normality, may NOT provide the best estimate of the mean.

Remark 1. *If* $Z \sim CD(\alpha, \beta, p)$, *then as* $\alpha \to 0$, *both*

$$\frac{Z - E(Z)}{\alpha\beta} \quad \textit{and} \quad \frac{1}{\alpha}\xi(Z/\beta) \rightsquigarrow \mathcal{N}(0, 1), \tag{10.45}$$

but the latter converges much more rapidly and is exact when $p = .5$.

Unbised Estimation

Let $X_1, \ldots, X_n$ be an iid sample from $\mathcal{W}(\alpha, \beta)$. If both α and β are unknown, the mles of (α, β) are

$$\hat{\alpha}^2 = \frac{\bar{X}_+}{\bar{X}_-} - 1, \qquad \hat{\beta} = \bar{X}_+. \tag{10.46}$$

While $\hat{\beta} \sim \mathcal{W}(\alpha/\sqrt{n}, \beta)$ is unbiased, one must modify $\hat{\alpha}^2$ to make it unbiased, namely,

$$A = \tilde{\alpha}^2 = \frac{n}{n-1}\left(\frac{\bar{X}_+}{\bar{X}_-} - 1\right) \sim \frac{\alpha^2 U V}{n-1}, \tag{10.47}$$

where here

$$U = \frac{\bar{X}_+}{\beta} \sim \mathcal{W}(\alpha/\sqrt{n}, 1) \perp V \sim \chi^2(n-1).$$

However, if α is known, the mle for β is *not* $\bar{X}_+$ but

$$\check{\beta} = \bar{X}_-\left[\frac{\alpha^2}{2} + \sqrt{\frac{\alpha^4}{4} + \frac{\bar{X}_+}{\bar{X}_-}}\right] = \hat{\beta}K(\hat{\alpha}^2) \tag{10.48}$$

where $\hat{\alpha}$ and $\hat{\beta}$ were defined in eqn (10.46) and we define the function

$$K(x) = \frac{\frac{\alpha^2}{2} + \sqrt{\frac{\alpha^4}{4} + 1 + x}}{1 + x} \quad \text{for } x > 0. \tag{10.49}$$

NB the function K can only be defined when α is presumed known. Does this yield a better estimate? One might consider whether $\mathrm{Var}(\hat{\beta}/\beta) > \mathrm{Var}(\check{\beta}/\beta)$ but since $\check{\beta}$ is not unbiased (and cannot be easily unbiased) the comparison might not be conclusive, moreover the investigation would presumably have to be done numerically.

We show instead that $\check{\beta}$ is closer to β than is $\hat{\beta}$ in expectation. Here "closer" means relative to the metric d which is defined by $d(x, y) = \xi^2(x/y)$ on the quadrant $\Re_2^+$. To check that d is a metric the only sticking point is the triangle inequality, viz.,

$$\xi^2(x/y) \le \xi^2(x/z) + \xi^2(z/y) \quad \text{for all } 0 < x < z < y.$$

But by setting $a = x/y$, $b = z/x$ and noting $-\xi(1/b) = \xi(b)$ the triangle inequality is equivalent with

$$\xi^2(a) \le \xi^2(b) + \xi^2(ab) \quad \text{for } 0 < a < 1 < b,$$

but this follows since for any $b > a$ we see

$$\xi^2(ab) + \xi^2(b) - \xi^2(a) \ge b - a + (1 + b - a)\xi^2[ab/(1 + b - a)] \ge 0.$$

Theorem 8. *Let $\tilde{\beta} = \hat{\beta}K(A)$, where A is the unbiased estimate of α^2 defined in eqn (10.47), then*

$$\mathrm{E}\,\xi^2(\tilde{\beta}/\beta) < \mathrm{E}\,\xi^2(\hat{\beta}/\beta) = \frac{\alpha^2}{n}.$$

PROOF. Let us set

$$H(x) := \frac{1}{K(x)} = -\frac{\alpha^2}{2} + \sqrt{\frac{\alpha^4}{4} + 1 + x}.$$

Then $K(x) = \frac{\alpha^2 + H(x)}{1+x}$ for $x > 0$. Thus, we have

$$\xi^2(\tilde{\beta}/\beta) = \frac{\hat{\beta}}{\beta}K(A) - 2 + \frac{\beta}{\hat{\beta}}H(A).$$

By using Taylor's theorem we find

$$K(A) = K(a) + (A - a)K'(a) + \int_a^A (A - x)K''(x)\,dx,$$

and so we see, setting $a = \alpha^2$, that $K(a) = H(a) = 1$. Hence we obtain

$$\xi^2(\tilde{\beta}/\beta) \sim \xi^2(U) + (A - a)\left[U \cdot K'(a) + \frac{H'(a)}{U}\right] + \int_a^A (A - x)\left[U \cdot K''(x) + \frac{H''(x)}{U}\right] dx.$$

Now taking expectations and realizing that

$$K'(a) = -H'(a) = \frac{-1}{2 + a} \quad \text{and} \quad \mathrm{E}\,[(A - a)U] = \mathrm{E}\,[(A - a)/U],$$

we find

$$\mathrm{E}\,\xi^2(\tilde{\beta}/\beta) = \mathrm{E}\,\xi^2(\hat{\beta}/\beta) + \mathrm{E}\left\{\int_a^A (A - x)\left[U \cdot K''(x) + \frac{H''(x)}{U}\right] dx\right\}. \tag{10.50}$$

We note that $\text{sgn}(A-a) = \text{sgn}(A-x)$ for all $x \in [a \wedge A, a \vee A]$. so the sign of the integral depends on the sign of the integrand in square brackets in eqn (10.50). Thus, it is sufficient to show the expectation of the integral is negative, which is implied,

$$U \cdot K''(x) + \frac{H''(x)}{U} \leq 0 \qquad \text{for all } x, U > 0. \tag{10.51}$$

We find that

$$K''(x) = \frac{H''(x) - 2K''(x)}{1+x}, \qquad H''(x) = \frac{-1}{4[H(x)]^3}.$$

Thus, asas we see eqn (10.51) is true iff for all $U, x > 0$,

$$\frac{(1+x)^2(U^2+1+x)}{4U^2} > -(1+x)H^2(x) + \alpha^2 H^3(x) - 2H^4(x). \tag{10.52}$$

But the rhs of eqn (10.52) is less than the polynomial $p(H)$, viz., $p(H) = -H^2 + \alpha^2 H^3 - 2H^4$ where $H \in \downarrow$ with $H(0) = (\alpha^4/4+1)^{-1/2}$ and $H(\infty) = 0$. But the maximum of this polynomial $p(H)$ is zero when $H(\infty) = 0$. Clearly then the lhs of eqn (10.52) is minimized at $x = 0$ with a minimum value of $[(U^2+1)/4U] > 0$. □

We infer, when α is small, any difference between the two populations, $\mathcal{W}(\alpha, \beta)$ and $\mathcal{N}\left(\beta, (\alpha\beta)^2\right)$, would not be detected from small samples. But too frequently in larger samples, assuming the latter model, the outliers would be regarded as spurious observations.

In those cases when the $\mathcal{W}$-distribution applies, rather than the normal, the mle of the mean, when α is known, is the appropriate function of the arithmetic and harmonic means as expressed in eqn (10.44).

In many instances practitioners may be deceived by ignoring their knowledge of the accuracy of measurement by assuming, in analogy with the normal distribution, that the sample mean is the mle estimator whether the coefficient of variation is known or not.

The problem of determining exact confidence intervals on the coefficient of variation, independent of the mean, has been of much interest in engineering and it has a long history. It has never been solved satisfactorily for any distribution of practical engineering interest, not even asympotically for the normal distribution.

But here the parameterization, using α and β in Wald's distribution, allows this to be accomplished, and done very simply when α is small.

10.7.2. Estimation for Incomplete Wald Samples

We now consider the procedure needed to obtain the mles for the parameters of the $\mathcal{W}$-density when the sample is censored. Denote the sample of size n by $x_1, x_2, \ldots, x_n$ where $x_1, \ldots, x_k$ are, wlog, the failure observations (dead-times) and $x_{k+1}, \ldots, x_n$ are right-censored (alive-times) for some $1 \leq k \leq n$. The log-likelihood is, from eqn (10.36),

$$\ell(\alpha, \beta|x_1, \ldots, x_n) = \sum_{i=1}^{k} \ln f_X(x_i;\alpha, \beta) - \sum_{j=k+1}^{n} H_X(x_i;\alpha, \beta)$$

with the hazard function $H_X = -\ln(1 - F_X)$ from which, for notational simplicity, we henceforth omit the parameters.

To obtain the stationary points we find, asas,

$$\frac{\partial \ell}{\partial \alpha} = \frac{1}{\alpha^3} \sum_{i=1}^{k} \left[-\alpha^2 + \frac{x_i}{\beta} - 2 + \frac{\beta}{x_i} \right] - \sum_{j=k+1}^{n} \frac{\partial H_X(x_j)}{\partial \alpha},$$

$$\frac{\partial \ell}{\partial \beta} = \frac{1}{2\alpha^2 \beta} \sum_{i=1}^{k} \left[\alpha^2 + \frac{x_i}{\beta} - \frac{\beta}{x_i} \right] - \sum_{j=k+1}^{n} \frac{\partial H_X(x_j)}{\partial \beta}.$$

Since β is a scale parameter, asas we find $\partial \ell/\partial \beta = 0$ is equivalent with

$$\alpha^2 + \frac{\bar{x}_+}{\beta} - \frac{\beta}{\bar{x}_-} = \frac{2\alpha^2}{k} \sum_{j=k+1}^{n} x_j h_X(x_j), \tag{10.53}$$

and we also find that $\partial \ell/\partial \alpha = 0$ is equivalent with

$$\alpha^2 = \frac{\bar{x}_+}{\beta} - 2 + \frac{\beta}{\bar{x}_-} + \frac{2\alpha^2}{k} \sum_{j=k+1}^{n} x_j h_X(x_j) - \frac{4}{k} \sum_{j=k+1}^{n} R(x_j). \tag{10.54}$$

NB here we have used $\bar{x}_+, \bar{x}_-$ for the arithmetic and harmonic means of the uncensored observations and defined

$$D(x;\alpha, \beta) = \frac{e^{2\alpha^{-2}} \Phi[-\psi(x/\beta)/\alpha]}{\Phi[-\xi(x/\beta)/\alpha] - e^{2\alpha^{-2}} \Phi[-\psi(x/\beta)/\alpha]},$$

where for notational simplicity, we have written $\psi(x) = 2x\xi'(x)$ and omitted the parameters from the function D in eqn (10.54).

Now we introduce the sample functions averaged over the censored observations depending only on the parameters.

$$\tilde{q}(\alpha, \beta) = \frac{1}{k} \sum_{j=k+1}^{n} x_j h_X(x_j;\alpha, \beta) \quad \text{and} \quad \tilde{D}(\alpha, \beta) = \frac{1}{k} \sum_{j=k+1}^{n} D(x_j;\alpha, \beta). \tag{10.55}$$

Using them we obtain the mles as the solutions of the two stationary equations in α, β,

$$\alpha^2 = \frac{\frac{\beta}{\bar{x}_-} - \frac{\bar{x}_+}{\beta}}{1 - 2\tilde{q}(\alpha, \beta)} \quad \text{and} \quad \beta = \frac{\bar{x}_+}{1 + 2\tilde{D}(\alpha, \beta)}. \tag{10.56}$$

Note that if $k = n$ then $\tilde{q} = \tilde{R} = 0$ and we have the usual mles for complete samples.

The method of computation is an iterative one; set

$$\beta_0 = \bar{x}_+ \quad \text{and} \quad \alpha_0^2 = \frac{\bar{x}_+}{\bar{x}_-} - 1,$$

then define for $j = 0, 1, 2 \ldots$:

$$\alpha_{j+1}^2 = \frac{\frac{\beta_j}{\bar{x}_-} - \frac{\bar{x}_+}{\beta_j}}{1 - 2\tilde{q}(\alpha_j, \beta_j)}, \quad \text{and} \quad \beta_{j+1} = \frac{\bar{x}_+}{1 + 2\tilde{D}(\alpha_j, \beta_j)}. \tag{10.57}$$

Then we have

$$\hat{\alpha} = \lim_{j\to\infty} \alpha_j, \qquad \hat{\beta} = \lim_{j\to\infty} \beta_j.$$

Convergence is rapid. This may be compared with the corresponding procedure for the $\mathcal{FL}$-distribution.

10.8. Estimation for the $\mathcal{FL}$-Distribution

10.8.1. Complete Samples

Now let us examine the estimator for β for the $\mathcal{FL}$-distribution when α is known. From [18, 1969] it is seen that the mle, $\hat{\beta}$, is the solution to the equation

$$\frac{\alpha^2}{n} \sum_{i=1}^{n} \frac{\beta - x_i}{\beta + x_i} = \alpha_0^2 \xi(\beta/\beta_0), \quad \text{with } \beta_0 = \sqrt{\bar{x}_+ \bar{x}_-},\ \alpha_0^2 = \sqrt{\frac{\bar{x}_+}{\bar{x}_-}}, \tag{10.58}$$

where $x_1, \ldots, x_n$ is a complete iid sample. We check that the rhs of eqn (10.58) is a rapidly increasing function of $\beta > 0$ with a range of $\Re$ while the lhs is a slowly increasing function of β only from $-\alpha^2$ to α^2.

But as noted *loc. cit.* the mle, $\hat{\beta}_n$, based on a sample of size n converges quickly (about $n = 5$ when $\alpha < .5$) to the mean-mean, since the lhs of eqn (10.58) converges quickly to 0. This is implied by the reciprocal property of the $\mathcal{FL}$-variates, i.e., $(X/\beta) \sim (\beta/X)$. Hence we conclude that in this case a knowledge of α makes little difference in the estimation of β when sample sizes are large.

We note that the mean-mean has the reciprocal property, namely that

$$\frac{\overline{\overline{X}}_n}{\beta} \sim \frac{\beta}{\overline{\overline{X}}_n}$$

but in addition we find that its asymptotic distribution falls again in the $\mathcal{FL}$–family.

Theorem 9. *If $X_1, \ldots, X_n$ is an iid sample from $\mathcal{FL}(\alpha, \beta)$, then*

$$\overline{\overline{X}}_n \rightsquigarrow \mathcal{FL}(\delta/\sqrt{n}, \beta) \qquad \text{as } n \to \infty$$

where

$$\delta = \frac{\alpha\sqrt{4+3\alpha^2}}{2+\alpha^2} < \alpha. \tag{10.59}$$

PROOF. We must show that

$$Z_n := \frac{\sqrt{n}}{\delta}\xi(\overline{\overline{X}}_n/\beta) \rightsquigarrow \mathcal{N}(0,1) \quad \text{as } n \to \infty,$$

that is, $\mathrm{P}[Z_n \le z] \to \Phi(z)$ for any $z \in \Re$ as $n \to \infty$. Now we have

$$\mathrm{P}[Z_n \le z] = \mathrm{P}\left[\overline{\overline{X}}_n/\beta \le \xi^{-1}(\delta z/\sqrt{n})\right]$$

where from the identities in Exercise Set 10.B we see

$$\left(\overline{\overline{X}}_n/\beta\right)^2 = \frac{\sum_1^n (X_i/\beta)}{\sum_1^n (\beta/X_i)} \sim \frac{\sum_1^n (U_i)}{\sum_1^n (1/U_i)},$$

where $U \sim \mathcal{FL}(\alpha, 1)$. Thus,

$$U \sim \frac{1}{U}, \quad \mathrm{E}U = 1 + \frac{\alpha}{2}, \quad \mathrm{Var}(U) = \alpha^2\left(1 + \frac{5\alpha^2}{4}\right).$$

Now let $\theta_n = \xi^{-1}(\delta z/\sqrt{n})$ and, omitting affixes for notational convenience, set

$$
\begin{aligned}
Y &= \frac{U}{\theta} - \frac{\theta}{U} \quad \text{so there follows:}\\
\mathrm{E}[Y] &= \left(\frac{1}{\theta} - \theta\right)\mathrm{E}[U],\\
\mathrm{Var}[Y] &= \left(\frac{1}{\theta^2} + \theta^2\right)\mathrm{Var}[U] + \alpha^2[2 + (\alpha^2/2)]\\
&= [\xi^2(\theta) + 2]^2\mathrm{Var}[U] - 2\alpha^4.
\end{aligned}
$$

Thus, we see $\mathrm{Var}[Y] \to \alpha^2(4+3\alpha^2)$ as $n \to \infty$. Note that

$$\mathrm{P}[Z_n \le z] = \mathrm{P}\left[\frac{\sum_1^n (Y_i - EY)}{\sqrt{n\mathrm{Var}(Y)}} \le \frac{-\sqrt{n}\mathrm{E}[Y]}{\sqrt{\mathrm{Var}[Y]}}\right].$$

To complete the proof we must show that

$$\frac{-\sqrt{n}\mathrm{E}(Y)}{\sqrt{\mathrm{Var}(Y)}} \longrightarrow \frac{\delta z(2+\alpha^2)}{\alpha\sqrt{4+3\alpha^2}} \quad \text{as } n \to \infty.$$

Since $-\sqrt{n}\mathrm{E}[Y] = \sqrt{n}[\theta_n - (\theta_n)^{-1}][1 + (\alpha^2/2)]$ we see by recalling $\psi(z) := z^{1/2} + z^{-1/2} = \sqrt{\xi^2(z) + 4}$ that

$$\xi^{-1}(x) - \frac{1}{\xi^{-1}(x)} = \xi[\xi^{-1}(x)] \cdot \psi[\xi^{-1}(x)] = x\sqrt{x^2+4}\,;$$

hence

$$-\sqrt{n}\mathrm{E}[Y] \to 2\delta z\left(1 + (\alpha^2/2)\right)\sqrt{1 + \frac{(\delta z)^2}{4n}},$$

and by definition of δ in eqn (10.59) we have the result. □

In the case when both α and β are unknown we have the MLEs, $\hat{\alpha}$, $\hat{\beta}$, as the solution of the pair of equations, namely, eqn (10.58) and the following:

$$\alpha^2 = \alpha_0^2\xi^2(\beta/\beta_0) + 2(\alpha_0^2 - 1). \tag{10.60}$$

10.8.2. Incomplete Samples of Fatigue-life Distribution

We now derive the estimates, using censored data, for the parameters of this cumulative damage distribution, given in eqn (10.9), with density given by

$$f_X(y) = \frac{1}{\alpha\beta}\xi'(y/\beta)\varphi[\frac{1}{\alpha}\xi(\beta/y)] \quad \text{for} \quad y > 0. \tag{10.61}$$

Here we have used the notation

$$\Phi' = \varphi \quad \text{and} \quad \xi'(x) = \frac{x+1}{2x^{3/2}}. \tag{10.62}$$

Let us assume that at each observation we record the minimum of the service-life X and some nonspecified but independent censoring time $T \sim F_T$, namely the couple (Y, Z), where $Y = \min(X, T)$ and $Z = \mathrm{I}(X < T)$. Here $\mathrm{I}(\cdot)$ is the indicator function. Thus, the joint density of the couple (Y, Z) is the same as in eqn (6.17), viz.,

$$f(y, z) = [f_X(y)S_T(y)]^z[f_T(y)S_X(y)]^{1-z} \quad \text{for } z = 0, 1; 0 < y < \infty.$$

We denote the corresponding survival distribution $S = 1 - F$ for any affix. Given an iid sample, say $(\boldsymbol{y}, \boldsymbol{z}) = (y_1, z_1, \ldots, y_n, z_n)$, the log-likelihood is

$$\ell(\alpha, \beta|\boldsymbol{y}, \boldsymbol{z}) = \sum_{i=1}^{n} \ln f(y_i, z_i) \cong \sum_{i=1}^{n} z_i \ln f_X(y_i) + \sum_{i=1}^{n}(1 - z_i)\ln S_X(y_i), \tag{10.63}$$

where $\cong$ denotes the reduced log-likelihood obtained by omitting all terms independent of the parameters. Now, without loss of generality, assume that

$$\sum_1^n z_i = k \quad \text{and} \quad z_1 = \cdots = z_k = 1 \quad \text{with} \quad z_{k+1} = \cdots = z_n = 0,$$

where $y_1 < \cdots < y_k$ are the *failure-times* and $y_{k+1} < \cdots < y_n$ are the *alive-times*. Thus, from eqn (10.63) the reduced log-likelihood for the censored data is given by

$$\ell(\alpha, \beta) = \sum_{i=1}^{k}\{-\ln(\alpha\beta) + \ln \xi'(y_i/\beta) - \frac{1}{2\alpha^2}\xi^2(\beta/y_i)\} + \sum_{j=k+1}^{n} \ln \Phi[\frac{1}{\alpha}\xi(\beta/y_j)]. \tag{10.64}$$

If we define the hazard rate, say $\hbar$, for the standard normal distribution by

$$\hbar(x) = \frac{\varphi(x)}{\bar{\Phi}(x)} \quad \text{and set} \quad q(x) = x\,\hbar(x), \quad \text{for any} \quad x \in \Re, \tag{10.65}$$

we can write

$$\frac{\partial \ln \Phi[\frac{1}{\alpha}\xi(\beta/y)]}{\partial \alpha} = \frac{1}{\alpha} q[\frac{1}{\alpha}\xi(y/\beta)].$$

Hence,

$$\frac{\partial \ell}{\partial \alpha} = \sum_{i=1}^{k}\{-\frac{1}{\alpha} + \frac{1}{\alpha^3}\xi^2(y_i/\beta)\} + \frac{1}{\alpha}\sum_{i=k+1}^{n} q[\frac{1}{\alpha}\xi(y_i/\beta)], \tag{10.66}$$

where $q(\cdot)$ is the function defined in eqn (10.65). Let us introduce statistics, based only on the fixed k failure observations, namely, the arithmetic and harmonic means,

$$\bar{y}_{\pm} = \left(\frac{1}{k}\sum_{i=1}^{k} y_i^{\pm 1}\right)^{\pm 1}$$

and the averaged value, based on the run-outs, of the function $q(\cdot)$, defined in eqn (10.65), namely,

$$q^\star(\alpha, \beta) = \frac{1}{k}\sum_{i=k+1}^{n} q[\frac{1}{\alpha}\xi(y_i/\beta)]. \tag{10.67}$$

By setting $\frac{\alpha^3}{k} \cdot \frac{\partial \ell}{\partial \alpha} = 0$, we then obtain from eqn (10.66)

$$\alpha^2 = \frac{(\bar{y}_+/\beta) - 2 + (\beta/\bar{y}_-)}{1 - q^\star(\alpha, \beta)}. \tag{10.68}$$

We again take the values for the initial estimates of the parameters, based on the arithmetic and harmonic means of failure data, with $\beta_0 = \sqrt{\bar{y}_+\bar{y}_-}$ and $\alpha_0^2 = \sqrt{\bar{y}_+/\bar{y}_-}$. Then we have, from eqn (10.68), the first equation that is needed, namely,

$$\alpha^2 = \frac{\alpha_0^2\xi^2(\beta/\beta_0) + 2(\alpha_0^2 - 1)}{1 - q^\star(\alpha, \beta)}. \tag{10.69}$$

By using the identities, namely, $2x\xi(x)\xi'(x) = x - x^{-1}$ and $-x\xi''(x)\xi'(x) = 2^{-1} + (x+1)^{-1}$, we obtain asas

$$\frac{\partial \ell}{\partial \beta} = \sum_{i=1}^{k}\left[\frac{\beta - y_i}{2\beta(y_i+\beta)}\right] - \frac{k}{2\alpha^2\beta}\left[\frac{\beta}{\bar{y}_-} - \frac{\bar{y}_+}{\beta}\right] + \sum_{i=k+1}^{n} \frac{\partial \ln \Phi[\frac{1}{\alpha}\xi(\beta/y_i)]}{\partial \beta}. \tag{10.70}$$

If we also define

$$\frac{\partial \ln \Phi[\frac{1}{\alpha}\xi(\beta/y)]}{\partial \beta} = \frac{\{\sqrt{\beta/y} + \sqrt{y/\beta}\}}{2\alpha\beta} \cdot h\left[\frac{1}{\alpha}\xi(y/\beta)\right] := \frac{g(y/\beta;\alpha)}{2\alpha\beta}, \tag{10.71}$$

and as in the preceeding case, we compute the averaged value over the run-outs

$$g^\star(\alpha,\beta) = \frac{1}{k}\sum_{i=k+1}^{n} g(y_i/\beta;\alpha), \tag{10.72}$$

then by setting $\frac{2\alpha^2\beta}{k} \cdot \frac{\partial \ell}{\partial \beta} = 0$ we have asas from eqns (10.70) and (10.71) the relation

$$\alpha_0^2\xi[(\beta/\beta_0)^2] = \frac{\alpha^2}{k}\sum_{i=1}^{k}\frac{\beta - y_i}{\beta + y_i} + \alpha g^\star(\alpha,\beta). \tag{10.73}$$

One can verify that $\ln[\xi^{-1}(2x)] = 2\sinh^{-1}(x)$ so eqn (10.73) can be written as

$$\beta = \beta_0 \exp\left\{\sinh^{-1}\left[\frac{\alpha^2}{2\alpha_0^2 k}\sum_{i=1}^{k}\frac{\beta - y_i}{\beta + y_i} + \frac{\alpha g^\star(\alpha,\beta)}{2\alpha_0^2}\right]\right\}. \tag{10.74}$$

This gives us the second equation sought.

A practical method of computation of these mles is by simple iteration. If we write eqn (10.69) formally as $\alpha = \psi_1(\alpha,\beta)$ and eqn (10.74) as $\beta = \psi_2(\alpha,\beta)$, then we can proceed iteratively as follows:

Given α_j, β_j for any $j = 0, 1, 2, \ldots$ we compute successive values

$$\alpha_{j+1} = \psi_1(\alpha_j,\beta_j) \qquad \beta_{j+1} = \psi_2(\alpha_j,\beta_j)$$

recursively, with the mles ultimately defined by

$$\hat{\alpha} = \lim_{j\to\infty}\alpha_j \qquad \hat{\beta} = \lim_{j\to\infty}\beta_j. \tag{10.75}$$

NB in the complete-sample case $k = n$ and so $q^\star = g^\star \equiv 0$. Hence eqns (10.69) and (10.73) become simply

$$\alpha^2 = \frac{\bar{y}_+}{\beta} - 2 + \frac{\beta}{\bar{y}_-}, \qquad \alpha_0^2\left(\frac{\beta}{\beta_0} - \frac{\beta_0}{\beta}\right) = \frac{\alpha^2}{n}\sum_{i=1}^{n}\frac{\beta - y_i}{\beta + y_i}. \tag{10.76}$$

Eliminating α^2, we obtain asas, the single equation in β, the solution of which is the mle $\hat{\beta}$, namely,

$$\beta^2 - \beta_0^2 = (\beta_0^2 - 2\beta\bar{y}_- + \beta^2)\left[-1 + \frac{2\beta}{n}\sum_{i=1}^{n}\frac{1}{\beta + y_i}\right]. \tag{10.77}$$

This equation can be easily solved numerically, by *Mathematica* or other such programming languages, using an initial value of β_0. It is easily shown that the term in square brackets is a monotone increasing function of β and

$$(1/n)\sum_{i=1}^{n}(\beta - Y_i)(\beta + Y_i)^{-1} \xrightarrow{p} 0 \text{ as } n \to \infty.$$

So for values of β near $\hat{\beta}$ the square-brackets factor in eqn (10.77) should be close to zero. The lhs of eqn (10.77) is an increasing quadratic for $\beta > 0$ with a zero at β_0 while the rhs at β_0 is $2\beta_0(\beta_0 - \hat{y}_-)[\cdots]$, which is small but can be either positive or negative. After computing $\hat{\beta}$ numerically we then set

$$(\hat{\alpha})^2 = \frac{\bar{y}_+}{\hat{\beta}} - 2 + \frac{\hat{\beta}}{\bar{y}_-}.$$

10.9. Estimation for Tweedie's Distribution

We now consider briefly the situation when we have observations from Tweedie's distribution. If $Y_i \sim \mathcal{T}(\alpha, \beta)$ for $i = 1, \ldots, n$ is an iid sample, then since $\frac{1}{Y_i} \sim \mathcal{W}(\alpha, 1/\beta)$ straightforwardly the mles are, in the case both parameters are unknown,

$$\hat{\beta} = \bar{Y}_-, \qquad \hat{\alpha}^2 = \frac{\bar{Y}_+}{\bar{Y}_-} - 1. \tag{10.78}$$

In the case that α is known, the mle of β is

$$\breve{\beta} = \bar{Y}_-\left[-\frac{\alpha^2}{2} + \sqrt{\frac{\alpha^4}{4} + \frac{\bar{Y}_+}{\bar{Y}_-}}\right] \asymp \bar{\bar{Y}} + \bar{Y}_-\left[-\frac{\alpha^2}{2} + \frac{\alpha^4}{8}(\bar{Y}_-/\bar{Y}_+)^{1/2} + O(\alpha^8)\right].$$

From the definition it follows that

$$\frac{\bar{X}_+}{\beta} \sim \frac{\beta}{\bar{Y}_-} \quad \text{and} \quad \frac{\beta}{\bar{X}_-} \sim \frac{\bar{Y}_+}{\beta}, \tag{10.79}$$

so that by direct analogy with eqn (9.26) we know that $\bar{Y}_-$ and $(\bar{Y}_+ - \bar{Y}_-)$ are independent. Since $\bar{Y}_- \sim \mathcal{T}(\alpha/\sqrt{n}, \beta)$ we find from eqn (10.78) that

$$\mathrm{E}\hat{\beta} = \beta(1 + \frac{\alpha^2}{n}) \quad \text{and} \quad \mathrm{E}\hat{\alpha}^2 = \frac{n-1}{n}\alpha^{2.}$$

Thus, we could define unbiased estimates of both parameters, as before.

Again by the reciprocal relationship as given in eqn (10.6), we see

$$U \sim \frac{\sqrt{n}}{\alpha}\xi(\bar{Y}_-/\beta) \quad \text{and} \quad V \sim \frac{\bar{Y}_-}{\alpha^2\beta}\sum_{i=1}^{n}\xi^2[Y_i/\bar{Y}_-] = \frac{n}{\alpha^2}\left[\frac{\bar{Y}_+}{\beta} - \frac{\bar{Y}_-}{\beta}\right]$$

are independent. Since it is an algebraic identity that

$$U^2 + V \sim \frac{1}{\alpha^2}\sum_{i=1}^{n}\xi^2[Y_i/\beta] \sim \chi^2(n),$$

we have the same distribution for confidence intervals as before.

However, neither $\hat{\alpha}^2$ nor $\tilde{\alpha}^2$ is pivotal for α since the distribution of both ratios depends on α. This deficiency is in contrast to the case when the λ parameterization for the IG-distribution is used. But it can largely be surmounted by the availability today of very good pseudo-random number generators and computational power.

Exercise Set 10.E

1. Prove the reproductive property of the Wald distribution using its cgf.
2. Prove that the mle $\hat{\beta}$ for the Wald distribution is both unbiased and consistent.
3. Derive the mles of α and β for the Wald distribution by using the results given in [26, 1989] for $\hat{\mu}$ and $\hat{\lambda}$ and the transformations in eqn (9.4) along with the invariance property of mles.
4. Show the exact discrepancy between the distribution of U and the standard normal distribution is

$$\sup_{u\in\Re}|F_U(u) - \Phi(u)| = e^{2n/\alpha^2}\bar{\Phi}(2\sqrt{n}/\alpha) \asymp \frac{\alpha}{\sqrt{8\pi n}}\left\{1 - \frac{\alpha^2}{4n} + \frac{1\cdot 3\alpha^4}{4n^2} - \cdots\right\},$$

where the asymptotic expansion is by Mill's ratio. Compute some values when $\sqrt{n}/\alpha$ is large. For $n = 9$ and $\alpha = .2$ what is the the maximum discrepancy?

5. Using Tweedie's parameterization compute some exact one-sided confidence intervals on the parameter $\lambda = \beta/\alpha^2$ by using the pivotal statistic $\hat{\lambda} = \frac{\bar{X}_+\cdot\bar{X}_-}{\bar{X}_+ - \bar{X}_-}$.
6. Verify that the mles $\hat{\alpha}, \hat{\beta}$ for a complete sample $\{x_1, \ldots, x_n\}$ from the $\mathcal{FL}$-distribution are the joint solution to the two equations in α, β, which from eqn (10.58) and (10.60), can be written

$$\frac{\alpha^2}{n}\sum_{i=1}^{n}\frac{\beta - x}{\beta + x_i} = \alpha_0^2\xi[(\beta/\beta_0)^2], \quad \left(\frac{\alpha}{\alpha_0}\right)^2 = [\xi(\beta/\beta_0)]^2 + \varepsilon^2, \qquad (10.80)$$

where the constants are

$$\beta_0 = \sqrt{\bar{x}_+\bar{x}_-}, \quad \alpha_0^2 = \sqrt{\frac{\bar{x}_+}{\bar{x}_-}}, \quad \varepsilon^2 = 2(1 - \alpha_0^{-2}).$$

Finally, compare the Equations in (10.80) above with those for an incomplete sample.

7. Using *Mathematica* verify for the Yokobori data (Dataset XII in Table 9.2), and the $\mathcal{FL}$-distribution that $\hat{\alpha} = 1.274$, $\hat{\beta} = 372.042$.

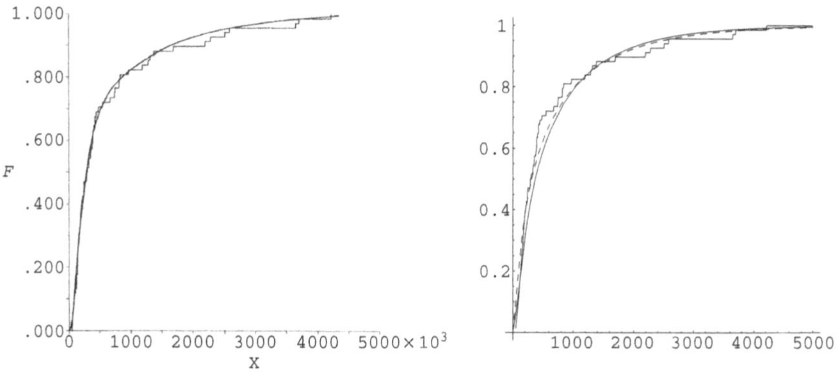

Figure 10.5. Yokobori Data A: Markov Chain fit B: $\mathcal{FL}$ fit by MLE, and MOM.

10.10. Cases of Misidentification

10.10.1. When the $\mathcal{FL}$-Distribution Is Unknown

One of the more daunting sets of fatigue-life data was obtained for carbon steel by Yokobori in 1951. It was discussed at length in [9, p. 221] where a seven-parameter Markov chain model was fitted and then smoothed for visual comparison with the edf. This comparison is reproduced in plot A in Figure 10.5, q.v. The difficulty with the data is that in the later stage of service-life, when only about 30% of the specimens survive, the failure rate decreases! (This was caused by nonuniform surface finish, here and many other fatigue samples, a fit is impossible for even an IHRA distribution, which excludes Weibull, Galton, gamma and others – a point made in [8].) The sample mean and signal-to-noise ratio were given, respectively, loc. cit., for the 68 observations as as $\hat{m} = 657$ and $\hat{m}/\hat{\sigma} = .726$.

If one merely calculates the $\mathcal{FL}(\alpha, \beta)$ parameters using the MOME,[1] by utilizing eqns (10.17) and (10.18) one obtains $\tilde{\beta} = 309.6971$ and $\tilde{\alpha} = 1.4976$. If we substitute them into a $\mathcal{FL}$-distribution and compare it with the edf a reasonable fit is obtained, see the dashed-line in Figure 10.4B. But if we compute the MLE, which weighs the influence of each observation equally, using eqn (10.80) we obtain a comparable fit as shown in the solid line in Figure 10.4B.

10.10.2. When the $\mathcal{CD}$-Distributions Are Unknown

Many datasets have been collected at the National Bureau of Standards (now the National Institute for Standards and Technology) which are pedagogical examples

[1] The method-of-moments estimators were introduced by Karl Pearson in the late 19th century but they have lost favor since Sir Ronald Fisher's papers on the MLEs appeared in mid 1920s.

of statistical technique, e.g. some are cited in the excellent textbook *Statistics* by Friedman et al. The data sets there are chosen purportedly to show that even if experimental technique is punctillious in high precision weighing protocols sometimes outliers occur that cannot be explained by Normal (Gaussian) distribution theory. Here an outlier is an observation, assuming Normality, which lies more than three standard deviations from the mean and hence should have a probability of occurring not more than $2\Phi(-3) \doteq 0.002$. The occurrence of such an observation in a sample of 10 observations is a "small miracle" and is regarded as an anomoly; see the discussion *loc, cit.*

The supposition for the repeated occurrence of such an implausible event must be some intermittent influence on an observation caused by some unknown variation in the experimental procedure or laboratory conditions. We posit another explanation. Despite the ubiquity of the normal distribition in such circumstances it may not be the right distribution to use! The weighing process is analogous to the addition of weight-units in small increments, nominally deterministic but in reality with stochastic variation, until the total weight "balances" with the object being weighed. This simple analogy seems closer to the generation of a $\mathcal{FL}$-distribution than to the observed variation being due to the sum of independent random influences, justifying the Gaussian distribution. And as we now show a $\mathcal{FL}$-distribution gives a much closer fit to the data.

To see the difference between these two assumptions one need only compare their densities; the Gaussian density falls off from its maximum as the exponential of a square while the $\mathcal{FL}$-distribution falls off linearly. For the Gaussian distribution the probability of getting an observation outside the 3σ limits in a sample of n observations is $1-(1-2\Phi(-3))^n \doteq .02$ for $n = 10$. But for the Wald-distribution (chosen for ease of comparison) the probability of obtaining an observation outside $\beta \pm 3\alpha\beta$ is, asas, independent of β and given by

$$\Phi(\tfrac{-3}{\sqrt{1-3\alpha}}) + \Phi(\tfrac{-3}{\sqrt{1+3\alpha}}) + e^{2/\alpha^2}[\Phi(\tfrac{3\alpha-2}{\alpha\sqrt{1-3\alpha}}) - \Phi(\tfrac{-3\alpha-2}{\alpha\sqrt{1+3\alpha}})].$$

At $\alpha = 0.1$ this evaluates to 0.005; since $1-(.995)^{10} \doteq .05$ (much larger than .002) so a "small miracle" would be expected to occur 25 times as frequently.

At $p = 1, \alpha = 0.15$ the value is .006388 so a "small miracle" would occur one time in 16! At $p = .5, \alpha = 0.15$ the value is .005076 so a "small miracle" would occur one time in 20; At $p = 0, \alpha = 0.15$ the value is .00417 so a "small miracle" would occur one time in 24. See Figure 10.6.

10.10.3. Weibull Distribution Contrasted with the $\mathcal{FL}$-Distribution

Carefully conducted experiments yielding large complete samples of fatigue-life data for even small structural components are inordinately expensive and so very rare. Weibull, himself, at one time collected almost a thousand fatigue observations using relatively inexpensive wire repeatedly twisted. To fit this dataset he adopted what would today be called a "change-point Weibull distribution" as a

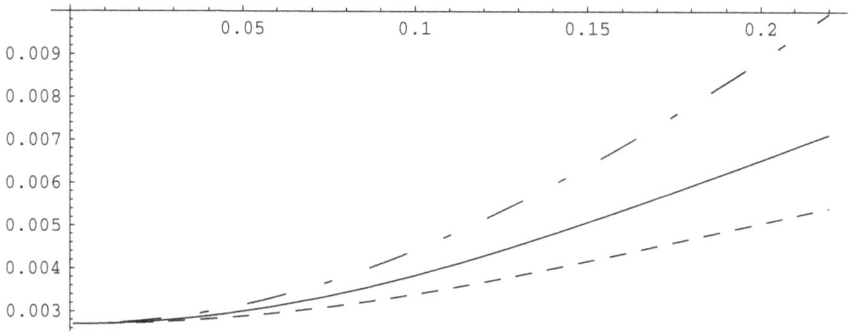

Figure 10.6. Probability outside $\mu \pm 3\sigma$ for the $\mathcal{CD}$-distributions.

model. This was likely, as we now show in Figure 10.7, only the consequence of misidentifying this dataset when, in fact, it was from the $\mathcal{FL}$-distribution.

The usual method of identifying a Weibull data set is to plot the double logarithm of an esd, suitably modified to avoid dificulty at the endpoints. It is based on the ordered observations, say $\{t_{1:n} < \cdots < t_{n:n}\}$, plotted vs. the logarithm of time. If $\hat{S}_n(t)$ is the K-M estimator we may examine, to ascertain if the sample is Weibull, the linearity of the plot of

$$\ln\left\{-\ln\left[\frac{n\hat{S}_n(e^x)+(1/2)}{n+1}\right]\right\} \quad \text{for} \quad \ln t_{1:n} \le x \le \ln t_{n:n}. \tag{10.81}$$

If the sample is from a $\mathcal{W}ei(\alpha, \beta)$ distribution, say $\bar{F}(t) = \exp\{-(t/\beta)^\alpha\}$ then the plot should approximate the straight line $\alpha(x - \ln\beta)$ when the abscissa is marked in $x = \ln t$. But it would show a characteristic concave nonlinearity when the data is from the $\mathcal{FL}$-distribution, i.e., $F(t) = \Phi[\frac{1}{\alpha}\xi(t/\beta)]$ for $t > 0$. In Figure 10.7 a graph of $\ln\left\{-\ln\bar{\Phi}[\frac{2}{\alpha}\sinh(x/2)\right\}$, where $x = \ln t - \ln\beta$, is presented for three values of α. NB the steepest curve when $\alpha = 0.1$ as in a fatigue sample, shows a bilinear behavior anomalous to the Weibull assumption.

10.10.4. Galton Distribution Mistaken for $\mathcal{FL}$-Distribution

Let us now examine another case of misidentification when the Galton (log-normal) distribution is falsely assumed for a $\mathcal{FL}$-distribution. In this case the comparison is between the distribution functions $\Phi[\ln(te^{-\mu})/\sigma]$ vs. $\Phi[\xi(t/\beta)/\alpha]$ for $t > 0$. $\tilde{F}_n(x) = (n\hat{F}_n(x) + .5)/(n+1)$ for $x = \ln t$, as the edf of the logarithms of the observations modified so as to avoid difficulties at the largest and smallest observations. For a large (complete) sample plotted on normal probability paper, i.e., $\Phi^{-1}\tilde{F}_n(x)$ on the ordinate and $x = \ln t$ on the abscissa, the data should, by the presumption, $\mathcal{LN}(\mu, \sigma^2)$ as in eqn (3.32), tend to the straight line $\frac{1}{\sigma}(x - \mu)$. But if the sample is from an $\mathcal{FL}(\alpha, \beta)$ distribution, the data should

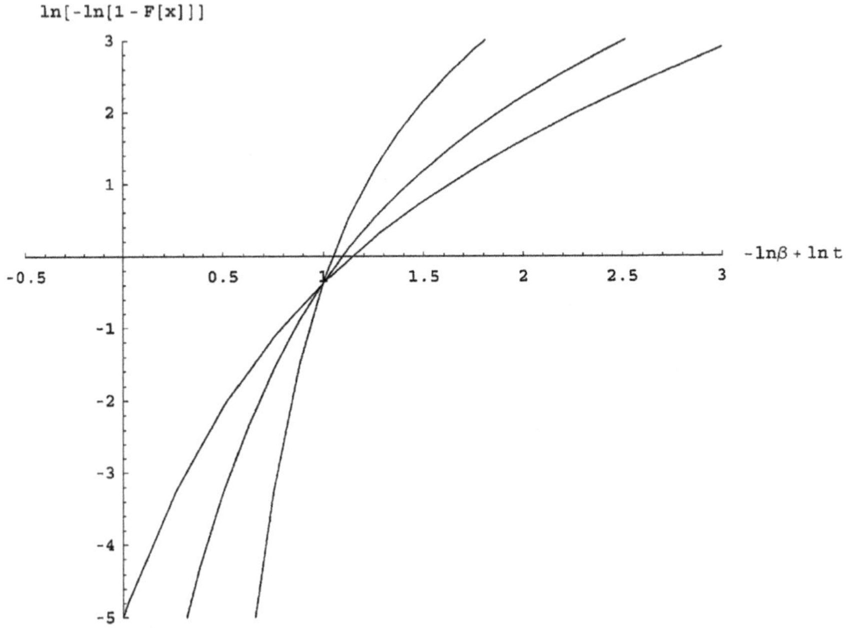

Figure 10.7. The curvature on a Weibull plot of $\mathcal{F}L(\alpha, \beta)$-data for $\alpha = 0.1, 0.2, 0.3$.

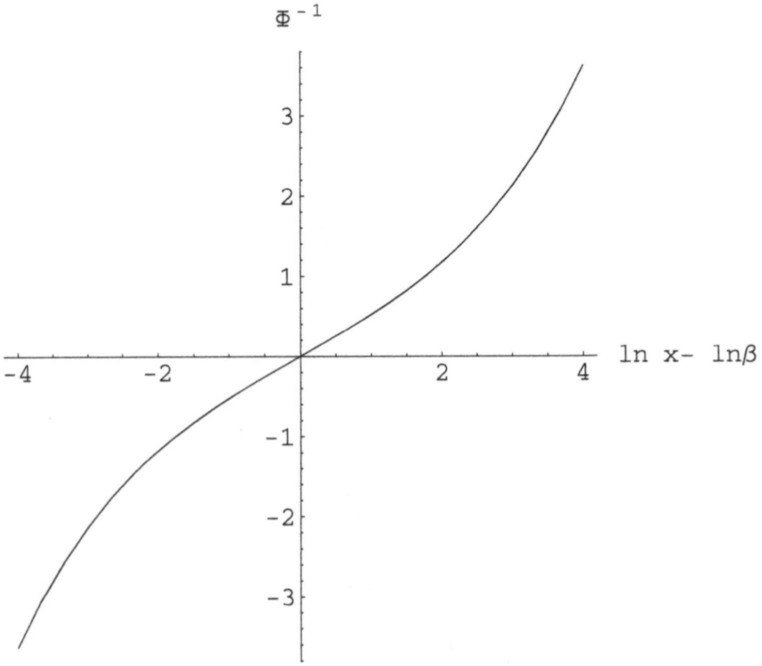

Figure 10.8. Graph of $\mathcal{F}L$-distribution on a log-normal plot.

tend to the curve $(2/\alpha)\sinh[(x - \ln\beta)/2]$. These two distributions can be easily mistaken for certain parameter values, if one looks only for linearity.

In astronomy the errors in parallax-estimation of stellar distances, traditionally, have always been assumed to be log-normal, with the occasional case of anomalous behavior. One such case is presented in Figure (10.8) of a large data set showing hyperbolic sine behavior.

Exercise Set 10.F

1. Which, if any, of the failure data sets in VIII, IX, X, XI, pp. 156,7, can be reasonably described using one of the CD-distributions? What are the reasons for your answer?
2. By comparing the ecdfs formed from these data sets on p. 156,7 with the Weibull or Galton distributions fitted with mle parameters can you make a better choice. How about a competing-risks Weibull model? What would you need to know? Explain why.

We must stop finding the exact answer to the approximate problem and start finding the approximate answer to the exact problem. Far better is the appoximate answer, which perhaps vague can always be made precise later, than any answer to the wrong problem

John Tukey

The imaginative statistician, far from being a narrow specialist, is often the direct opposite – a generalist; a man who can walk into almost any scientific or commercial operation, find a numerical way to analyze what is going on, and to increase output . . . The gains to the economy and productivity of a country – any country, at any level of industrialization – from such mathematized use of intelligence are so great compared to the cost involved that a sound and imaginative statistical generalist ranks high among heavy intellectuals.

The Nobel Physicist John A. Wheeler
(after getting to know John Tukey)

In Statistics we need to learn not only technical capability and applicable theory but some expertise in both mathematics and computation, along with strategic vision, i.e., the ability to discern where best to apply our science at the frontiers of knowledge.

Emanuel Parzen

CHAPTER 11

Analysis of Dispersion

11.1. Applicability

The analysis of variance is one of the most highly developed areas of mathematical statistics. Not coincidentally, it is one of the most widely utilized. But many of the deductions therefrom in the analysis and interpretations of experiments depend upon the accuracy of the assumption of normality of the observations.

Few textbooks present the general linear model and mention that departure from normality entails an error of unknown magnitude and then attempt to deal with this error. One of the exceptions is that of Kempthorne [52], which suggests some transformations which may be useful in "reshaping" certain datasets so they are sufficiently close to normality that inferences are 'good enough for agricultural work.' Historically, skewness and kurtosis were used for determining how close datasets were to normality.

These suggested functional transformations include:

$$\ln(x), \quad \sqrt{x}, \quad \sin^{-1}\sqrt{x}, \quad \lambda^{-1}\sinh(\lambda\sqrt{x}), \quad \frac{x^{\lambda}-1}{\lambda}, \qquad \text{etc.},$$

where an arbitrary location parameter can also be introduced. It is the careless statistician who applies to all data the techniques that pertain only to Gaussian theory. Even studies of robustness cannot completely allay the error, and culpability is compounded when every multiplicative effect is first transformed by logarithm.

Analysis of variance depends on algebraic decomposition, such as

$$\frac{1}{n}\sum_{i=1}^{n}\frac{(x_i-\mu)^2}{\sigma^2} = \frac{1}{n}\sum_{i=1}^{n}\frac{(x_i-\bar{x}_+)^2}{\sigma^2} + \frac{(\bar{x}_+-\mu)^2}{\sigma^2}$$

(letting $\bar{x}_+$ denote the sample mean) and the distributional theory that follows from

$$X \sim \mathcal{N}(\mu,\sigma^2) \quad \text{implying} \quad \left(\frac{X-\mu}{\sigma}\right)^2 \sim \chi^2(1).$$

Of course, the results of ANOVA cannot be duplicated for any distribution other than normal since only for Gaussian rv's X, Y, dependent or not, do we have, for any real numbers a, b, that $aX + bY$ also Gaussian. Thus, Cochran's theorem only applies in its generality under such circumstances. Nevertheless there are many situations when an exact test, for small samples, would be useful when the datasets representing cumulative damage, are skewed and nonnegative and hence not normal.

Since the $\mathcal{CD}$ family has the Chi-Square property, what we propose in this chapter is an exact analysis for the $\mathcal{W}(\alpha, \beta)$-family of distributions in (10.1), and frequently others in the $\mathcal{CD}$ class, which is analogous to ANOVA for the normal case. Sometimes this yields a related approximate analysis for, say, the $\mathcal{FL}$-family, under certain tests of hypothesis which involve the dispersion. In these analyses are used such decompositions as

$$\frac{1}{n\alpha^2}\sum_{i=1}^{n}\xi^2(x_i/\beta) = \frac{\beta}{n\alpha^2\bar{x}_+}\sum_{i=1}^{n}\xi^2(x_i/\bar{x}_+) + \frac{1}{\alpha^2}\xi^2(\bar{x}_+/\beta),$$

with its associated distributional theory.

11.2. Schrödinger's Distribution

In his original derivation of the random time T for a particle under Brownian moton to cover a fixed distance d, when moving at a mean velocity of ν, we recall the density derived by Schrödinger involves three parameters; the additional parameter is δ. This gives us eqn (5.16),

$$\frac{d}{\sqrt{2\pi\delta t^3}}\exp\left\{-\frac{(\nu t - d)^2}{2\delta t}\right\} \quad \text{for } t > 0. \tag{11.1}$$

Here δ is the *diffusion-rate* constant. To transform this density into the reduced parameterization of the $\mathcal{W}$-density involving only α, β, we note

$$\left[\frac{1}{\alpha}\xi(t/\beta)\right]^2 = \frac{(t-\beta)^2}{\alpha^2\beta t} \equiv \frac{(\nu t - d)^2}{\delta t} \quad \text{implies} \quad \beta = \frac{d}{\nu}, \;\; \alpha^2 = \frac{\delta}{\nu d}.$$

Thus $\alpha^2\beta = \delta\nu^{-2}$, which is independent of the distance d. If the dimensions chosen were say, cm-sec, then we have $\dim(\delta) = \text{cm}^2 \times \text{sec}^{-1}$ and correspondingly $\dim(\alpha^2\beta) = \dim(\beta) = \text{sec}$, so the dimension of $\alpha^2\beta$ is *time* only, with α being always a pure (dimensionless) number.

11.3. Sample Distributions under Consonance

Let us extend our notation and write for any indexed set, say $\{x_1, \ldots, x_n\}$, when n is fixed:

$$x_{(r)} := \left(\sum_{i=1}^{n} x_i^r\right)^{1/r}, \quad \text{and} \quad \bar{x}_{(r)} := \left(\frac{1}{n}\sum_{i=1}^{n} x_i^r\right)^{1/r} \quad \mathit{for} \quad r \in \Re, r \neq 0. \tag{11.2}$$

Note we will sometimes write simply $\bar{x}_+ = \bar{x}_{(1)}$ and $\bar{x}_- = \bar{x}_{(-1)}$.

Definition. For any $\mathcal{CD}(\alpha, \beta, p)$ population, i.e., the distribution given in (10.12), we call α^2 the dispersion and β the *scale*. If $Z_i \sim \mathcal{CD}(\alpha_i, \beta_i, p)$ for $1 \leq i \leq n$, there is *consonance* among the $\mathcal{CD}$ rvs whenever

$$\alpha_1^2\beta_1 = \cdots = \alpha_n^2\beta_n. \tag{11.3}$$

This means, for the Wald distribution, when $p = 1$, the *scaled dispersion*, $\alpha_i^2\beta_i$, is constant for $i = 1, \ldots, n$, viz., that the variance divided by the mean remains the same.

Lemma 1. *Let $X_i \sim \mathcal{W}(\alpha_i, \beta_i)$ for $i = 1, 2, \cdots, n$ be independent. Then their sample sum satisfies $X_+ \sim \mathcal{W}(\alpha_{(-2)}, \beta_+)$ iff they are consonant, that is they satisfy eqn (11.3), with*

$$\frac{1}{\alpha_{(-2)}^2} = \sum_{i=1}^{n} \alpha_i^{-2} \quad and \quad \beta_+ = \sum_{i=1}^{n} \beta_i. \tag{11.4}$$

Proof. This is immediate if we consider the cumulant generating function of $X_1 + \cdots + X_n$, namely,

$$\sum_{i=1}^{n} \frac{1}{\alpha_i^2}[1 - \sqrt{1 - 2t\alpha_i^2\beta_i}] = \left(\sum_{i=1}^{n} \frac{1}{\alpha_i^2}\right)\left[1 - \sqrt{1 - 2t\alpha_{(-2)}^2\beta_+}\right],$$

since it follows from eqn (11.3) that

$$\frac{1}{\alpha_{(-2)}^2} = \frac{\beta_1}{\alpha_1^2\beta_1} + \cdots \frac{\beta_n}{\alpha_n^2\beta_n} = \frac{\beta_+}{\alpha_i^2\beta_i}$$

and so $\alpha_i^2\beta_i = \alpha_{(-2)}^2\beta_+$. □

Remark. *Since β is a scale parameter, under the assumptions of this lemma, the sample-mean distribution also lies in the same family, viz., $\bar{X}_+ \sim \mathcal{W}(\alpha_{(-2)}, \bar{\beta}_+)$.*

We state the corresponding result for Tweedie variates:

Corollary 1. *If $Y_i \sim \mathcal{T}(\alpha_i, \beta_i)$ for $i = 1, \ldots, n$ with $Y_1 \perp \cdots \perp Y_n$, then the sample harmonic mean satisfies*

$$\bar{Y}_- \sim \mathcal{T}(\alpha_{(-2)}, \bar{\beta}_-) \text{ iff the populations are consonant.} \tag{11.5}$$

Proof. Since $1/Y_i \sim \mathcal{W}(\alpha_i, 1/\beta_i)$ for $i = 1, \ldots, n$, from eqn (11.3) and Lemma 1 we have necessary and sufficient conditions that $Y_- = \sum_{i=1}^n (1/Y_i) \sim \mathcal{W}(\alpha_0, \beta_-)$. This is equivalent with eqn (11.5). □

We know for two Wald variates satisfying eqn (11.3) we have not only $\sum_{i=1}^2 \frac{1}{\alpha_i^2}\xi^2(X_i/\beta_i) \sim \chi^2(2)$ but also from Lemma 1 that

$$Z = \frac{X_1}{\alpha_1^2\beta_1} + \frac{X_2}{\alpha_2^2\beta_2} = \frac{X_1 + X_2}{\alpha_{(-2)}{}^2(\beta_1 + \beta_2)} \sim \mathcal{W}(\alpha_{(-2)}, \alpha_{(-2)}{}^{-2}),$$

and hence $\alpha_{(-2)}{}^{-2}\xi^2(\alpha_{(-2)}{}^2 Z) \sim \chi^2(1)$. Let us set

$$V = \sum_{i=1}^{2} \frac{1}{\alpha_i^2}\xi^2(X_i/\beta_i) - \frac{1}{\alpha_{(-2)}{}^2}\xi^2\left(\frac{X_1+X_2}{\beta_1+\beta_2}\right)$$
$$= \frac{\beta_1}{\alpha_1^2 X_1} + \frac{\beta_2}{\alpha_2^2 X_2} - \frac{1}{\alpha_{(-2)}{}^2}\left(\frac{\beta_1+\beta_2}{X_1+X_2}\right), \tag{11.6}$$

and ask under what assumptions, if any, is it true that $Z \perp V$ with $V \sim \chi^2(1)$?

Following tedious, but standard, transformation-of-variables techniques, the joint density of (V, Z) is found to be

$$f_{V,Z}(v,z) = \frac{\exp\{-\frac{1}{2}[v + \alpha_{(-2)}{}^{-2}\xi^2(\alpha_{(-2)}{}^2 z)]\}}{\pi\sqrt{vz}\left(\frac{x_1}{\beta_1}+\frac{x_2}{\beta_2}\right)}, \tag{11.7}$$

where the term in round brackets, in the denominator above, can be expressed as a function of z alone iff $\beta_1 = \beta_2$. But in that circumstance under consonance from eqn (11.3) we must have $\alpha_1 = \alpha_2$ and so:

Corollary 2. *Iff $X_1 \perp X_2 \sim \mathcal{W}(\alpha, \beta)$ are the arithmetic and harmonic means related so that*

$$\bar{X}_+ \perp V, \quad \textit{where} \quad V := \frac{2\beta}{\alpha^2}\left[\frac{1}{\bar{X}_-} - \frac{1}{\bar{X}_+}\right],$$

with $\bar{X}_+ \sim \mathcal{W}(\frac{\alpha}{\sqrt{2}}, \beta)$ and $V \sim \chi^2(1)$. Moreover $V \perp \frac{2}{\alpha^2}\xi^2(\bar{X}_+/\beta)$.

These consequences are not new; they are a restatement of results given originally by Tweedie in his parameterization.

From this follows the decomposition theorem of Seshadri, which we state without proof since it now follows easily by induction.

Remark 1. *If $X_i \sim \mathcal{W}(\alpha, \beta)$ are iid for $i = 1, \ldots, n = 2^k$, for k any natural number, with $\bar{X}_+$ the sample mean, then setting*

$$U = \frac{n}{\alpha^2}\xi^2(\bar{X}_+/\beta) \quad \textit{and} \quad V = \frac{1}{\alpha^2}\sum_{i=1}^{n}\xi^2(X_i/\beta) - \frac{n}{\alpha^2}\xi^2(\bar{X}_+/\beta)$$

then there exists a decomposition of $V = \sum_{i=1}^{n-1} V_i$ where $U \perp V_i$ for all $i = 1, \ldots, n-1$ with $U \sim V_i \sim \chi^2(1)$ and $V \sim \chi^2(n-1)$.

Note that this remark allows a decomposition of V, when $n = 2^k$, into a sum of $(n-1)$ of $\chi^2(1)$ variates. For example, for $n = 4$ set

$$V_1 = \frac{\beta}{\alpha^2}\left[\frac{1}{X_1} + \frac{1}{X_2} - \frac{4}{X_1 + X_2}\right], \quad V_2 = \frac{\beta}{\alpha^2}\left[\frac{1}{X_3} + \frac{1}{X_4} - \frac{4}{X_3 + X_4}\right],$$

$$V_3 = \frac{\beta}{\alpha^2}\left[\frac{4}{X_1 + X_2} + \frac{4}{X_3 + X_4} - \frac{16}{X_1 + X_2 + X_3 + X_4}\right],$$

then $V = V_1 + V_2 + V_3$ and the V_i are iid $\chi^2(1)$.

It seems surprising that such a decomposition is possible for sums of variates in number which are powers of two but not for other numbers of observations! This was a point of contention among early investigators, see the comments of Chhikara and Folks [26, 1987]. We now state the key result to be proved later in this chapter.

Theorem 1. *Let $X_i \sim \mathcal{W}(\alpha_i, \beta_i)$ for $i = 1, 2$ with $X_1 \perp X_2$, where*

$$\alpha_1^2\beta_1 = \alpha_2^2\beta_2 \quad \textit{and we set} \quad \frac{1}{\alpha_{(-2)}^2} := \frac{1}{\alpha_1^2} + \frac{1}{\alpha_2^2}, \quad \beta_+ = \beta_1 + \beta_2, \tag{11.8}$$

then

$$\alpha_i^2\beta_i = \alpha_{(-2)}^2\beta_+ \quad \textit{for} \ \ i = 1, 2.$$

If we define

$$U = \frac{1}{\alpha_{(-2)}{}^2}\xi^2\left(\frac{X_1+X_2}{\beta_1+\beta_2}\right), \tag{11.9}$$

and

$$V = \frac{\alpha_{(-2)}^2}{(\alpha_1\alpha_2)^2}\frac{\beta_1+\beta_2}{X_1+X_2}\xi^2\left(\frac{X_2/\beta_2}{X_1/\beta_1}\right), \tag{11.10}$$

which is an identity of eqn (11.6), then $U \perp V$ and

$$U + V \equiv \sum_{i=1}^{2}\frac{1}{\alpha_i^2}\xi^2(X_i/\beta_i) \sim \chi^2(2) \quad \textit{so that} \quad U, V \sim \chi^2(1).$$

From eqn (11.10) we see that despite V having one degree of freedom it cannot be expressed as being proportional to $\xi^2(Y)$ where Y is some function of X_1, X_2.

Of course Theorem 1 has an analgous result for Tweedie variates which we, for didactic reasons, also present for two variates.

We state this as

Corollary 3. *Let $Y_i \sim \mathcal{T}(\alpha_i, \beta_i)$ for $i = 1, 2$ be independent rvs with parameters that satisfy*

$$\frac{\alpha_1^2}{\beta_1} = \frac{\alpha_2^2}{\beta_2}, \tag{11.11}$$

then for $i = 1, 2$

$$\frac{\alpha_i^2}{\beta_i} = \alpha_{(-2)}^2 \beta_- \quad \textit{where} \quad \frac{1}{\alpha_0{}^2} = \frac{1}{\alpha_1^2} + \frac{1}{\alpha_2^2}, \quad \frac{1}{\beta_-} = \frac{1}{\beta_1} + \frac{1}{\beta_2}.$$

If we set

$$U = \frac{1}{\alpha_{(-2)}{}^2} \xi^2(Y_-/\beta_-) \quad \textit{and} \quad V = \frac{Y_1}{\alpha_1^2 \beta_1} - \frac{Y_1}{\alpha_2^2 \beta_2} - \frac{\beta_-}{\alpha_{(-2)}^2 Y_-},$$

then $U + V \sim \chi^2(2)$, $U \perp V$ *so that* $U, V \sim \chi^2(1)$.

Proof. We know by eqn (11.5) and Theorem 1 that $U \sim \chi^2(1)$ and since $V = \sum_{i=1}^{2} \frac{1}{\alpha_i^2} \xi^2(Y_i/\beta_i) - U$ we see that $U + V \sim \chi^2(2)$. To show $U \perp V$ it is sufficient to produce an identification between these Tweedie rv's with parameters satisfying eqn (11.11) and independent Wald rvs having parameters satisfying eqn (11.8). Consider

$$\frac{Y_-}{\alpha_{(-2)}^2 \beta_-} = \frac{\beta_1 \beta_2 \left(\frac{1}{Y_1} + \frac{1}{Y_2}\right)}{\alpha_{(-2)}^2 (\beta_1 + \beta_2)} = \frac{\beta_1}{\alpha_1^2 Y_1} + \frac{\beta_2}{\alpha_2^2 Y_2}.$$

The variates

$$X_1 = \frac{\beta_1 \beta_2}{Y_1} \sim \mathcal{W}(\alpha_1, \beta_2) \quad \text{and} \quad X_2 = \frac{\beta_1 \beta_2}{Y_2} \sim \mathcal{W}(\alpha_2, \beta_1)$$

have parameters satifying eqn (11.8), and therefore $X_1 \perp X_2$, but this conditon is equivalent with eqn (11.11). □

The characterization theorem of Tweedie, [115, 1957], for inverse-Gaussian variates is as follows:

Remark 2. *If the* $X_i > 0$ *are iid for* $i = 1, \ldots, n$ *with both* EX_i^2 *and* EX_i^{-1} *positive and finite, then they are inverse-Gaussian variates iff*

$$\bar{X}_+ \perp T \quad \textit{where} \quad T := \frac{1}{\bar{X}_-} - \frac{1}{\bar{X}_+}.$$

It is known, *vide supra,* that if the assumption of identical distribution is relaxed, then the independence between $\bar{X}_+$ and T is no longer true. Nevertheless, Theorem 1 says, despite the variates X_1 and X_2 not being identically distributed, that for certain combinations of the parameters specified in eqn (11.8), we have $\xi^2[(X_1 + X_2)/(\beta_1 + \beta_2)]$ independent of V even though $X_1 + X_2$ is not!

From Theorem 1 we see if $\alpha_1 = \alpha_2$, then by the restriction of eqn (11.8), we must have $\beta_1 = \beta_2$. Thus, $X_1 \sim X_2$, in which case

$$U = \frac{2}{\alpha^2} \xi^2(\bar{X}_1/\beta) \quad \text{and} \quad V = \frac{2\beta T}{\alpha^2}.$$

What does the consonance restriction (11.3) imply for the the usual IG-parameters? Recall $\mu = \beta$ and $\lambda = \beta/\alpha^2$ so that the condition in eqn (11.8) becomes

$$\frac{\mu_1^2}{\lambda_1} = \frac{\mu_2^2}{\lambda_2} = \frac{(\mu_1+\mu_2)^2}{\bar{\lambda}} \quad \text{where} \quad \bar{\lambda} = \left[\frac{\lambda_1}{\mu_1} + \frac{\lambda_2}{\mu_2}\right](\mu_1+\mu_2).$$

The independence of $\bar{X}_+$ and T, is of didactic interest since it implies that $V = \frac{n\beta}{\alpha^2}T \sim \chi^2(n-1)$. Consider the moment generating functions; if $\bar{X}_+ \perp T$,

$$M_T(t) = \frac{\mathrm{E}\exp\left\{[t/\bar{X}_-] - [(t-s)/\bar{X}_+]\right\}}{M_{1/\bar{X}_+}(s)}.$$

But this holds in particular for $s = t$, so there follows

$$\mathrm{E}\exp\{tT\} = \frac{\mathrm{E}[\exp\{t/\bar{X}_-\}]}{\mathrm{E}[\exp\{t/\bar{X}_+\}]}. \tag{11.12}$$

Now the claim follows since eqn (11.12), setting $\tau = (\alpha^2 t)/(n\beta)$, implies

$$\mathrm{E}[\exp\{t/\bar{X}_-\}] = \mathrm{E}[\exp\{\tau\beta/X\}]^n = \exp\{\frac{n}{\alpha^2}(1-\sqrt{1-2\tau}) + n\ln(1-2\tau)^{-1/2}\}.$$

But we also know that

$$\mathrm{E}[\exp\{t/\bar{X}_+\}] = \exp\{\frac{n}{\alpha^2}(1-\sqrt{1-2\tau}) + \ln(1-2\tau)^{-1/2}\},$$

and so by eqn (11.12) we obtain the mgf of T.

Moreover we have an identity, for any positive variates $X_1, \ldots, X_n$ and any positive numbers α and β, namely,

$$\frac{n}{\alpha^2}\xi^2[\bar{X}_+/\beta] + \frac{\beta}{\alpha^2\bar{X}_+}\sum_{i=1}^{n}\xi^2[X_i/\bar{X}_+] = \frac{1}{\alpha^2}\sum_{i=1}^{n}\xi^2[X_i/\beta]. \tag{11.13}$$

This identity corresponds to the usual decomposition of sums of squares using the mean when applied to normal variates. The question is how close are the two terms on the lhs of eqn (11.13) to being independent and what choice of parameters, if any, makes this independence true?

The answer to this question is given in

Corollary 4. *Let $X_i \sim \mathcal{W}(\alpha, \beta)$ be iid for $i = 1, \ldots, n$ and define for $j = 1, \cdots, n-1$*

$$\bar{X}_j = \tfrac{1}{j}\textstyle\sum_{i=1}^{j} X_i \quad \textit{for} \quad j = 1, \ldots, n-1, \textit{with}$$

$$V_j = \frac{j}{\alpha^2}\xi^2(\bar{X}_j/\beta) + \frac{1}{\alpha^2}\xi^2(X_{j+1}/\beta) - \frac{j+1}{\alpha^2}\xi^2(\bar{X}_{j+1}/\beta).$$

Now set

$$U = \frac{n}{\alpha^2}\xi^2[\bar{X}_+/\beta] \quad \text{and} \quad V := \frac{1}{\alpha^2}\sum_{i=1}^{n}\xi^2(X_i/\beta) - U = \frac{\beta}{\alpha^2}\sum_{i=1}^{n}\left(\frac{1}{X_i} - \frac{1}{\bar{X}_+}\right)$$

then $U \sim \chi^2(1)$ *and* $V_1 \perp V_2 \perp \cdots \perp V_{n-1} \sim \chi^2(1)$ *with* $U \perp V$ *and*

$$V = \sum_{i=1}^{n-1} V_i \sim \chi^2(n-1).$$

PROOF. We merely exhibit the identity, setting $S_j = \sum_{i=1}^{j} X_i$, for any integer $j = 1, 2, \ldots$:

$$\begin{aligned} V = &\left[\frac{1}{\alpha^2}\xi^2(X_1/\beta) + \frac{1}{\alpha^2}\xi^2(X_2/\beta) - \frac{2}{\alpha^2}\xi^2(S_2/2\beta)\right] \\ &+ \left[\frac{2}{\alpha^2}\xi^2(S_2/2\beta) + \frac{1}{\alpha^2}\xi^2(X_3/\beta) - \frac{3}{\alpha^2}\xi^2(S_3/3\beta)\right] \\ &\cdots\cdots\cdots\cdots\cdots\cdots\cdots\cdots \\ &+ \left[\frac{n-1}{\alpha^2}\xi^2(S_{n-1}/(n-1)\beta) + \frac{1}{\alpha^2}\xi^2(X_n/\beta) - \frac{n}{\alpha^2}\xi^2(S_n/n\beta)\right], \end{aligned}$$

where by Theorem 1 each of the $(n-1)$ summands has a Chi-square distribution with one degree of freedom. □

NB this supplies a positive answer, in the case of Wald variates, to the question about decomposing $V \sim \chi^2(n-1)$ into the sum of $(n-1)$ iid chi-square summands for any positive integer n rather than, as in Remark 1, only for $n = 2^k$.

We now state a lemma for $\mathcal{CD}$ variates, which contains Theorem 1 as a special case, and we provide an elementary proof. This might be compared with the proof of Samantha and Schwarz [102, 1991] which shows that for Wald variates when the sample is iid that $X_+ \perp V$.

Lemma 2. *Let* $X_1 \perp X_2$ *with* $X_i \sim \mathcal{CD}(\alpha_i, \beta_i, p)$ *for* $i = 1, 2$, *where the parameters satisfy eqn (11.8), and* U *and* V *are defined as in eqns (11.9) and (11.10), respectively. Then* U, V *have joint density given by*

$$f_{U,V}(u, v) = \frac{e^{-(u+v)/2}}{2\pi\sqrt{uv}}\left[1 + q^2\varepsilon(u, v)\right] \quad \text{for} \quad u, v > 0,$$

where $q = 1 - p$ *and*

$$\varepsilon(u, v) = \frac{\alpha_{(-2)}{}^2[1 + \alpha_{(-2)}{}^2 v](u - v)}{(1 + \alpha_{(-2)}^2 v)^2 + \alpha_{(-2)}^4 uv}.$$

While $\varepsilon(u, v) \not\equiv 0$, *it is negligible over the region of probabilistic interest when* $\alpha_{(-2)} \ll 1$.

Proof. By the usual transformation of variables we obtain $f_{U,V}$, the joint density of U, V, from g_{X_1,X_2}, the joint density of X_1, X_2. We write, omitting subscripts on the densities,

$$f(u, v) = g(x_1, x_2) \frac{1}{\left| \frac{\partial(u, v)}{\partial(x_1, x_2)} \right|_+}, \tag{11.14}$$

where the affix + here denotes the absolute value of the Jacobian, and the variables x_i must yet be expressed in terms of u and v. We find, asas,

$$g(x_1, x_2) = \frac{\alpha_1 \alpha_2 \, e^{-(u+v)/2}}{2\pi(\beta_1 + \beta_2)^2 \alpha_{(\text{-}2)}^4} \prod_{j=1}^{2} \sqrt{\frac{\beta_j}{x_j}} \left(\frac{p\beta_j}{x_j} + q \right) \tag{11.15}$$

and

$$\left| \frac{\partial(u, v)}{\partial(x_1, x_2)} \right|_+ = \frac{\left|1 - (\bar{\beta}_+/\bar{x}_+)^2\right| \cdot \left|(\beta_1/x_1)^2 - (\beta_2/x_2)^2\right|}{(2\alpha_0{}^2\bar{\beta}_+)^{-2}}. \tag{11.16}$$

Thus, the Jacobian changes signs on both lines $x_1 + x_2 = \beta_1 + \beta_2$ and $x_2/\beta_2 = x_1/\beta_1$, in the (x_1, x_2)-plane. These regions generate the four domains defined by the two relations in the new variates u, v, from eqns (11.9) and (11.9), viz.,

$$u = \frac{1}{\alpha_{(\text{-}2)}{}^2} \xi^2(\bar{x}/\bar{\beta}) \quad \text{and} \quad v = \frac{\alpha_{(\text{-}2)}{}^2 \bar{x}}{\bar{\beta}(\alpha_1\alpha_2)^2} \xi^2(x_2\beta_1/x_1\beta_2).$$

For convenience let us write $a = \frac{\alpha_1\alpha_2}{\alpha_{(\text{-}2)}}$ and $y_i = x_i/\beta_i$ for $i = 1, 2$ and set $t = \xi^{-1}(\alpha_{(\text{-}2)}\sqrt{u})$. Then the boundaries of the four regions are given by

$$\frac{\bar{\beta}}{\bar{x}} = t, \qquad \pm a\sqrt{\frac{v}{t}} = \xi(y_2/y_1); \tag{11.17}$$

$$\frac{\bar{x}}{\bar{\beta}} = t, \qquad \pm a\sqrt{vt} = \xi(y_2/y_1). \tag{11.18}$$

Thus, from eqns (11.14), (11.15), (11.16), we have asas

$$f(u, v) = \frac{e^{-(u+v)/2}}{2\pi\sqrt{uv}} \cdot \sum_{k=1}^{4} \left[\frac{1}{1 + (\bar{\beta}/\bar{x})} \left(\frac{p^2}{y_1 + y_2} + pq + \frac{q^2}{\frac{1}{y_1} + \frac{1}{y_2}} \right) \right] \Bigg|_{D_k},$$

where the last factor is the sum of contributions, within the enclosed brackets, from the Jacobian over each of the four domains; label them D_k for $k = 1, \ldots, 4$.

Write, for short, $r_i = \beta_i/(\beta_1+\beta_2)$ for $i = 1, 2$ and set $z = \xi^{-1}(a\sqrt{\frac{v}{t}})$ and $w = \xi^{-1}(a\sqrt{vt})$, then since $y = \xi(x)$ iff $-y = \xi(1/x)$ say:

$$D_1: \quad \frac{\bar{\beta}}{\bar{x}} = t, \quad y_2 = zy_1 \quad \text{gives} \quad y_1 = \frac{1}{t(r_1 + r_2 z)};$$
$$D_2: \quad \frac{\bar{\beta}}{\bar{x}} = t, \quad y_1 = zy_2 \quad \text{gives} \quad y_2 = \frac{1}{t(r_2 + r_1 z)};$$
$$D_3: \quad \frac{\bar{x}}{\bar{\beta}} = t, \quad y_2 = wy_1 \quad \text{gives} \quad y_1 = \frac{t}{r_1 + r_2 w};$$
$$D_4: \quad \frac{\bar{x}}{\bar{\beta}} = t, \quad y_1 = wy_2 \quad \text{gives} \quad y_2 = \frac{t}{r_2 + r_1 w}.$$

The sum of the coefficients of p^2 over the four regions, asas using $r_1 + r_2 = 1$, is found to be unity. Likewise the sum of the coefficients of pq is two. But, alas, the sum of the coefficients of q^2, call it C, is not unity; tidpat it is

$$C = \frac{1}{1+t}\left[\frac{1}{t+\alpha_{(-2)}^2 v} + \frac{t^2}{1+\alpha_{(-2)}^2 vt}\right] = \frac{1+\alpha_{(-2)}^2 v + \xi^2(t)}{(1+\alpha_{(-2)}^2 v)^2 + \alpha_{(-2)}^2 v\xi^2(t)}. \quad (11.19)$$

Then since $\xi(t) = \alpha_{(-2)}\sqrt{u}$, we define $\varepsilon = C - 1$ and simplify to obtain the result. □

Of course with $q = 0$ this basic lemma becomes Theorem 1. However, with this result we can check the robustness of this decomposition for all CD variates. We now have another consequence of this basic lemma, which relaxes the iid assumption yet obtains a decomposition into chi-square variates.

Theorem 2. *Let $X_i \sim \mathcal{W}(\alpha_i, \beta_i)$ for $i = 1, \ldots, n \geq 2$ be independent with consonant parameters that satisfy eqn (11.3). Then with $X_+ = \sum_{i=1}^n X_i$ and the notation of eqn (11.4) we have*

$$X_+ \sim \mathcal{W}(\alpha_{(-2)}, \beta_+) \quad \textit{and} \quad U = \frac{1}{\alpha_{(-2)}^2}\xi^2(X_+/\beta_+) \sim \chi^2(1). \quad (11.20)$$

Moreover, $U \perp V$ where

$$V = \sum_{i=1}^n \frac{1}{\alpha_i^2}\xi^2(X_i/\beta_i) - \frac{1}{\alpha_{(-2)}^2}\xi^2(X_+/\beta_+) \quad (11.21)$$
$$= \sum_{i=1}^n \frac{\beta_i}{\alpha_i^2 X_i} - \frac{\beta_+}{\alpha_{(-2)}^2 X_+} \sim \chi^2(n-1). \quad (11.22)$$

Proof. We proceed by induction. For $n = 2$ the result is Theorem 1. Assume it is true for $n - 1 \geq 2$ so that we have $S_{n-1} = \sum_{i=1}^{n-1} X_i \sim \mathcal{W}(\alpha_{(-2)}, \beta_+)$ where we set

$$\alpha_{(-2)}^2\beta_+ = \alpha_1^2\beta_1 = \cdots = \alpha_{n-1}^2\beta_{n-1} \quad \text{with} \quad 1/\alpha_{(-2)}^2 = \sum_{i=1}^{n-1}\alpha_i^{-2}, \quad \beta_+ = \sum_{i=1}^{n-1}\beta_i.$$

Let $X_n \sim \mathcal{W}(\alpha_n, \beta_n)$ for which we have the equality holding $\alpha_{(-2)}^2\beta_+ = \alpha_n^2\beta_n = (\beta_n + \beta_+)/(\alpha_n^{-2} + \alpha_{(-2)}^{-2})$. Since eqn (11.20) holds, replacing X_+ by S_{n-1}, consider

$$V = \sum_{i=1}^{n-1}\frac{1}{\alpha_i^2}\xi^2(X_i/\beta_i) - \frac{1}{\alpha_{(-2)}^2}\xi^2(S_{n-1}/\beta_+) \tag{11.23}$$

$$+ \frac{1}{\alpha_{(-2)}^2}\xi^2(S_{n-1}/\beta_+) + \frac{1}{\alpha_n^2}\xi^2(X_n/\beta_n) - \frac{1}{\alpha_{(-2)}^2}\xi^2\left(\frac{S_{n-1}+X_n}{\beta_+ + \beta_n}\right). \tag{11.24}$$

By the induction hypothesis the right-hand side of expression (11.23) is chi-square with $(n-2)$ d.f. and is independent of $\frac{1}{\alpha_{(-2)}^2}\xi^2(S_{n-1}/\beta_+)$. By assumption X_n is independent of $X_1, \ldots, X_{n-1}$, so expression (11.24) is independent of expression (11.23). But again by Theorem 1, the expression (11.23) is chi-square distributed with one degree of freedom. □

The practical implications of this decomposition are apparent. NB there is an analagous result for Tweedie variates, which we do not bother to state.

11.3.1. And Student's Distribution?

Arguably, the most used test in Statistical Science is Student's t-test. The test-statistic with the t-distribution applies for statistically independent $X_i \sim \mathcal{N}(\mu, \sigma^2)$ for $i = 1, \ldots, n$ because then

$$T := \frac{\bar{X} - \mu}{\sqrt{\frac{\sum_1^n (X_i - \bar{X})^2}{n-1}}} \quad \text{has a t-distribution with } (n-1) \text{ d.f.} \tag{11.25}$$

The ratio of $Z \div \sqrt{U/(n-1)}$, where $Z \sim \mathcal{N}(0,1)$ and $U \sim \chi^2(n-1)$ with $U \perp Z$, has a t-distribution. The independence of numerator and denominator in eqn (11.25), in fact, characterizes the Gaussian distribution and for no other distribution is there a statistic of such convenience.

However, by utilizing modern computing power we can obtain analogous results. For example, when dealing with the Weibull distribution, or one that can be so transformed, as in Chapter 6.4.4, with unknown parameters α, β. In this case even for type II censored samples the mles, $\hat{\alpha}, \hat{\beta}$, have the property that $\hat{\alpha}/\alpha$ and $\hat{\beta}/\beta$ have a joint distribution independent of α and β. Thus, the distribution of the statistic W where $W = \hat{\alpha}\ln[\hat{\beta}/\beta]$ must also be independent of both α and β. Here the logarithm merely transforms the sample space from $\Re_+$ to $\Re$.

So this property would allow a test of H:$\beta = \beta_0$ with α unknown if the distribution of W were tabulated for the particular sample size and the censoring specification. Today the appropriate values can be generated within a few minutes using the method previously given for generating order observations (in Chapter 6) by simulating the sample distribution.

Now consider the situation for Wald variates. If $X_i \sim \mathcal{W}(\alpha, \beta)$ are iid for $i = 1, \ldots, n$, then $U \perp V$ where

$$U = \frac{n}{\alpha^2}\xi^2[\bar{X}_1/\beta] \sim \chi^2(1) \quad \text{and} \quad V = \frac{n\beta}{\alpha^2}\left(\frac{1}{\bar{X}_{-1}} - \frac{1}{\bar{X}_1}\right) \sim \chi^2(n-1).$$

Here a test of $H_0 : \beta = \beta_0$ can be performed using the F-test, since $(n-1)U/V \sim F_{1,n-1}$, to determine if H_0 is true. But note from eqn (11.25) that $T^2 \sim F_{1,n-1}$; yet for $X \sim \mathcal{W}(\alpha, \beta)$ the variate $\xi(X/\beta)/\alpha$ does not have a standard normal distribution! However, if $Y \sim \mathcal{FL}(\alpha, \beta)$, then $\xi(Y/\beta)/\alpha \sim \mathcal{N}(0, 1)$. Could this be exploited to obtain a valid t-test?

Exercise Set 11.A

1. The great utility of Gaussian variates is they are, uniquely, closed under arbitrary linear combinations. Because of this property they are sometimes utilized as an approximation for nonnegative variates, which incurrs some error. What about linear combinations of $\mathcal{CD}$-variates?
   (a) Let $X_i \sim \mathcal{W}(\alpha_i, \beta_i)$ for $i = 1, 2$ with $X_1 \perp X_2$ and $X_1 + X_2 \sim F$. How closely does $G \simeq \mathcal{W}(\alpha_0, \beta_{(1)})$ where

   $$\beta_{(1)} = \beta_1 + \beta_2, \quad \alpha_0 = \alpha_{(-2)}\sqrt{1 + [\xi(r)/\psi(s)]^2},$$

   with $r = \alpha_1^2\beta_1/(\alpha_2^2\beta_2)$, $\quad s = \beta_1/\beta_2$, and $\frac{1}{\alpha_{(-2)}^2} = \frac{1}{\alpha_1^2} + \frac{1}{\alpha_2^2}$?
   (b) Using *Mathematica* or a similar program, compute some values of

   $$\sup_{y>0} |F_Y(y) - G(y)| \quad \text{for instructive values of } r \text{ near unity.}$$

   (c) Let $X_i \sim \mathcal{W}(\alpha_i, \beta_i)$ be independent for $i = 1, \cdots, n$ with $\varsigma_{i,j}$ satisfying $\alpha_i^2\beta_i\varsigma_{i,j} = \alpha_j^2\beta_j$ for all $i, j = 1, \cdots, n$ then $Y_j := \sum_{i=1}^n \varsigma_{i,j}X_i \sim \mathcal{W}(\alpha_{(-2)}, \beta_j^*)$ for each $j = 1, \cdots, n$ where $\alpha_{(-2)} = (\sum_1^n \alpha_i^{-2})^{-1/2}$ and $\beta_j^* = \sum_{i=1}^n \varsigma_{i,j}\beta_i$. If

   $$\alpha_k^2\beta_k = \min_{i=1}^n \alpha_i^2\beta_i, \quad \text{and} \quad \alpha_\ell^2\beta_\ell = \max_{i=1}^n \alpha_i^2\beta_i$$

   then follows $Y_k \preceq X_+ \preceq Y_\ell$. Thus, we have bounds on the distribution of X_+ and if β_k^* and β_ℓ^* are close, then the distribution of X_+ is well approximated by $\mathcal{W}(\alpha_{(-2)}, \beta_+)$.
2. State and prove a corollary to Theorem 2 for Tweedie variates.
3. Determine the maximum percentage error for the density in Lemma 2 for $0 < u, v < 20$ and calculate the corresponding cumulative probability $G(20, 20)$, where

   $$\frac{f_{U,V}(u, v) - g(u, v)}{g(u, v)} = q^2\varepsilon(u, v)$$

   with $g(u, v) = \frac{e^{-(u+v)/2}}{2\pi\sqrt{uv}}$ for $0 < u, v < \infty$. Would you use this lemma as a suitable approximation for $\mathcal{FL}$ distribution, i.e., when $q = .5$, for $\alpha \leq .1$?

4. Using *Mathematica*, or similar software, generate a program for determining the sample distribution of $W = \hat{\alpha}\ln[\hat{\beta}/\beta]$ from a Weibull distribution, where $\hat{\alpha}$, $\hat{\beta}$ are the mles from the smallest k-order observations out of n.

5. Let $X_i \sim \mathcal{FL}(\alpha, \beta)$ be statistically independent for $i = 1, \ldots, n$ and then set $Z_i = \frac{1}{\alpha}\xi(X_i/\beta)$, and define $U = \bar{Z}/\sqrt{n}$ and $V = \sum_1^n (Z_i - \bar{Z})^2$.
   (a) What is the distribution of $T = U\sqrt{\frac{n-1}{V}}$?
   (b) Let

$$\bar{X}_{(r)} := \left(\frac{1}{n}\sum_{i=1}^{n} X_i^r\right)^{1/r} \quad \text{for } r \neq 0,$$

   and then define

$$A = \frac{1}{\alpha\sqrt{n}}\sqrt{\frac{\bar{X}_{(1/2)}}{\bar{X}_{(-1/2)}}}, \quad \tilde{\beta} = \sqrt{\bar{X}_{(1)}\bar{X}_{(-1)}}, \text{ and } \quad \beta^* = \sqrt{\bar{X}_{(1/2)}\bar{X}_{(-1/2)}},$$

   and show that

$$U = 2A\sinh[\ln\sqrt{\beta^*/\beta}],$$
$$V = \frac{n}{\alpha^2}\left(\frac{\bar{X}_{(1)}}{\beta} - 2 + \frac{\beta}{\bar{X}_{(-1)}}\right) - A^2\left(\frac{\beta}{\beta^*} - 2 + \frac{\beta^*}{\beta}\right).$$

   (c) Can this be used as a t-test? What are the advantages and disadvantages?

11.4. Classifications for Dispersion Analysis

Let's say, for short, the *dispersion* of an rx X is the square of the coefficient of variation and write

$$\text{Dis}[X] := \frac{\text{Var}[X]}{\{\text{E}[X]}^2.$$

For the Wald distribution, the dispersion is exactly the square of the coefficient of variation, α, and in this study it takes the place of the variance. So we shall refer to the corresponding type of tests of hypotheses which when addressed, assuming a Gaussian distribution, are called "analysis of variance," as *analyis of dispersion.* Tweedie [116, 1957] called the similar analyses he performed the "analysis of reciprocals." But analysis of dispersion (ANODI) seems just as alliterative and here describes more correctly what is being analysed.

We also find it necessary to introduce an alternate notation for the *arithmetic* and *harmonic* means, namely, the affixes, H, A, to be used for multiple subscripts $x_{i,j}$ for $i = 1, \cdots, n$.

$$x_{\text{H},j} = \frac{n}{x_{-,j}} = \bar{x}_{-,j}, \quad x_{\text{A},j} = \frac{x_{+,j}}{n} = \bar{x}_{+,j}, \quad \text{where } x_{\pm,j} := \sum_{i=1}^{n} x_{i,j}^{\pm 1}. \quad (11.26)$$

11.4.1. A Single Classification

Let $X_{i,j} \sim \mathcal{W}(\alpha, \beta_i)$ for $j = 1, \ldots, n_i$ be an iid sample from each of the $i = 1, \ldots, k$ categories. We want to test the hypothesis that the means of each category are the same, i.e.,

$$H_0 : \beta_1 = \cdots = \beta_k = \beta,$$

when both mean β and dispersion α^2 are unknown.

For each category we know

$$X_{i,+} = \sum_{j=1}^{n_i} X_{i,j} \sim \mathcal{W}(\alpha/\sqrt{n_i}, n_i\beta) \qquad \text{for each } \ i = 1, \ldots, k.$$

We now must verify that the hypothesis of Theorem 2 is satisfied. Note consonance signifies the dispersion×mean is constant for each component, being $\alpha^2/n_i \times n_i\beta$, and for the sum it is $n_+\beta \times \alpha^2/n_+$. Thus, by Theorem 2 we see

$$U = \frac{n_+}{\alpha^2}\xi^2(X_{+,+}/\beta n_+) \sim \chi^2(1) \quad \text{and} \quad U \perp V_\star,$$

where

$$\begin{aligned} V_\star &= \sum_{i=1}^{k} \frac{n_i}{\alpha^2}\xi^2(X_{i,+}/\beta n_i) - \frac{n_+}{\alpha^2}\xi^2(X_{+,+}/\beta n_+), \\ &= \frac{\beta}{\alpha^2}\left[\sum_{i=1}^{k} \frac{n_i^2}{X_{i,+}} - \frac{n_+^2}{X_{+,+}}\right] \sim \chi^2(k-1). \end{aligned}$$

This corresponds to the usual SSM (sum of squares between categories) under the common Gaussian assumptions. But we also have for each category, say the ith for any $i = 1, \ldots, k$,

$$\begin{aligned} V_i &= \sum_{j=1}^{n_i} \frac{1}{\alpha^2}\xi^2(X_{i,j}/\beta_i) - \frac{n_i}{\alpha^2}\xi^2(X_{i,\mathrm{A}}/\beta_i) \\ &= \frac{n_i\beta_i}{\alpha^2}\left[\frac{1}{X_{i,\mathrm{H}}} - \frac{1}{X_{i,\mathrm{A}}}\right] \sim \chi^2(n_i - 1), \end{aligned}$$

where we have now utilized the notation of eqn (11.26), namely,

$$X_{i,\mathrm{H}} = \left[\frac{1}{n_i}\sum_{j=1}^{n_i} X_{i,j}^{-1}\right]^{-1} \quad \text{and} \quad X_{i,\mathrm{A}} = \left[\frac{1}{n_i}\sum_{j=1}^{n_i} X_{i,j}\right]$$

for the corresponding harmonic and arithmetic means over the appropriate index.

Thus, under the null hypothesis H_0, we have

$$V_+ = \sum_{i=1}^{k} V_i = \frac{\beta}{\alpha^2}\sum_{i=1}^{k} n_i\left[\frac{1}{X_{i,\mathrm{H}}} - \frac{1}{X_{i,\mathrm{A}}}\right] \sim \chi^2(n_+ - k).$$

This corresponds to the usual SSE (the within-groups sum of squares). Hence under H_0 and by Theorem 2 we have

$$V = \sum_{i=1}^{k}\left[\sum_{j=1}^{n_i}\frac{1}{\alpha^2}\xi^2(X_{i,j}/\beta)\right] - \frac{n_+}{\alpha^2}\xi^2(X_{+,+}/\beta n_+) \sim \chi^2(n_+ - 1).$$

Hence we have that $V = V_+ + V_\star$ with $V_+ \perp V_\star$.

By using the corresponding test statistic we have asas

$$\frac{V_\star/(k-1)}{V_+/(n_+ - k)} = \frac{\left[\frac{n_+}{X_{H_2,A_1}} - \frac{n_+}{X_{+,+}}\right]}{\sum_{i=1}^{k} n_i\left[\frac{1}{X_{i,H}} - \frac{1}{X_{i,A}}\right]} \cdot \frac{n_+ - k}{k-1} \sim \mathcal{F}(k-1, n_+ - k), \tag{11.27}$$

which is independent of both α and β.

As remarked, Tweedie [116], in his analysis of reciprocals for a single-classification used the usual parameterization of the inverse Gaussian distribution; see Chikkara and Folks [26]. That was the initial extent of the theory.

11.4.2. A Two-Way Classification for Multiplicative Effects

Let us now consider a special two-way classification having the same number of observations, say n, in each cell with each observation having the same dispersion; assume there are $i = 1, \ldots, r$ rows and $j = 1, \ldots, c$ columns.

We assume a multiplicative row and column effect so that for the kth observation in the (i, j) cell we have

$$X_{i,j,k} \sim \mathcal{W}(\alpha, \beta_i\gamma_j) \quad \text{consequently} \quad X_{i,j,+} = \sum_{k=1}^{n} X_{i,j,k} \sim \mathcal{W}(\frac{\alpha}{\sqrt{n}}, n\beta_i\gamma_j),$$

for any whole number n and $i = 1, \ldots, r$ and $j = 1, \ldots, c$.

We now consider three null hypotheses:

- $H_0 : \ \beta_i\gamma_j \equiv \delta$ for all (i, j)
- $H_1 : \ \gamma_1 = \cdots = \gamma_c \equiv \gamma$
- $H_2 : \ \beta_1 = \cdots = \beta_r \equiv \beta$

where the common values β, γ, δ are all unknown.

No Row or Column Effects

If we consider H_0 true, then we have the unbiased estimate for δ

$$\hat{\delta} = \frac{X_{+,+,+}}{rcn} \sim \mathcal{W}(\alpha/\sqrt{nrc}, \delta).$$

We set

$$V_\bullet = \frac{n}{\alpha^2}\sum_{i,j=1}^{r,c}\xi^2\left(\frac{X_{i,j,+}}{n\delta}\right) - \frac{nrc}{\alpha^2}\xi^2\left(\frac{X_{+,+,+}}{nrc\delta}\right)$$

$$= \frac{n^2\delta}{\alpha^2}\left[\sum_{i,j=1}^{r,c}\frac{1}{X_{i,j,+}} - \frac{r^2c^2}{X_{+,+,+}}\right] \sim \chi^2(rc-1). \quad (11.28)$$

Since $X > 0$ *a.s.* implies for any $c > 0$ that $\text{Dis}\,(cX) = \text{Dis}\,(X)$, then

$$\frac{X_{i,j,+}}{n} \sim \mathcal{W}(\tfrac{\alpha}{\sqrt{n}}, \beta_i\gamma_j) \quad \text{for all } i = 1, \cdots, r \text{ and } j = 1, \cdots, c, \quad (11.29)$$

so we have the result claimed in eqn (11.28).

Now consider

$$V_+ = \frac{1}{\alpha^2}\sum_{i,j,k=1}^{r,c,n}\xi^2\left(\frac{X_{i,j,k}}{\delta}\right) - \frac{nrc}{\alpha^2}\xi^2\left(\frac{X_{+,+,+}}{nrc\delta}\right)$$

$$= \frac{\delta}{\alpha^2}\sum_{i,j,k=1}^{r,c,n}\left[\frac{1}{X_{i,j,k}} - \frac{r^2c^2n^2}{X_{+,+,+}}\right] \sim \chi^2(rcn-1). \quad (11.30)$$

By Theorem 2 we know

$$\tfrac{nrc}{\alpha^2}\xi^2\left(\tfrac{X_{+,+,+}}{nrc\delta}\right) \perp V_+.$$

Thus, if the value of the mean δ were hypothesized, the test statistic should be

$$\frac{\frac{nrc}{\alpha^2}\xi^2\left(\frac{X_{+,+,+}}{nrc\delta}\right)}{\frac{V_+}{rcn-1}} \sim \mathcal{F}(1, rcn-1), \quad (11.31)$$

which is functionally dependent on both α and δ. For both δ and α unknown, use the statistic

$$\frac{(nrc-1)V_\bullet}{(rc-1)V_+} \sim \mathcal{F}(rc-1, rcn-1), \quad \text{for } n > 1, \quad (11.32)$$

after showing $V_\bullet \perp V_+$, which is independent of both δ and α. The computing expressions for $V_\bullet$ and V_+ are given in eqns (11.28) and (11.30).

No Column Effects

Now consider the second hypothesis:

$$H_1 : \gamma_1 = \cdots = \gamma_c = \gamma \quad \text{with } \gamma \text{ unknown.}$$

From eqn (11.29) we fix i and set

$$V_{i,+} = \sum_{j=1}^{c}\frac{n}{\alpha^2}\xi^2\left(\frac{X_{i,j,+}}{n\beta_i\gamma}\right) - \frac{cn}{\alpha^2}\xi^2\left(\frac{X_{i,+,+}}{cn\beta_i\gamma}\right)$$

$$= \sum_{j=1}^{c}\left(\frac{n}{\alpha^2}\cdot\frac{n\beta_i\gamma}{X_{i,j,+}}\right) - \frac{nc}{\alpha^2}\cdot\frac{nc\beta_i\gamma}{X_{i,+,+}}$$

$$:= \frac{nc\beta_i\gamma}{\alpha^2}\left[\frac{1}{X_{i,\mathrm{H}_2,\mathrm{A}_1}} - \frac{1}{X_{i,\mathrm{A}_2,\mathrm{A}_1}}\right] \sim \chi^2(c-1), \quad (11.33)$$

where subscripts H, A denote the harmonic and arithmetic means, respectively, as first defined in eqn (11.26), and their subscripts now indicate the order of their computation.

We check that for each fixed i, that the mean×dispersion remains constant, namely, for all $j = 1, \ldots, c$ we see that

$$\left(\frac{\alpha^2}{n}\right)(n\beta_i\gamma) \equiv \frac{\alpha^2}{nc}(nc\beta_i\gamma) \qquad \text{for each fixed } i = 1, \ldots, r.$$

Alternatively, under the same assumption, we have

$$\begin{aligned} V^\star_{i,+} &= \sum_{j=1}^{c}\left[\sum_{k=1}^{n}\frac{1}{\alpha^2}\xi^2\left(\tfrac{X_{i,j,k}}{\beta_i\gamma}\right) - \frac{n}{\alpha^2}\xi^2\left(\tfrac{X_{i,j,+}}{n\beta_i\gamma}\right)\right] \\ &= \sum_{j=1}^{c}\left[\sum_{k=1}^{n}\frac{1}{\alpha^2}\frac{\beta_i\gamma}{X_{i,j,k}} - \frac{n}{\alpha^2}\frac{n\beta_i\gamma}{X_{i,j,+}}\right] \\ &:= \frac{nc\beta_i\gamma}{\alpha^2}\left[\frac{1}{X_{i,\text{H}_2,\text{H}_1}} - \frac{1}{X_{i,\text{H}_2,\text{A}_1}}\right] \sim \chi^2[c(n-1)], \end{aligned} \tag{11.34}$$

where, to be explicit, $X_{i,j,\text{H}} = [\frac{1}{n}\sum_{k=1}^{n}(1/X_{i,j,k})]^{-1}$.

Since

$$\begin{aligned} V_{i,+} + V^\star_{i,+} &= \sum_{j=1}^{c}\sum_{k=1}^{n}\frac{1}{\alpha^2}\xi^2\left(\tfrac{X_{i,j,k}}{\beta_i\gamma}\right) - \sum_{j=1}^{c}\frac{n}{\alpha^2}\xi^2\left(\tfrac{X_{i,j,+}}{n\beta_i\gamma}\right) \\ &\quad + \sum_{j=1}^{c}\frac{n}{\alpha^2}\xi^2\left(\tfrac{X_{i,j,+}}{n\beta_i\gamma}\right) - \frac{cn}{\alpha^2}\xi^2\left(\tfrac{X_{i,+,+}}{cn\beta_i\gamma}\right), \\ V_{i,+} + V^\star_{i,+} &= \frac{nc\beta_i\gamma}{\alpha^2}\left[\frac{1}{X_{i,\text{H}_2,\text{H}_1}} - \frac{1}{X_{i,\text{A}_2,\text{A}_1}}\right] \sim \chi^2(nc-1). \end{aligned}$$

By Theorem 2 we have for each fixed i that $V_{i,+} \perp V^\star_{i,+}$. Therefore, the test statistic is

$$T_i = \frac{c(n-1)V_{i,n}}{(c-1)V^\star_{i,+}} \sim \mathcal{F}[(c-1), c(n-1)] \quad \text{for each } i = 1, \ldots, r,$$

which are iid and functionally independent of $\beta_i, \gamma, \alpha^2$.

Let $f_{1-\epsilon}(m_1, m_2)$ be the $100(1-\epsilon)$ percentile of the $\mathcal{F}$-distribution with (m_1, m_2) degrees of freedom. Hence a critical region to reject H_2 is if any $T_i > f_{1-\epsilon}[c-1, c(n-1)]$; the test is of level

$$\Pr[\max_{i=1}^{r} T_i < f_{1-\epsilon}(c-1, cn-c)] = (1-\epsilon)^r.$$

Proposition 1. *In this case the unbiased estimator of the dispersion* α^2 *is*

$$\check{\alpha}^2 = \frac{1}{r}\sum_{i=1}^{r}\tilde{\alpha}_i^2 \quad \textit{with } \operatorname{Dis}[\check{\alpha}^2] = \frac{2 + (\alpha^2/n)(c+1)}{\sqrt{r}(c-1)},$$

where

$$\tilde{\alpha}_i^2 = \frac{n X_{i,A_2,A_1}}{c-1}\left[\frac{1}{X_{i,H_2,A_1}} - \frac{1}{X_{i,A_2,A_1}}\right] \qquad \textit{for } i = 1.\ldots, r.$$

PROOF. We see that we can write

$$\tilde{\alpha}_i^2 = \frac{\alpha^2 X_{i,A_2,A_1}}{c\beta_i\gamma} \cdot \frac{V_{i,+}}{c-1},$$

where

$$X_{i,A_2,A_1} \sim \mathcal{W}(\alpha/\sqrt{cn}, c\beta_i\gamma) \perp V_{i,+} \sim \chi^2(c-1)$$

is itself unbiased for α^2. Now since for any constant b and rvs $X \perp V$, we have

$$\text{Dis}[b \cdot X \cdot V] = \text{Dis}[X] \cdot \text{Dis}[V] + \text{Dis}[X] + \text{Dis}[V],$$

the result follows. □

The actual distribution of a product $W = X \cdot V$ where $X \perp V$ with $X \sim \mathcal{W}(\alpha, 1)$ and $V \sim \chi^2(\nu)$ can be found using numerical integration from

$$\Pr[W \le 2w] = \int_0^\infty \left\{\bar{\Phi}[\frac{1}{\alpha}\xi(y/w)] + e^{2/\alpha^2}\bar{\Phi}[\frac{1}{\alpha}\sqrt{4+\xi^2(y/w)}]\right\} \times \frac{y^{\nu/2-1}e^{-y}}{\Gamma(\nu/2)}\, dy.$$

No Row Effects

The corresponding test for $H_2 : \beta_1 = \cdots = \beta_r = \beta$ can be simply obtained by symmetry in the indices i, j and r, c,

When Does Consonance Occur?

Consonance arises frequently when one thinks in terms of multiplicative effects. Consider a comparison of duration-of-life tests for polymeric coatings. The basic assumption is that the coating, independent of accidental chipping or abrasion, degrades due to environmental exposure from ultraviolet radiation. The damage from UV is, by accepted theory, proportional to the *absorbed dosage*. The absorbed dosage is obtained by integrating the product of the measured spectral irradiance, at a given site, and the spectral absorption of the particular coating, e.g., acrylic melamine, over the range of wavelengths impinging on the specimen for the duration of exposure at the specific humidity and temperature.

Thus to predict the mean service life at various geographical locations the expected absorbed dosage for each coating must be calculated over the duration of its duty cycle (which may be annually) at the average temperature and humidity at each location. These calculations need not concern us here. Assume the annual expected absorbed dosages calculated at the k service locations are $\rho_1, \cdots, \rho_k$. Since all absorbed dosages are to be compared with one distinguished site, wlog let $\rho_k = 1$.

Assume we have n_j test specimens measured at the jth site so we have iid

$$X_{i,j} \sim \mathcal{W}(\alpha, \beta_j) \text{ for } i = 1, \cdots, n_j, \text{ and } j = 1, \cdots, k.$$

Then the total life statistic at the ith site is

$$X_{+,j} := \sum_{i=1}^{n_j} X_{i,j} \sim \mathcal{W}(\tfrac{\alpha}{\sqrt{n_j}}, n_j\beta_j). \tag{11.35}$$

Let the working hypothesis here be that the jth test site environment is ρ_j times as severe as that at the distinguished site, that is the ratio of the two expected absorbed dosages is ρ_j with $\rho_j\beta_j = \beta$ for $j = 1, \cdots, k$ where $\rho_k = 1$. Then by eqn (11.35) we have for $i = 1, \cdots, k$

$$Y_j := \rho_j X_{+,j} \sim \mathcal{W}(\alpha_j, \beta_j') \quad \text{where } \alpha_j = \tfrac{\alpha}{\sqrt{n_j}}, \quad \text{and } \beta_j' = n_j\rho_j\beta_j.$$

We conclude that we have consonance for this series of tests since $\alpha_j^2\beta_j' = \alpha^2\beta$ for $j = 1, \ldots, k$, and we need not know either α or β.

Exercise Set 11.B

1. Prove that the rv V as defined in eqn (11.10) is equivalent with the expression given in eqn (11.6).

2. Verify the joint density of V, Z is as given in eqn (11.7).

3. Verify the expression for eqn (11.19) as the sum of the contributions over each of the quadrants D_1, D_2, D_3, D_4. Hint: Remember that: $y = \xi(x) := \sqrt{x} - \frac{1}{\sqrt{x}} = 2\sinh(\ln\sqrt{x})$ implies $x = \xi^{-1}(y) = \exp\{2\sinh^{-1}(y/2)\} = \left(\frac{y}{2} + \sqrt{1 + \frac{y^2}{4}}\right)^2$.

4. If $Z_i \sim \mathcal{CD}(\alpha_i, \beta_i, p_i)$ are independent and consonant, i.e., $\alpha_1^2\beta_1 = \cdots = \alpha_n^2\beta_n$, show using the mgfs as given in eqn (10.13), that the cdf of $Z_+ = \sum_1^n Z_i$ is given by,
   (a) $\sum_{k=0}^n \binom{n}{k} p^{n-k}q^k F * G_k$, when $p_i = p$ for $i = 1, \cdots, n$ and $q = 1 - p$.
   (b) $\prod_{i=1}^n q_i \sum_{k=0}^n \varsigma_k F * G_{n-k}$, when $q_i > 0$ for $i = 1, \cdots, n$, and the ς_k are the symmetric polynomials in $r_i = p_i/q_i$, viz.,
   $\varsigma_0 = 1, \quad \varsigma_1 = \sum_{i=1}^n r_i, \quad \varsigma_2 = \sum_{i<j} r_i r_j, \quad \cdots, \quad \varsigma_n = \prod_{i=1}^n r_i.$
   Here $F \simeq \mathcal{W}(\alpha_{(\text{-}2)}, \beta_+)$ and $G_k \simeq \mathcal{G}am(k/2, 2\alpha_0^2\beta_+)$ is a Gamma distribution for $k = 1, \cdots, n$ with G_0 being the identity distribution with a saltus of 1 at zero.

5. Let $X_i \sim \mathcal{W}(\alpha_i, \beta_i)$ for $i = 1, \cdots, n$ be independent and consonant. Show that not only do we have $\alpha_i^2\beta_i \equiv \alpha_{(\text{-}2)}^2\beta_+$ but also $\alpha_i^2\beta_i \equiv \alpha_{(2)}^2\beta_-$ where $\alpha_{(2)} = \sqrt{\sum_1^n \alpha_i^2}$. Does this imply an alternate theory for dispersion analysis for Tweedie variates?

6. State and prove the results for Tweedie variates analgous to Theorem 2.

7. If $X_1 \perp X_2$ with $X_i \sim \mathcal{CD}(\alpha_i, \beta_i, p)$ for $i = 1, 2$ but $\alpha_1^2\beta_1 \neq \alpha_2^2\beta_2$ show that a good approximation for the distribution of $X_1 + X_2$ is given by the distribution of $Z_1 + Z_2$ where for $i = 1, 2$ with $Z_i \sim \mathcal{CD}[\alpha_i, (1 + (-1)^{i-1}\rho)\beta_i, p]$, by setting

$$\nu_i = \alpha_i^2/(\alpha_1^2 + \alpha_2^2) \quad \text{and} \quad \beta' = \frac{2\beta_1\beta_2}{\beta_1\nu_1 + \beta_2\nu_2},$$

we obtain $Z_1 + Z_2 \sim \mathcal{CD}(\alpha_{-2}, \beta', p)$,

An education is what remains after what you've memorized has been forgotten.

B.F. Skinner

The law of conservation of difficulty in an academic discipline implies that obscurantism will expand usage of nebulous terminology to fill any vacuum of intrinsic simplicity.

The spread of secondary and tertiary education has created a large population of people, often with well-developed literary and scholary tastes, who have been educated far beyond their capacity to undertake analytical thought.

Richard Dawkins

CHAPTER 12

Damage Processes

There is a marked difference between damage that occurs only as the number of duty cycles, or service hours, accumulate and when damage also can occur by some means extrinsic to time spent in service, such as random peaks in strain (earthquakes, clear-air turbulence, or tornados), bursts of ultraviolet radiation, or sudden fluctuations in humidity (like tsunamis). Consequently, an accurate prediction of damage during such service experience requires more inclusive models and accounting practices.

12.1. The Poisson Process

Consider a counting process where $N_{(0,t]}$ = the number of random events in the interval $(0, t]$. The following are equivalent definitions of a *Poisson process* with intensity rate $\lambda > 0$. We shall denote this stochastic process by $N_{(0,t]} \sim \mathcal{P}P(\lambda)$.

Definition I. The interoccurrence times between events are iid with the time between the $(i-1)$st and ith events being $T_i \sim \mathrm{E}p(\lambda)$.

Definition II. $N_{(0,t]}$ has stationary, independent increments. For all $s, t > 0$.

$$(i): N_{(0,t]} \sim N_{(s,t+s]} \quad (ii): N_{(0,s]} \perp N_{(s,t+s]} \quad (iii): \mathrm{E}N_{(0,t]} = \lambda t.$$

Definition III. (Feller) We have three conditions of incremental behavior:

$$(a): \ \Pr[N_{(t,t+\epsilon]} = 1] = \lambda\epsilon + o(\epsilon),$$
$$(b): \ \Pr[N_{(t,t+\epsilon]} \geq 2] = o(\epsilon),$$
$$(c): \ \Pr[N_{(0,0]} = 0] = 1.$$

From any one of these definitions, now writing $N(t) := N_{(0,t]}$, there follows

$$\Pr[N(t) = k] = \frac{e^{-\lambda t}(\lambda t)^k}{k!} \quad \text{for } k = 0, 1, 2, \ldots.$$

This corresponds to the exponential waiting time between events, since if $T_1 \sim \mathrm{Ep}(\lambda)$ is the waiting time to the first event, then

$$\Pr[T_1 > t] = \Pr[N(t) = 0] = e^{-\lambda t} \quad \text{for any } t > 0.$$

Let $S_n = \sum_{i=1}^{n} T_i$ be the time of the nth occurrence of the (failure) event. The basic *duality* between the counting process and waiting time until failure is the equivalence of the events relating partial sums of 'times between events' with counts of failures, namely,

$$[S_n \leq t] = [N(t) \geq n]. \tag{12.1}$$

Since $T_i \sim G$, say, are iid, then it follows that $S_n \sim G^{n*} = G * G * \cdots * G$, the n-fold convolution of G with itself. Thus, for all $x > 0$, $n = 1, 2, \cdots$, we have the cdf given by

$$G^{n*}(x) := \int_0^x G^{(n-1)*}(x-y)\, dG(y) = \int_0^{\lambda x} \frac{y^{n-1} e^{-y}}{\Gamma(n)}\, dy, \tag{12.2}$$

where $G^{0*} = \mathrm{I}(x \geq 0)$ for all $x \in \Re$. NB that $2\lambda S_n \sim \chi^2(2n)$ with the corresponding density

$$\frac{d\,G^{n*}(x)}{dx} := g^{n*}(x) = \frac{\lambda(\lambda x)^{n-1} e^{-\lambda x}}{\Gamma(n)}.$$

It can be verified, using repeated integration by parts, that

$$\overline{G^{n*}}(x) = \sum_{j=1}^{n-1} \frac{(\lambda x)^j e^{-\lambda x}}{j!}. \tag{12.3}$$

From these relationships we can verify the duality relationship of eqn (12.1) directly as

$$\Pr[S_n > t] = \overline{G^{n*}}(t) = \Pr[N(t) < n].$$

We now note the

Theorem 1. *The conditional joint density of* $S_1, \ldots, S_k | S_{k+1}$ *is*

$$f_{S_1,\ldots,S_k|S_{k+1}}(s_1, \ldots, s_k|s) = \frac{k!}{s^k} \quad \textit{for} \quad 0 < s_1 < \cdots < s_k < s.$$

Proof. We use the definition of conditional density on the simplex $0 < s_1 < s_2 < \cdots < s_k < s$ to obtain

$$\begin{aligned} f_{S_1,\ldots,S_k|S_{k+1}}(s_1, \ldots, s_k|s) &= \frac{f_{S_1,\ldots,S_{k+1}}(s_1, \ldots, s_k, s)}{f_{S_{k+1}}(s)} \\ &= \frac{\lambda e^{-\lambda s_1} \lambda e^{-\lambda(s_2 - s_1)} \cdots \lambda e^{-\lambda(s - s_k)}}{g^{(k+1)*}(s)}, \end{aligned}$$

which gives the result by eqns (12.2) and (12.3). □

This means that knowing only the event $[N(s) = k + 1]$ implies the times of the preceding occurrences $S_1, \ldots, S_k$ can be thought of as being jointly distributed as the order statistics in a sample of size k from a Uni $(0, s)$ population.

12.1.1. The Superposition of Poisson Processes

Immediately we obtain the following:

Remark 1. *If $N_i(t) \sim \mathcal{P}P(\lambda_i)$ for $i = 1, \ldots, k$ are the counting functions for independent Poisson processes, then*

$$N(t) = \sum_{i=1}^{k} N_i(t) \sim \mathcal{P}P(\sum_{i=1}^{k} \lambda_i).$$

Of course this generalizes. If $N_\lambda(t) \sim \mathcal{P}P[g(\lambda)]$ for all $\lambda \in \Lambda$ represents the counting process for the arrival of radiation (photons) at each wavelength λ, not to be confused with Poisson intensities λ_i, over the spectrum Λ (the range of possible wavelengths of radiation), then

$$N_\Lambda(t) = \int_{\lambda \in \Lambda} N_\lambda(t)\, d\lambda \sim \mathcal{P}P[\textstyle\int_{\lambda\in\Lambda} g(\lambda) d\lambda].$$

12.1.2. The Decomposition of Poisson Processes

Let $N(t) \sim \mathcal{P}P(\lambda)$ be the counting process for certain events having iid interarrival times $T_i \sim \text{Ep}(\lambda)$ for $i = 1, 2, \ldots$. But each event is *damaging* only if, at arrival, a chance circumstance occurs simultaneously and such circumstances are indicated by a sequence of Bernoulli variates $X_i \sim \mathcal{B}(1, p)$ for $i = 1, 2, \ldots$. Let us calculate the interoccurrence times for these damaging events. Let D denote the random number of successive events until the first damaging event occurs. Then since $\text{E}X_i = p$,

$$\Pr[D = n] = pq^{n-1} \quad \text{for } n = 1, 2, \ldots; q = 1 - p.$$

Let $T^\star$ be the waiting time until the first damaging event and $S_n = \sum_{i=1}^{n} T_i$ be the time of the nth event. Then $T^\star := S_D = T_1 + \cdots + T_D$ so that by eqn (12.2), and the theorem of total probability, we obtain

$$\Pr[T^\star > t] = \textstyle\sum_{n=1}^{\infty} \Pr[S_n > t | D = n] \cdot pq^{n-1} = \sum_{n=1}^{\infty} \int_t^\infty \frac{p\lambda(q\lambda x)^{n-1} e^{-\lambda x}}{\Gamma(n)}\, dx = e^{-p\lambda t}.$$

Hence the times between damaging events are themselves iid Epstein variates with hazard rate λp and it follows that the counting process for damaging events is Poisson with an intensity rate reduced by the fraction p.

12.2. Damage Due to Intermittant Shocks

Consider that damage occurs intermittantly according to a *renewal process* with interarrival times, $T_i > 0$, which are iid with $T_i \sim G$ for $i = 1, 2, \ldots$, which need not be Epstein. Then the nth arrival occurs at $S_n = \sum_{i=1}^{n} T_i \sim G^{n*}$, which is the n-fold convolution of G; denote the associated counting process by $N(t)$, for which duality holds with the partial sums, namely,

$$[N(t) \geq n] = [S_n \leq t].$$

Remark 2. *The counting process has distribution*

$$\Pr[N(t) = n] = G^{n*}(t) - G^{(n+1)*}(t) \quad \textit{for } n = 0, 1, 2 \cdots \quad \textit{and } t > 0, \tag{12.4}$$

where $G^0(t) = 1$. *The associated* renewal function, *defined as the expected number of arrivals of incremental damage by time* $t > 0$, *is given by*

$$R(t) := EN(t) = \sum_{n=1}^{\infty} G^{n*}(t). \tag{12.5}$$

Proof. We see that for exactly n arrivals to have occurred at time t we must have had exactly n before time t with the next arrival occurring after, that is, $[N(t) = n] = [S_n \leq t, S_{n+1} > t]$, proving eqn (12.4). Then there follows

$$\mathrm{E}N(t) = \sum_{n=0}^{\infty} n[G^{n*}(t) - G^{(n+1)*}(t)] = \sum_{n=0}^{\infty} G^{(n+1)}(t). \qquad \square$$

When the damage resulting from the ith arrival (shock or load sustained during the duty cycle) is also random, label it $0 < X_i \sim F$, then the total damage sustained by the structure after the n shock arrival (or duty cycle) will be denoted, accordingly as the X_i are iid or not, by

$$Y_n = \sum_{i=1}^{n} X_i \sim F^{n*} \quad \text{or} \quad Y_n = \sum_{i=1}^{n} X_i \sim F^{(n)}.$$

If the shocks themselves occur at random according to some counting process, then we have a more complicated distribution.

Theorem 2. *Let* $N(t)$ *count the shocks arriving before time* $t > 0$ *that occur according to a Poisson process with intensity* λ. *At the nth shock the cumulative damage to the structure is* $S_n = \sum_{i=1}^{n} X_i \sim F^{n*}$, *where the* $X_i > 0$ *are iid with cdf F. Thus, at time* $t > 0$ *the cumulative damage to the structure is* $S_{N(t)}$. *If the structure will fail when for the first time its cumulative damage exceeds a given critical level, say* ω, *then the waiting time until failure,* $W \sim F_W$, *is* IHRA *with sdf*

$$\bar{F}_W(t) = \Pr[S_{N(t)} \leq \omega] = \sum_{n=0}^{\infty} \frac{(\lambda t)^n e^{-\lambda t}}{n!} F^{n*}(\omega). \tag{12.6}$$

PROOF. To obtain eqn (12.6) note by duality that $[W > t] = \left[\sum_{i=1}^{N(t)} X_i < \omega\right]$ (where a vacuous sum is null), and then apply the theorem of total probability using the Poisson definition of $N(t)$ and that incremental damages are iid. The proof of F_W being *IHRA* follows from the total positivity of the Poisson density and the fact that $[F^{n*}]^{1/n} \in\downarrow$ in n. See pp. 93, 160 in Barlow and Proschan [7]. □

Sometimes this distribution may have an asymptotic form that is simple, yet maintains all the relevant properties. As the first step to obtain the limiting distribution for the cdf as given in eqn (12.6) we state Wald's lemma:

Lemma 1. *If $Y_n = \sum_{i=1}^{n} X_i$, where X_i are iid for $i \in \aleph$ and N is an rv on $\aleph$ such that the event $[N = n]$ is always a "stopping event," i.e., one that depends only on $\{X_1, \ldots, X_n\}$, then the random sum S_N has mean and variance given by*

$$E[S_N] = EN \cdot EX, \qquad Var[S_N] = EN \cdot Var[X] + [EX]^2 \cdot Var[N]. \quad (12.7)$$

PROOF. Write $S_N = \sum_{i=1}^{\infty} X_i \mathrm{I}(N \geq i)$ and so the expectation is $\mathrm{E}[S_N] = \sum_{i=1}^{\infty} \mathrm{E}\{X_i[1 - \mathrm{I}(N < i)]\}$, but by definition we have $[N < i] \perp X_i$ so that $\mathrm{E}\{X_i[1 - \mathrm{I}(N < i)]\} = \mathrm{E}[X] \cdot \Pr[N \geq i]$ and there follows $\mathrm{E}[S_N] = \mathrm{E}[X] \cdot \mathrm{E}[N]$. The result for the variance is proved similarly. □

Theorem 3. *Under the assumptions of Theorem 2, the waiting time until failure from cumulative damage exceeding a critical limit has, asymptotically, a fatigue-life distribution.*

PROOF. Set $\mathrm{E}X = \nu$, $\mathrm{Var}[X] = \delta^2$ and label the cumulative damage by time t as $Y_t = \sum_{i=1}^{N(t)} X_i$. By Wald's lemma we have that $\mathrm{E}[Y_t] = t\lambda\nu$, and $\mathrm{Var}[Y_t] = t\lambda(\delta^2 + \nu^2)$. For notational convenience set $\mu = \lambda\nu$, $\sigma^2 = \lambda(\nu^2 + \delta^2)$, thus the standardized cumulative damage Z_t satisfies the CLT, namely,

$$Z_t = \frac{t\mu - Y_t}{\sqrt{t}\sigma} \rightsquigarrow \mathcal{N}(0, 1) \text{ as } t \to \infty. \quad (12.8)$$

We thus obtain the cdf of the waiting time, W_λ, which depends on the arrival rate λ,

$$\begin{aligned} \Pr[W_\lambda > t] = \Pr[Y_t \leq \omega] &= \Pr\left[Z_t \geq \sqrt{t}\frac{\mu}{\sigma} - \frac{\omega}{\sqrt{t}\sigma}\right] \\ &= \Pr\left[Z_t \geq \frac{\sqrt{\mu\omega}}{\sigma}\xi(t\mu/\omega)\right], \end{aligned} \quad (12.9)$$

where as before $\xi(x) = \sqrt{x} - \dfrac{1}{\sqrt{x}}$.

We now consider the limiting distribution as the time between shock occurrences becomes small. We can do this by setting

$$\alpha^2 = \frac{\sigma^2}{\mu\omega} = \frac{\nu^2+\delta^2}{\nu\omega}, \quad \text{and} \quad \beta = \frac{\omega}{\nu},$$

where neither α nor β are functions of λ. Now set $s = \lambda t$ to obtain from eqn (12.9)

$$\Pr[\lambda W_\lambda > s] = \Pr\left[Z_{(s/\lambda)} \geq \frac{1}{\alpha}\xi(s/\beta)\right] \tag{12.10}$$

Letting λ approach 0, by eqn (12.8), we obtain the asymptotic waiting time as

$$\lim_{\lambda \downarrow 0} \Pr[\lambda W_\lambda > s] = \bar{\Phi}[\xi(s/\beta)/\alpha]. \tag{12.11}$$

Thus, the limiting distribution is $\mathcal{F}L(\alpha, \beta)$. □

12.3. Renewal Processes

We now consider the interarrival times as iid with $T_i > 0$ but $T_i \sim G$ where the cdf G is not necessarily exponential. Thus, for each renewal process with counting function $N(t)$, we have the associated *renewal function* defined by eqn (12.5). We see that we can write

$$R(t) = G(t) + \sum_{k=1}^{\infty}[G^{k*}*G](t) = G(t) + \sum_{k=1}^{\infty}\int_0^t G^{k*}(t-x)\,dG(x),$$

$$R(t) = G(t) + \int_0^t R(t-x)\,dG(x). \tag{12.12}$$

Equation (12.12) is known as the *fundamental renewal equation.* If a density $F' = f$ exists, then there is a *renewal density*, namely, $r(t) = R'(t)$, which satisfies the integral equation

$$r(t) = f(t) + \int_0^t r(t-x)f(x)\,dx \quad \text{for } t > 0.$$

The *Laplace–Stieltjes transform* of any distribution G, on $[0, \infty)$, see eqn (1.13), is denoted by

$$G^\dagger(s) = \int_0^\infty e^{-ts}\,dG(t),$$

and it plays a useful role because we see from eqn (12.5) that

$$R^\dagger(s) = \sum_{k=1}^{\infty}[G^\dagger(s)]^k = \frac{G^\dagger(s)}{1-G^\dagger(s)}, \tag{12.13}$$

or equivalently

$$G^{\dagger}(s) = \frac{R^{\dagger}(s)}{1 + R^{\dagger}(s)}. \tag{12.14}$$

EXAMPLE 1. Let the interarrival time be Epstein, i.e., $T \sim EP(\lambda)$, so that $\bar{G}(t) = e^{-\lambda t}$ and $G^{\dagger}(s) = \lambda/(\lambda + s)$ and hence $R^{\dagger}(s) = \lambda/s$. Upon inverting $R^{\dagger}$ we find $R(t) = \lambda t$, which we know is $\mathrm{E}[N(t)]$.

EXAMPLE 2. Let the iid interarrival time T be the sum of two different, independent Epstein variates so that

$$T \sim G \quad \text{where } \bar{G}(t) = \frac{be^{-at} - ae^{-bt}}{b - a} \qquad \text{for } t > 0; 0 < a < b.$$

The Laplace–Stieltjes transform then becomes

$$G^{\dagger}(s) = \frac{ab}{(s + a)(s + b)},$$

so that substitution into eqn (12.13) yields

$$R^{\dagger}(s) = \frac{ab}{s^2 + s(a + b)} = \frac{ab}{a + b}\left[\frac{1}{s} - \frac{1}{s + a + b}\right],$$

which upon inversion yields

$$r(t) = \frac{ab}{a + b}[1 - e^{-(a+b)t}].$$

By integration we obtain an explicit representation for the renewal function,

$$R(t) = \frac{ab}{a + b}\left[t - \frac{1 - e^{-(a+b)t}}{a + b}\right] \qquad \text{for } t > 0. \tag{12.15}$$

We can now state the important asymptotic average count.

Theorem 4. *Let the interarrival time for a renewal process be T with $\mathrm{E}T = \mu$. Then*

$$\frac{N(t)}{t} \longrightarrow \frac{1}{\mu} \quad \text{a.s. as } t \to \infty.$$

NB if $\mu = \infty$, then $\frac{1}{\mu} = 0$.

PROOF. Consider case(1) where $0 < \mu < \infty$. We have by definition

$$[S_{N(t)} \le t \le S_{N(t)+1}]$$

from which follows

$$\frac{S_{N(t)}}{N(t)} \le \frac{t}{N(t)} \le \frac{S_{N(t)+1}}{N(t)+1} \cdot \frac{N(t)+1}{N(t)}.$$

Now as $t \to \infty$ we have $N(t) \to \infty$ a.s. so that by the strong law of large numbers we conclude that $t/N(t) \to \mu$ a.s.

In case (2) when $\mu = \infty$ we truncate T at some large value and then let it approach infinity. □

This elegant but easily proved result allows us to state the *elementary renewal theorem* as

Corollary 1. *If the interarrival time has* E$T = \mu$, *then*

$$\frac{R(t)}{t} \to \frac{1}{\mu} \quad \text{as } t \to \infty.$$

There are other related results of importance.

Definition 1. An rv X, or its distribution, is said to be *lattice* (or its sample space is *periodic*) iff there exists an $\epsilon > 0$ such that $\sum_{n=0}^{\infty} \Pr[X = n\epsilon] = 1$. That is the support of X is contained in some lattice $\{0, \epsilon, 2\epsilon \ldots\}$.

We now state Blackwell's theorem.

Theorem 5. *If the interarrival time between events in a renewal process is non-lattice with mean* μ, *then for any* $\epsilon > 0$

$$\lim_{t \uparrow \infty} [R(t + \epsilon) - R(t)] = \frac{\epsilon}{\mu}.$$

A proof is given in Feller [33].

We also have the key renewal theorem of Smith, which is equivalent with the Blackwell theorem.

Theorem 6. *If the interarrival time between events in a renewal process is non-lattice with mean* μ *and if* Q *is a Riemann-integrable function on* $(0, \infty)$, *then*

$$\lim_{t \uparrow \infty} \int_0^t Q(t - x) dR(x) = \frac{1}{\mu} \int_0^{\infty} Q(t)\, dt.$$

12.3.1. Renewal Function for the Wald Distribution

There are only a few distributions where a closed form can be given for the renewal function, the Epstein distribution being the most familiar. We now consider this evaluation for the Wald distribution.

Let $G \simeq \mathcal{W}(\alpha, \beta)$ then $R(y) = \sum_{n=1}^{\infty} G^{(n*)}(y)$ where $G^{(n*)} \simeq \mathcal{W}(\alpha/\sqrt{n}, n\beta)$. Hence, asas, we find

$$G^{(n*)}(\beta y) = \int_0^y \frac{n}{\alpha t^{3/2}\sqrt{2\pi}} \exp\{-\tfrac{n}{2\alpha^2}\xi^2(n/t)\}\, dt \quad \text{for } y > 0.$$

With fixed t, we set

$$f_t(x) := \frac{x}{\sqrt{2\pi}} \exp\{-\tfrac{x}{2\alpha^2}\xi^2(x/t)\} = x\varphi[\tfrac{x-t}{\alpha\sqrt{t}}] \quad \text{for all } x > 0.$$

Thus, we can write

$$R(\beta y) = \sum_{n=1}^{\infty} \int_0^y \frac{1}{\alpha t^{3/2}} f_t(n)\, dt = \int_0^y \frac{t^{-3/2}}{\alpha} \sum_{n=1}^{\infty} f_t(n)\, dt.$$

We evaluate this sum using Euler's summation formula (Problem 1 in EXERCISE SET 16.B)

$$\sum_{n=1}^{\infty} f_t(n) = \int_0^{\infty} f_t(x)\, dx + \int_{x=0}^{\infty} B(x)\, df_t(x),$$

since clearly $f_t(0) = f_t(\infty) = 0$; here $B(x)$ is the trignometric sum given *loc. cit.* These two integrals above converge. The first integral is

$$\begin{aligned} I(t) &:= \int_0^{\infty} f_t(x)\, dx = \int_0^{\infty} x\varphi[\tfrac{x-t}{\alpha\sqrt{t}}]\, dx = \int_{z=-\sqrt{t}/\alpha}^{\infty} \alpha\sqrt{t}(t + \alpha\sqrt{t}z)\varphi(z)\, dz, \\ &= \alpha t^{3/2}\Phi(\sqrt{t}/\alpha) + \alpha^2 t \int_{z=\sqrt{t}/\alpha}^{\infty} z\varphi(z)\, dz = \alpha t^{3/2}\Phi(\sqrt{t}/\alpha) + \tfrac{\alpha^2 t}{\sqrt{2\pi}} e^{-t/2\alpha^2}. \end{aligned}$$

The second step follows considering the odd and even nature of the integrands.

By changing the variable of integration the second integral is

$$\begin{aligned} II(t) &:= \int_{x=0}^{\infty} B(x)\, df_t(x) = \int_{x=0}^{\infty} B(x)\, d\Big[x\varphi[\tfrac{x-t}{\alpha\sqrt{t}}]\Big], \\ &= \int_{z=-\sqrt{t}/\alpha}^{\infty} B(t + \alpha\sqrt{t}z)\, d[(t + \alpha\sqrt{t}z)\varphi(z)], \\ &= \int_{z=-\sqrt{t}/\alpha}^{\infty} B(t + \alpha\sqrt{t}z)[-zt - \alpha\sqrt{t}z^2 + \alpha\sqrt{t}]\varphi(z)\, dz. \end{aligned}$$

By eliminating the scale factor the renewal function can be written as

$$\begin{aligned} R(\beta y) &= \int_0^y \frac{I(t) + II(t)}{\alpha t^{3/2}}\, dt \\ &= 2\alpha^2 \int_0^{\sqrt{y}/\alpha} u\Phi(u)\, du + \frac{\alpha^2}{\sqrt{\pi}} \int_0^{y/2\alpha^2} u^{-1/2} e^{-u}\, du + \varepsilon(y, \alpha). \end{aligned} \tag{12.16}$$

Here the error term, $\int_0^y \frac{t^{-3/2}}{\alpha} II(t)\, dt$, can be expressed by the double integral

$$\begin{aligned} \varepsilon(y, \alpha) &= \int_{t=0}^{y} \frac{t^{-3/2}}{\alpha} \int_{-\sqrt{t}/\alpha}^{\infty} B(t + \alpha\sqrt{t}z)[\alpha\sqrt{t} - zt - \alpha\sqrt{t}z^2]\varphi(z)\, dz\, dt \\ &= 2\int_{u=0}^{\sqrt{y}/\alpha} \int_{z=-u}^{\infty} B[\alpha^2(u^2 + uz)] \left(\tfrac{1-z^2}{u} - z\right) \varphi(z)\, dz\, du, \end{aligned} \tag{12.17}$$

with the last expression obtained by making the change of variable $\alpha u = \sqrt{t}$ for $t > 0$.

12.3.2. Negligible Replacement Times for Units in Service

Let us consider a renewal process as the regular (instantaneous) replacement of failed components in service with nominally identical new ones. Then at any time $t > 0$ the rvs $A(t) = t - S_{N(t)}$ and $L(t) = S_{N(t)+1} - t$ are, respectively, the *age* and *life remaining* of the unit in service, with

$$\Pr[A(t) \geq x] = \bar{G}(x) \quad \text{and} \quad \Pr[L(t) > u] = \bar{G}(t+u) + \int_0^t \bar{G}(t-x+u)\, dR(x).$$

12.3.3. Tauberian Theorems for the Laplace Transform

A "Tauberian theorem" is a result that allows one to determine the asymptotic behavior of a function, or of its transform, from the other by knowing the relationships between them. For example, $A(t)$ at time $t \to \infty$ may often be found from the behavior of its Laplace transform, say $A^{\dagger}(s)$, as $s \downarrow 0$; it is named for the German mathematician A. Tauber, who first investigated such results in 1897. We restate, without proof and in our notation, a result from [120, p. 179] which is often applicable to renewal theory.

Theorem 7. *Let $A^{\dagger}(s)$ for $s > 0$ be the Laplace–Stieltjes transform of the function A on $[0, \infty)$. If there exist constants $\beta, \gamma > 0$ such that $\beta t^{\gamma} + A(t)$ is non-decreasing on $[0, \infty)$, then we have, respectively,*

$$A^{\dagger}(s) \asymp \frac{\alpha}{s^{\gamma}} \quad \textit{for} \quad \begin{cases} s \downarrow 0 \\ s \to \infty \end{cases}, \quad \textit{iff,}$$

$$A(t) \asymp \frac{\alpha t^{\gamma}}{\Gamma(\gamma+1)} \quad \textit{for} \quad \begin{cases} t \to \infty \\ t \downarrow 0 \end{cases}. \tag{12.18}$$

Exercise Set 12.A

1. Describe the behavior of the renewal function for a $\mathcal{W}(\alpha, \beta)$ distribution when α approaches zero. Show using eqn (12.17) that $\lim_{\alpha \to 0} \varepsilon(y, \alpha) = 0$.
   Hint: recall that $\varphi'(x) = -x\varphi(x)$ and $\varphi''(x) = (1 - x^2)\varphi(x)$.

2. In a manufacturing production run a certain machine tool component must be periodically replaced when sufficiently worn. Assume the service life of this component until "dull" is $T \sim G$. The high initial cost of a new component is c_0 and the cost of refurbishing a worn component in service, while initially low, increases the older it becomes. Can the renewal function be used to decide when the cost per unit of time is minimised to determine the optimal replacement period?

12.4. Shock Models with Varying Intensity

Consider a structural component sustaining a duty cycle that can be modeled by an intensity function $A(t)$ for $t > 0$ (often periodic) where $A(t) = a$ means the intensity of stress loading is a. We assume that at intensity level a,:

(a) The component is exposed to shocks having an interarrival time of Ep[$\lambda(a)$].
(b) The shock magnitudes generated are iid with cdf $G(\cdot; a)$; say $Y(a) \sim G(\cdot; a)$.

Thus, during the duty cycle when the loading intensity is a then $N_a(t)$, a Poisson process, counts the shocks during any time interval of length t. Let $Y_i(a) \sim G(\cdot; a)$ for $i = 1, \ldots, N_a(t)$ be the magnitude of shocks during the period of length $t > 0$ at a given intensity a. Let $X_a(t)$ be the maximum shock during the period of time of length t. Then

$$\begin{aligned}\Pr[X_a(t) > y] &= \sum_{n=0}^{\infty} \Pr[N_a(t) = n] \cdot \Pr[\bigwedge_{i=1}^{n} Y_i(a) > y]\\ &= \sum_{n=0}^{\infty} \frac{e^{-t\lambda(a)}(t\lambda(a))^n}{n!} \{1 - [G(y : a)]^n\} = 1 - e^{-t\lambda(a)\bar{G}(y;a)}.\end{aligned}$$

Thus, we have the

Lemma 2. *Under the assumptions* (a) *and* (b) *above, the waiting time between shocks of magnitude exceeding y, when the component is loaded at the constant intensity level a, is* Ep[$\lambda(a)\bar{G}(y; a)$].

From this we have

Theorem 8. *If Lemma 2 above holds and the duty cycle follows an intensity function* $A(\cdot)$ *governing shock frequency, then the maximum shock during* $(0, t]$ *the component will encounter, call it* $X^\star(t)$, *has cdf given by*

$$\Pr[X^\star(t) \le y] = \exp\left\{-\int_0^t \lambda[A(s)]\bar{G}[y; A(s)]\, ds\right\}.$$

Proof. Consider the interval $(0, t]$ subdivided into the intervals $(t_{i-1}, t_i]$ for $i = 1, \ldots, n+1$ with $t_0 = 0$ and $t_{n+1} = t$, and that during the ith time interval the activity level is $A(t_i^{'})$ for some $t_{i-1} \le t_i^{'} \le t_i$. Let $X(\Delta t_i)$ denote the maximum gust during $(t_{i-1}, t_i]$. These are independent for $i = 1, \ldots, n+1$. Thus, by the lemma

$$\Pr[X(\Delta t_i) \le y] = \exp\{-\int_{t_{i-1}}^{t_i} \lambda[A(s)]\bar{G}[y; A(s)]\, ds\}.$$

By the independence we have

$$\Pr[\max_{i=1}^{n} X(\Delta t_i) \le y] = \exp\left\{-\sum_{i=1}^{n+1} \int_{t_{i-1}}^{t_i} \lambda[A(s)]\bar{G}[y; A(s)]\, ds\right\}.$$

Thus, by duality the waiting time T until a shock of magnitude exceeding some critical threshhold, say ω, satisfies $[T \le t] = [X^\star(t) \ge \omega]$ so we have the result. □

Exercise Set 12.B

1. If the assumptions of Theorem 9 must hold, to which of the following situations might it apply?
   (a) To the gust-loading on a wing station of an airframe flying at an altitude a during its ground-air-ground cycle?
   (b) To the calculation of seasonal salt (de-icer) contamination to the re-bar (bars of steel reinforcement used in concrete) which may be exposed to sufficient loads to cause microscopic cracks in the tensile portions of prestressed concrete girders?
   (c) To the diurnal fluctuation of temperature, humidity and UV-radiation on a polymer coating?
2. Write out and carefully justify all the assumptions made in the above choices and any other application you might consider.

12.4.1. The Marshall–Olkin Distribution

The model proposed by Marshall and Olkin (see [70]) supposes that a coherent system has two components that are susceptible to shocks which in service encounters exactly three types of independent shocks that cause failure of the components.

Model I: Suppose that:

(a) A shock of type i causes failure only of component i, and the waiting time between such shocks is $U_i \sim \text{Ep}(\lambda_i)$ for $i = 1, 2$.
(b) A shock of type 3 causes failure of both components, and the waiting time between such shocks is $U_{1,2} \sim \text{Ep}(\lambda_{1,2})$

Thus, the life length of component 1 is $T_1 = U_1 \wedge U_{1,2}$ while the random life of component 2 is $T_2 = U_2 \wedge U_{1,2}$. Hence the joint sdf is, for $t_1 > 0, t_2 > 0$, given by

$$\bar{F}(t_1, t_2) = \Pr[T_1 > t_1, T_2 > t_2] = e^{-\lambda_1 t_1 - \lambda_2 t_2 - \lambda_{1,2} \max(t_1, t_2)}. \tag{12.19}$$

This is the *Marshall-Olkin bivariate distribition* and it has marginal distributions which are Epstein(exponential), namely, for $i = 1, 2$

$$\bar{F}_i(t) = \Pr[T_i > t] = e^{-(\lambda_i + \lambda_{1,2})t} \quad \text{for } t > 0.$$

NB that T_1, T_2 are associated rvs.

Model II: Suppose that the independent shocks that occur in Model I only cause failure with a given probability.

(a) A shock from source i causes failure of component i with probability q_i for $i = 1, 2$.
(b) A shock from source 3 causes failure of both components with probability $q_{1,1}$; failure of component 1 only with probability $q_{1,0}$; failure of component 2 only with probability $q_{0,1}$; of neither component with probability $q_{0,0}$, where $q_{1,1} + q_{1,0} + q_{0,1} + q_{0,0} = 1$.

Then one finds asas, see [7], p 136, that for $0 \leq t_1 \leq t_2$

$$\ln\{\Pr[T_1 > t_1, T_2 > t_2]\} = -t_1[\lambda_1 q_1 + \lambda_{1,2} q_{1,0}] - t_2[\lambda_2 q_2 + \lambda_{1,2}(1 - q_{0,0} - q_{1,0})],$$

and by symmetry for $0 \leq t_2 \leq t_1$

$$\ln\{\Pr[T_1 > t_1, T_2 > t_2]\} = -t_1[\lambda_1 q_1 + \lambda_{1,2}(1 - q_{0,0} - q_{0,1})] - t_2[\lambda_2 q_2 + \lambda_{1,2} q_{0,1}].$$

Combining the two we obtain the Marshall–Olkin bivariate sdf

$$-\ln\{\bar{F}(t_1, t_2)\} = \nu_1 t_1 + \nu_2 t_2 + \nu_{1,2} \max(t_1, t_2),$$

where $\nu_1 = \lambda_1 q_1 + \lambda_{1,2} q_{1,0}$, $\nu 2 = \lambda_2 q_2 + \lambda_{1,2} q_{0,1}$, and $\nu_{1,2} = \lambda_{1,2} q_{1,1}$.

Of course this immediately generalizes to a multivariate distribution with all the marginals having the Marshall–Olkin distribution;

$$-\ln[\bar{F}(t_1, \ldots, t_n)] = \sum_{i=1}^{n} \lambda_i t_i + \sum_{i<j} \lambda_{i,j} \max(t_i, t_j) + \sum_{i<j<k} \lambda_{i,j,k} \max(t_i, t_j, t_k) + \cdots + \lambda_{1,2,\cdots,n} \max_{i=1}^{n} t_i. \quad (12.20)$$

It is one of the few multivariate distributions in which all the marginal and conditional distributions can be determined.

Exercise Set 12.C

1. What models for risk would you consider if you were to provide insurance for a group of experienced mountaineers who were making an ascent of Mt. Everest but taking with them several inexperienced ‘thrill seekers’ (for \$65K each) but guaranteeing that all would make it to the summit?
2. What information would you need to know to answer (1)? What data could you reasonably obtain to estimate parametrs and what would you have to supply with Bayesian methods, i.e., ‘climbing experience’ (guesses)?
3. Who would be the loser if unisex premiums for individuals were applied separately to a man and his wife?
4. Do you think the M-O distribution would apply to flood and earthquake insurance? Why?

12.4.2. The Bivariate Poisson

From the Marshall–Olkin distribution one can obtain another related discrete multivariate distribution. Let us assume for Shock Model-II above with two components that upon failure of any component it is instantly replaced with a new one. Let $N_i(t)$ count the number of replacements in component i by time $t > 0$ for $i = 1, 2$. Let us consider the Poisson process $Z(t) = Z_1(t) + Z_2(t) + Z_{1,2}(t)$ with parameter $\lambda = \lambda_1 + \lambda_2 + \lambda_{1,2}$ that governs the sequence of shocks of the three types. Given that a shock has occurred in the $Z(t)$ process, the probability is

$p_{1,1} = (\lambda_{1,2}/\lambda)q_{1,1}$ that both components are replaced;
$p_{1,0} = (\lambda_1/\lambda)q_1 + (\lambda_{1,2}/\lambda)q_{1,0}$ that only component 1 is replaced;
$p_{0,1} = (\lambda_2/\lambda)q_2 + (\lambda_{1,2}/\lambda)q_{0,1}$ that only component 2 is replaced;
$p_{0,0} = (\lambda_1/\lambda)(1 - q_1) + (\lambda_2/\lambda)(1 - q_2) + (\lambda_{1,2}/\lambda)q_{0,0}$ that neither is replaced.

Check that $p_{1,1} + p_{1,0} + p_{0,1} + p_{0,0} = 1$.

By the calculus of probabilities one obtains the corresponding *bivariate Poisson process*, viz.,

$$\begin{aligned}\Pr[N_1(t) = n_1, N_2(t) = n_2] \\ = e^{-\lambda t(p_{1,1}+p_{1,0}+p_{0,1})} \sum_{j=0}^{n_1 \wedge n_2} \frac{(\lambda p_{1,1})^j (\lambda p_{1,0})^{n_1-j} (\lambda p_{0,1})^{n_2-j}}{j!(n_1 - j)!(n_2 - j)!}.\end{aligned} \tag{12.21}$$

12.5. Stationary Renewal Processes

There are other inequalities for the renewal function that involue the NBU and NWU classes for component life.

Definition 2. The sequence of independent variates $\{\hat{T}_i : i = 1, 2, \ldots\}$ is a *stationary renewal process* when all $\{\hat{T}_i : i = 2, 3, \ldots\}$ are iid with mean μ and cdf G but $\hat{T}_1 \sim G_1$ where

$$G_1(t) = \frac{1}{\mu} \int_0^t \bar{G}(x)\, dx. \tag{12.22}$$

Thus, if we have any renewal process $\{T_i : i = 1, 2, \ldots\}$ with iid interarrival times having cdf G and $ET_i = \mu$ we can define the *associated renewal process* as $\{\hat{T}_i = T_{i-1} : i = 2, 3, \ldots\}$ where $\hat{T}_1 \sim G_1$, with G_1 defined as in eqn (12.22) where the counting process satifies

$$[\hat{N}(t) = n] = \left[\sum_{i=1}^{n} \hat{T}_i \le t, \sum_{i=1}^{n+1} \hat{T}_i > t \right].$$

The reason it is called *stationary* is as follows:

Theorem 9. *Any stationary renewal process satisfies*

$$E\hat{N}(t) := R_1(t) = \frac{t}{\mu} \quad \textit{for all} \quad t > 0.$$

Definition 3. The rv $X > 0$ is NBUE (new better than used in expectation) iff

$$\mathrm{E}[X] \geq \mathrm{E}[X - t | X > t] = \frac{\int_t^\infty \bar{F}(x)\,dx}{\bar{F}(t)} \quad \text{for all} \quad t > 0.$$

An rv $X > 0$ is NWUE (new worse than used in expectation) iff the inequality above is reversed.

Theorem 10. *Let* $0 < T_i \sim F$ *for* $i = 1, 2, \ldots$ *be the interarrival times for a renewal process for which* $\mathrm{E}T_1 = \mu < \infty$. *Then*

$$R(t) \geq \frac{t}{\mu} - 1 \qquad \textit{for } t \geq 0. \tag{12.23}$$

If F is NBUE(NWUE) *then, respectively,*

$$R(t) \leq (\geq) \frac{t}{\mu} \qquad \textit{for} \quad t \geq 0. \tag{12.24}$$

Proof. To prove eqn (12.23) we note that for all $t > 0$ we have $S_{N(t)} \leq t < S_{N(t)+1}$. Thus, we have

$$0 \leq \mathrm{E}\left(S_{N(t)+1} - t\right) = \mu[R(t) + 1] - t$$

so this case is true since by Wald's lemma

$$\mathrm{E}[S_{N(t)+1}] = \sum_{n=1}^{\infty} \mathrm{E}[S_{n+1} | N(t) = n] \cdot \Pr[N(t) = n] = \sum_{n=0}^{\infty} (n+1)\mu \Pr[N(t) = n].$$

To prove eqn (12.24), assume F is NBUE, then $F_1(t) := \frac{1}{\mu} \int_0^t \bar{F}(x)\,dx \geq F(t)$. So

$$\begin{aligned}\mathrm{E}N(t) = \frac{t}{\mu} &= \sum_{n=1}^{\infty} \int_0^t F^{(n-1)*}(t - x)\,dF_1(x) \geq \sum_{n=1}^{\infty} \int_0^t F^{(n-1)*}(t - x)\,dF(x),\\ &= R(t),\end{aligned}$$

completing the proof. □

Exercise Set 12.D

1. Prove the equivalence of the three definitions of a Poisson process.
2. Show that the fatigue-life distribution can be obtained as the limiting distribution if we let the critical damage level ω become large and the mean incremental damage per shock ν become small at the same rate so that $\omega \cdot \nu$ remains fixed in eqn (12.10).
3. Find the renewal function when the interarrival time is the sum of three different Epstein variates.

4. Prove that if G_1 is defined by eqn (12.22) with $G_n = G$ for $n = 2, 3, \ldots$ that $\mathrm{E}\hat{N}(t) = t/\mu$. What is $\mathrm{Var}\hat{N}(t) = t/\mu$?

5. If $\hat{X}_1, \hat{X}_2, \cdots$ is a stationary renewal process with $\mathrm{E}\hat{X}_2 = \mu$, find $\mathrm{E}\hat{X}_1$ and show that $2\mathrm{E}\hat{X}_1 \geq \mathrm{E}\hat{X}_2$.

6. Show that if $\hat{X}_1 \sim \hat{X}_2$ for a stationary renewal process, then their common distribution is Epstein.

7. Let R be the renewal function for a stationary renewal process. Then $R^\dagger = F^\dagger/(1 - F^\dagger)$. Use this to show that $R_1(t) = t/\mu$, where $\mu = \int_0^\infty F(x)\,dx$.

8. Suppose a renewal process with interarrival time distribution G began at time $t = -\infty$. Show that if observation begins at time $t = 0$ the waiting time until observing the first renewal has the distribution $\frac{1}{\mu}\int_0^t \bar{G}(x)\,dx$.

12.6. The Miner–Palmgren Rule and Additive Damage

> A successful rule in engineering practice is often merely a mathematical theorem the hypothesis of which is being ignored because it is either unknown or it has been forgotten.
>
> Z.W. Birnbaum

Near the end of WWII a Douglas engineer published a paper [79, 1945] describing a method of calculating the fatigue life of aluminum airframe components which were being subjected to a spectrum of loads in service. The rule he proposed was the same as one previously published by Palmgren, [88, 1922], a Swedish engineer, for calculating the fatigue life of steel ball-bearings.

It was in fact the same solution as to the Algebra I problem: using a push lawnmower John can, by himself working steadily, cut the lawn in two and a half hours; his brother Dale can, under the same conditions, cut it in three hours while Stan takes three and a quarter hours. Without knowing how big the lawn is or how many sq-ft each boy can cut in one hour calculate how long it would take them if they could all mow simultaneously? The answer, in hours, is the reciprocal of $\frac{2}{5} + \frac{1}{3} + \frac{4}{13}$, obtained by a trivial application of the harmonic mean, but the result is not trivial.

To see this same relationship holds for a metallic component sustaining duty cycles, suppose that during each duty cycle a load of category i occurs k_i times, each load causing damage of μ_i, and there are $i = 1, \ldots, \ell$ load categories. If the component will fail when a critical amount of damage has occurred, how many duty cycles will it survive? How many loads? These questions seem at first to be impossible to answer. But suppose we can estimate how many loads of each type i it takes to fail the component, say the number of loads to failure is ν_i for each $i = 1, \ldots, \ell$. Now an answer can be estimated!

Let us suppose the unknown critical cumulative damage level is ω. Then by the linear cumulative damage hypothesis we must have $\nu_i = \omega/\mu_i$ and thus the number of duty cycles to generate the damage ω, using all types of load categories, will be

$$\nu = \frac{\omega}{\sum_{i=1}^{\ell} k_i \mu_i} = \frac{\omega}{\sum_{i=1}^{\ell} k_i \omega / \nu_i} = \frac{1}{\sum_{i=1}^{\ell} k_i / \nu_i}.$$

Hence we have an answer based only on knowing only k_i and ν_i for $i = 1, \ldots, \ell$, which are the quantities that can be estimated, under the linear damage accumulation, by testing a sample of the relatively inexpensive specimens of the component material at known (increased) stress levels to determine the life ν_i. How many loads of all categories can be sustained to accumulate the unknown critical damage level? Since we can encounter exactly $\sum_{i=1}^{\ell} k_i$ loads per duty cycle the answer is

$$\nu \sum_{i=1}^{\ell} k_i = \left[\sum_{i=1}^{\ell} \frac{k_i}{\nu_i \sum_{j=1}^{\ell} k_j} \right]^{-1} = \frac{1}{\sum_{i=1}^{\ell} \frac{\lambda_i}{\nu_i}}, \tag{12.25}$$

where $\lambda_i = k_i / \sum_{j=1}^{\ell} k_j$ denotes the relative frequency of occurence of the ith type of load per duty cycle.

Of course this rule to compute fatigue life under cycles of various stresses was only an analogy, made by deterministic presupposition, and other more specific rules were proposed during the next decade. In the early 1960s the Air Force sponsored an exhaustive series of fatigue tests on airframe components using more than a score of then extant "fatigue rules" to predict lifetimes under representative stress regimes. None was sufficiently accurate but "Miner's rule was the best on average." This would seem to mean that Miner's rule is only true in expectation.

12.6.1. Miner's Rule as an Expectation

For each fixed $j = 1, \ldots \ell$ let $X_{1,j}, X_{2,j}, \ldots$ be an iid sequence of NBUE rv's representing incremental damage under the jth type of load. Let $W \sim G$, independent of all $\{X_{i,j}\}$, denote the critical damage level to the structure. Let $S_{n,j} = \sum_{i=1}^{n} X_{i,j}$ be the total cumulative damage due to the jth category of load in the spectrum. Define failure under the jth type of load as

$$[N_j(W) = n] = [S_{n-1,j} \leq W, S_{n,j} > W].$$

Setting $\omega = \mathrm{E}[W]$, $\mathrm{E}N_j(W) = \nu_j$, $\mathrm{E}[X_{i,j}] = \mu_j$, we obtain, by the renewal theorem, the inequality

$$\frac{\omega}{\mu_j} - 1 \leq \nu_j \leq \frac{\omega}{\mu_j} \qquad \text{for } j = 1, \ldots, \ell. \tag{12.26}$$

Let $T_i = \sum_{j=1}^{\ell} X_{i,j}$ for $i = 1, 2, \ldots$ be the total damage due to the ith repetition of the spectrum. Set $\mu = \mathrm{E}T_i = \sum_{j=1}^{\ell} \mu_j$, then by duality we express the event

$$[N(W) = n] = \left[\sum_{j=1}^{n-1} T_j \leq W, \sum_{j=1}^{n} T_j > W \right].$$

Again applying the renewal theorem and conditional expectation we have

$$\frac{\omega}{\mu} - 1 \le \nu \le \frac{\omega}{\mu}, \qquad \text{setting } \nu = \mathrm{E}[N(W)]. \tag{12.27}$$

But from eqn (12.26) we find that

$$\frac{\omega}{\nu_j + 1} \le \mu_j \le \frac{\omega}{\nu_j} \qquad \text{for each } j = 1, \ldots, \ell.$$

Thus from eqn (12.27) we have asas

$$\frac{1}{\sum_{j=1}^{\ell} \frac{k_i}{\nu_i}} - 1 \le \nu = \mathrm{E}[N(W)] \le \frac{1}{\sum_{j=1}^{\ell} \frac{k_i}{\nu_i+1}}, \tag{12.28}$$

where the k_i represents the number of repetitions of the ith load in the spectrum. From this we conclude that under the assumptions made to obtain Miner's rule it will always give a value *of the expectation* lying between these two bounds which are, when the ν_i are large, very close. But note, one would also like to know the distribution of this estimate, not just its expectation.

Exercise Set 12.E

1. If $X \sim \mathcal{W}(\alpha, \beta)$, show that X is *infinitly divisible*, i.e., for any whole number n we have $X = \sum_{i=1}^{n} X_i^{(\frac{1}{n})}$ where $X_i^{(\frac{1}{n})} \sim \mathcal{W}(\sqrt{n}\alpha, \beta/n)$ are iid for $i = 1, \cdots, n$. Thus, $\mathrm{E}X = \beta$, $\mathrm{Var}(X) = (\alpha\beta)^2$, while $\mathrm{E}X_i^{(\frac{1}{n})} = \frac{\beta}{n}$ and $\mathrm{Var}(X_i^{(\frac{1}{n})}) = \frac{(\alpha\beta)^2}{n}$ for $i = 1, \cdots, n$.
2. Let $Y = X_1 + X_2 + X_3$ be the fatigue life of a component partitioned into three stages: $X_1 \sim \mathcal{W}(\alpha_1, \beta_1)$ is the time until crack initiation, $X_2 \sim \mathcal{FL}(\alpha_2, \beta_2)$ is the time after crack initiation until criticality is reached, $X_3 \sim \mathcal{W}(\alpha_3, \beta_3)$ is the time from criticality until component failure; but all are consonant with $\alpha_1^2\beta_1 = \alpha_2^2\beta_2 = \alpha_3^2\beta_3$.

   Prove that $Y \sim \mathcal{FL}(\alpha_{(-2)}, \beta_+)$ where $\alpha_{(-2)} = \left(\sum_{i=1}^{3} \alpha_i^{-2}\right)^{-\frac{1}{2}}$ and $\beta_+ = \sum_{i=1}^{3} \beta_i$.
   (a) Using the three-stage decomposition above with $\beta_1 = 10,000$, $\beta_2 = 100$, $\beta_3 = 1$, and $\alpha_1 = .02$, $\alpha_2 = .2$, $\alpha_3 = 2$, show that $\beta_+ = 10101$, yet $\alpha_0 \doteq \alpha_1$.
   (b) What is the reliability of a unit with low-stress but high-cycle service life for, say 24 hours at 15,000 rpm, when in duty cycles $\beta_1 = 2 \times 10^7$, $\beta_2 = 2 \times 10^5$, $\beta_3 = 20$, and $\alpha_1 = .01$, $\alpha_2 = .1$, $\alpha_3 = 10$.
   (c) What would you advise for high-strength materials having a critical crack-size which is microscopic in dimension so that fatigue crack intiation cannot be determined by field inspection?

12.6.2. How Applicable Is This Theory?

Deterioration occurs everywhere in Nature but in many instances damage increments, over the length of human life, can be assumed to have a stationary distribution. A current source of cumulative damage is the contaminants from industrialized nations which, dispersed by the winds, settle everywhere; then by various pathways they accumulate in the flesh of animals at the top of the food chain. This is noticed in the amount of potentially toxic methyl-mercury found in fish.

These levels of mercury in fish taken from US rivers and streams, ranging from 0.16ppm to 0.63ppm (when the EPA's maximum safety level is 0.3ppm), resulted in recommendations that women of child-bearing age should eat less than 8 oz. per month while children younger than 6 should eat no more than 4 oz. every two months. The pervasiveness of this problem is shown by both PCBs (polychlorinated biphenyls) and mercury appearing in the breast milk of Artic women who consume whale meat and other animals as their main protein source, see [27].

Cumulative damage affects the infrastructure of mankind; it can occur in metals from galling, abrasion, oxidation, fatigue or stress-corrosion. It may occur in the spalling of structural members due to the internal corrosion of steel re-bars from contaminated sand or perhaps from chemical additives used to hasten concrete curing. Highway surfaces deteriorate both from the traffic load and the corrosion of structural bridge supports from de-icing compounds.

Further, economic concern arises from the failure of protective polymer coatings from degradation due both to UV-dosage plus temperature/humidity fluctuation. Cumulative damage arises frequently in human health. This includes dental x-rays, the accumulation of PCBs and heavy metals (mercury in fish and lead or hexavalent chromium in water) as well as from hormone disruption due to the ingestion and retention of small amounts, of toxic pesticides in bodily tissue. Lastly, genetic damage in humans may, in part, be a consequence of synthetic chemical pollution, see [103].

12.7. Other Cumulative Damage Processes

We now wish to consider damage processes, in particular, situations where the passage of time (time of unit in service) may not be directly related to the amount of incremental damage sustained. This occurs when environmental conditions, extrinsic to service time such as humidity, ultraviolet exposure or elevated temperature ranges, contribute so directly to damage they must be accounted for. The same distinction can arise between "service hours" containing only intermittant "duty cycles."

12.7.1. Deterioration of Polymer Coatings

Consider a polymer coating as a film held together by a network of long, interlinked polymer chains (think of chicken wire). Call the points of connection "nodules" and determine the area, say a, of the nodule wherein a photon strike, of sufficient energy, will sever the chain. Assume there are ν such "distinquishable" nodules per square millimeter of exposed surface of the coating. Thus, $m = \nu a$ is the number of a-unit areas suseptible to photon strike (also the fraction of area) within each square millimeter on the exposed coating surface.

Assume the irradiance of photons is η per unit of time per mm^2, so the arrival rate of nodule-striking photons is $\lambda = m\eta$. Thus, the number of such nodule-damaging photons during a period of length t, say $N(t)$, has a Poisson distribution,

$$\mathrm{P}[N(t) = n] = \frac{e^{-\lambda t}(\lambda t)^n}{n!} \quad \text{for } n = 0, 1, 2, \cdots.$$

If $N(t) = n$ damaging photons have arrived, let K_n be the random number of the distinct nodules that will have been severed, and thus over a period of length t the random number of different severed nodules will be denoted by $K_N(t)$. From the theorem of total probability and eqn (3.22) we have,

$$\begin{aligned}
\mathrm{P}[K_N(t) = k] &= \sum_{n=0}^{\infty} \mathrm{P}[N(t) = n]\mathrm{P}[K_n = n] \\
&= \binom{m}{k} \sum_{j=0}^{k} (-1)^{k-j} \binom{k}{j} \sum_{n=0}^{\infty} \left(\frac{j}{m}\right)^n \frac{e^{-\lambda t}(\lambda t)^n}{n!} \\
&= e^{-\lambda t} \binom{m}{k} \sum_{j=0}^{k} (-1)^{k-j} \binom{k}{j} e^{j\lambda t/m} = e^{-\lambda t} \binom{m}{k} \left(e^{\lambda t/m} - 1\right)^k.
\end{aligned}$$

Let $p = 1 - q$ be the fraction of m distinquished nodules at which the total number of scissions becomes critical, then

$$\begin{aligned}
\mathrm{P}[K_N(t) \leq pm] &= \sum_{k=0}^{pm} \binom{m}{k} e^{-(m-k)\lambda t/m} \left(1 - e^{-\lambda t/m}\right)^k \\
&= \sum_{j=mq}^{mp} \binom{m}{k} e^{-j\lambda t/m} \left(1 - e^{-\lambda t/m}\right)^{m-j} \\
&= \frac{1}{B(mq, mp+1)} \int_0^{e^{-\lambda t/m}} x^{mq-1}(1-x)^{mp}\, dx,
\end{aligned}$$

where $B(a, b) := \frac{\Gamma(a)\Gamma(b)}{\Gamma(a+b)}$ is the Beta function of classical analysis.

Thus, if $T^\star$ is the random waiting time until the critical number of chain scissions have occurred on a square millimeter of coating surface, then by duality, it has the distribution, since $\lambda = mq$

$$\mathrm{P}[T^\star \geq t] = \frac{1}{B(mq, mp+1)} \int_0^{e^{-\eta t}} x^{mq-1}(1-x)^{mp}\, dx, \tag{12.29}$$

12.7.2. Varying Duty Cycles

Consider a renewal process with iid "interoccurrence" times $T_1, T_2, \ldots$, where,

$$T_i \sim G \quad \text{for } i = 1, 2 \cdots \quad \text{with} \quad S_n = \sum_{i=1}^{n} T_i \sim G^{n*} \quad \text{for } n \geq 1 \tag{12.30}$$

Here these times represent the varying duration of the duty cycles from the commencement of one, including the service and maintenance following, until the start of the next; of course sometimes they may be merely diurnal cycles. Let $N(t)$ be the number of duty cycles completed on or before time $t > 0$. Then by the duality theorem

$$P[N(t) \geq n] = G^{n*}(t) \quad \text{for } t > 0, \text{ and } n = 1, 2, \cdots,$$

where G^{n*} is the n-fold convolution of G, with renewal function given by $G^{\natural}(t) := EN(t) = \sum_{k=1}^{\infty} G^{k*}(t)$.

The shocks, radiation, humidity causing stress, or whatever type of damage is being sustained, occurs during the duty cycles but its distribution may not be stationary over time. After the start of the ith duty cycle and before the $(i+1)st$ duty cycle there occurs a random amount of damage, say X_i for $i = 1, 2, \cdots$. Thus, an amount of damage X_n accrues after S_n and before S_{n+1}, and we let $Y_n = \sum_{i=1}^{n} X_i \sim F_n$ for $n = 1, 2, \cdots$ be the total accumulated damage at time S_n, after the completion of the nth duty cyle. NB the X_i need not be iid but can be stochastically increasing so as to model an increase in wear during the life of the system.

Let $\omega > 0$ be the fixed and known critical damage threshold, i.e., failure occurs when it is first exceeded. Define W as the time at which the service life is ended because the critical damage is first exceeded at that time during some duty cycle. Let $W \sim H$ then its survival distribution, as a function of service-life, is

$$\begin{aligned}\bar{H}(t) &:= P[W > t] = P[Y_{N(t)} \leq \omega] = \sum_{n=1}^{\infty} P[N(t) = n, Y_n \leq \omega], \\ &= \sum_{n=1}^{\infty} F_n(\omega) P[N(t) = n] = \sum_{n=1}^{\infty} F_n(\omega)[G^{n*}(t) - G^{(n+1)*}(t)]. \qquad (12.31)\end{aligned}$$

Define the *fractional renewal function*, with ω given, as

$$K(t) := \sum_{n=1}^{\infty} F_n(w) G^{n*}(t); \quad \text{then } \bar{H}(t) = [\bar{G} * K](t), \qquad (12.32)$$

which can be seen, utilizing integration by parts, from

$$\begin{aligned}\bar{H}(t) &= K(t) - \sum_{n=1}^{\infty} F_n(w) \int_0^t G^{n*}(t-y)\, dG(y) \\ &= K(t) - \int_0^t K(t-y)\, dG(y) = \int_0^t \bar{G}(t-y)\, dK(y).\end{aligned}$$

Exercise Set 12.F

1. When any interruption due to malfunction during a duty cycle is to be avoided, as in airline service or a military mission, it becomes important to calculate life in terms of duty cycles and to schedule maintence appropriately. Let $N^{\natural}$ be the number of the duty cycle during which failure occurs from damage exceeding ω.

(a) Argue that we can obtain the distribution of $N^\natural$ by using $[N^\natural \geq n] = [Y_n \leq \omega]$.
(b) Show that $\mathrm{E}N^\natural = \sum_{n=1}^{\infty} F_n(\omega)$.
(c) Prove directly that the mean service life, $\mathrm{E}W$, satisfies Wald's lemma, viz.,

$$\mathrm{E}W = \int_{y=0}^{\infty} \int_{t=y}^{\infty} \bar{G}(t-y)\,dt\,d\mathrm{K}(y) = \mathrm{E}T_1 \cdot \mathrm{E}N^\natural.$$

2. Show that the expected remaining life for $W \sim H$, at given age t, where H is defined in eqn (12.32) is the ratio $\int_t^\infty \bar{H}(x)\,dx \div \bar{H}(t)$

3. Show $0 < X \sim F$ and $\mathrm{E}X^a < \infty$ implies $\mathrm{E}X^a = \int_0^\infty \bar{F}(t)\,dt^a$ for any $a \in \Re$, and use it to prove that if W has the distribution in eqn(12.32) then $\mathrm{E}\binom{W}{k} = \sum_{n=1}^{\infty} \mathrm{E}\left[\binom{S_{n+1}}{k} - \binom{S_n}{k}\right]$, where S_n is given in eqn(12.30) and then calculate $\mathrm{E}W^2$ and $\mathrm{Var}W$.

4. Suppose for a system produced having service life $T^\natural$, under a fixed duty cycle, that after service time t_0 a change to more severe duty is contemplated. How could you show what this more severe duty would do to the reliability? How could you show what this alteration of 'mission' would do to the expected life remaining by calculating what percentage the expected remaining life is decreased? Remember economics is important.

12.8. When Linear Cumulative Damage Fails

It was discovered, in the British DeHaviland "Comet" disasters, that the test overloads imposed during their fatigue investigations produced a longer estimated fatigue life than was encountered in service. Later investigations showed that even the same load spectra, when reordered within a duty cycle, can produce very different fatigue lives. The history of civilization, not just NASA, is replete with managers who, when making operational decisions, disregard the caution from groups of technical experts. This is because they know the great expense of added safety but they ignore the enormous cost of failure because its probability is unknown. This source of unreliability occurs more frequently than from situations when theory itself is deficient.

The incremental damage following each load fluctuation is not always proportional to the incremental stress intensity during that fluctuation! This fact vitiates the universal application of the Miner–Palmgren rule not only in the fatigue of materials but in other circumstances too. Damage accumulation, until ultimate failure under variable amplitude loads, is determined by the confluence of several mechanisms, and under certain situations the use of the linear damage rule can lead to erroneous predictions for the safe (or expected) life.

It may be counterintuitive but the M-P rule predicts too great a degree of fatigue damage from higher stress amplitudes during cyclic stresses. It is known that tensile overloads applied to metalic materials, with built in stress concentrations, temporarily reduce the rate of fatigue crack growth while high compressive stresses do the opposite. This unknown effect was the cause of the Comet disasters; for the metallurgical reasons see Suresh [109]. For a graphical depictions of the magnitude of this effect see Bogdanoff and Kozin [9].

It is the exceptional extreme loads that influence the damage rate, from subsequent loads, after a crack has initiated. Thus, the order of their occurrence during the duty cycle is important. In such circumstances it is not always sufficient to model the accumulative of damage only as a function of stress amplitude or the number of duty cycles. When a more accurate model must be developed to predict service life include situations when the load-order becomes significant in affecting the resulting strain on the system/structure or when the influence of an ancillary variable, such as environmental variation in humidity or temperature serves as a catalyst for chemical degradation which increases the rate of deterioration/damage. Using only life data obtained from sinusoidal fatigue testing at different maximum stresses, with the M-P rule, may be dangerous in such applications.

12.8.1. Load-Order Effects in Crack Propagation

We now consider a problem concerning the fatigue lives of metallic components which are subjected to the same spectra of stresses but the expected order of of load occurrence varies in duty cycles within different service environments. What we propose in this section is the construction of a simple damage function sensitive to load order which is applicable to fatigue crack propagation under high stresses. A quantitative, metallurgical investigation of load-order consequence has been done by McMillan and Pelloux [74], in which they measure, using an electron microscope, the rate of propagation of a fatigue crack under separate regimes. They exhibit four different phenomena which they classify as crack-arrest, crack deceleration, crack-acceleration, and crack-jump. However, the formulation of a model which accurately reflect these phenomena, using extensive data from a particular alloy, which give sufficient accuracy for fatigue life prediction is, not surprisingly, considered proprietary information.

Under programmed repetitions of the same load oscillations, say $\lambda_1, \ldots, \lambda_m$ representing stress fluctuation with periodicity $\lambda_{jm+i} := \lambda_i$ for $1 \le i \le m$ and $j \ge 1$, let us assume the incremental crack growth for the kth load oscillation, say Δd_k, is the proportionality

$$\Delta d_k \propto \left[\vartheta(\lambda_k) - \textstyle\sum_{j=0}^{k} \varsigma^j \varphi(\lambda_{k-j})\right]^+ \qquad \text{for } k = 1, 2, \cdots, \tag{12.33}$$

where we write $x^+ = \max\{x, 0\}$ for $x \in \Re$.

Here $\vartheta(\cdot)$ and $\varphi(\cdot)$ are material-specific functionals and $0 < \varsigma < 1$ is the parameter of influence from the preceding strain. A tensile stress overload causes plastic deformation at the crack tip "dulling" it, which inhibits advance while compressive stress "sharpens" the tip, which facilitates crack advance during the next load fluctuation.

Since fatigue damage is independent of the frequency of oscillation (unless the vibration is so rapid the material heats) we can represent a sequence of load oscillations as a set of points, say,

$$(l_0, u_0), (\ell_1, u_1), (\ell_2, u_2), (\ell_3, u_3), \cdots \quad \text{with} \quad \ell_i < u_i > \ell_{i+1} \text{ for } i = 1, 2, \cdots,$$

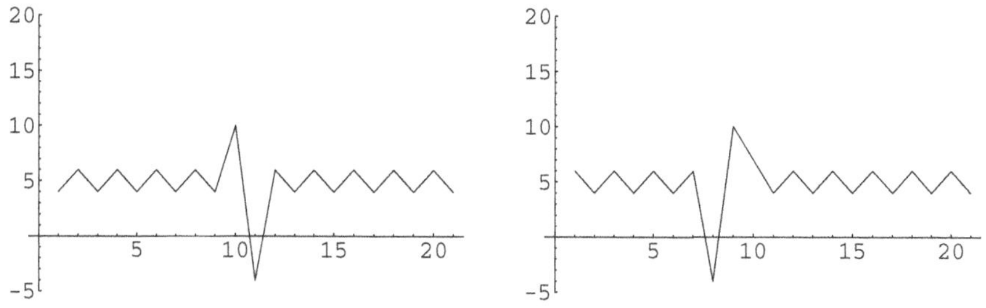

Figure 12.1. Programmed loads with damages showing load-order effect.

where $\sup \lambda_i = u_i$ and $\inf \lambda_i = \ell_i$. (Compressive stresses here take negative values.) A common model for incremental fatigue crack growth in a metallic alloy is $\Delta d \propto \max(\lambda)[\max(\lambda) - \min(\lambda)]^a$, for some $a > 0$ with a a characteristic of the material. However this model exhibits neither the lag nor acceleration effects cited above. Another model which does, and might be suitable for damage per duty cycle in some cases, is

$$\vartheta(\lambda_k) = u_k(u_k - \ell_k)^a \quad \text{with} \quad \varphi(\lambda_k) = bu_k \quad \text{for some } a, b > 0.$$

Some graphs exhibiting crack advance under this simple model are presented in Figures 12.1 and 12.2. NB that the parameters should be adjusted with any change of material or alteration of environmental conditions.

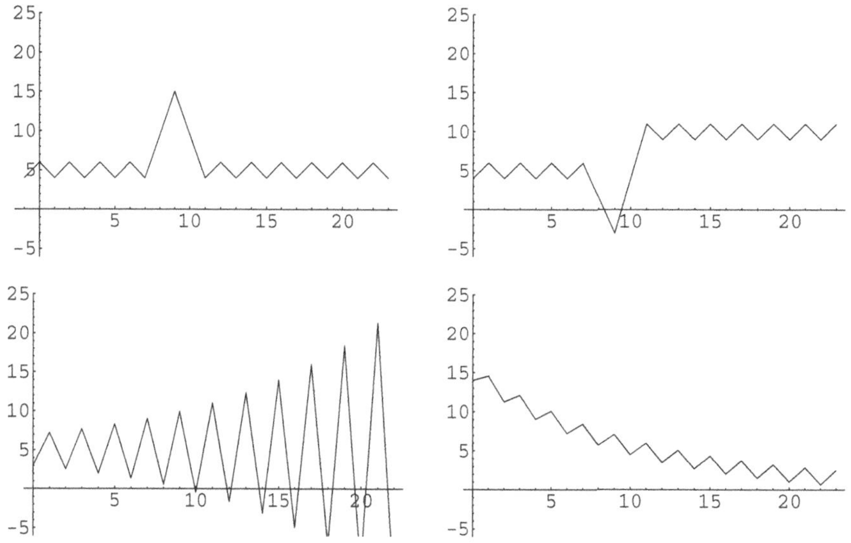

Figure 12.2. Programmed loads causing crack arrest, jump, acceleration and deceleration.

Theorem 11. *Under the incremental model in eqn (12.33) one concludes:*

1. *Crack advance will eventually occur under repetitions of a constant amplitude sinusoidal load fluctuation iff* $(1-\varsigma)\vartheta(\lambda) > \varphi(\lambda)$*, which is always true if* ς *is sufficiently near 1.*
2. *Crack arrest will occur during each duty cycle following the maximum load until the kth oscillation, for* $2 \le k \le m-1$*, when* $\lambda_0 := \lambda_m$ *is the maximum load and* $\lambda_1 = \cdots = \lambda_{m-1}$ *iff*
$$\frac{(1-\varsigma)\vartheta(\lambda_1)-\varphi(\lambda_1)}{(1-\varsigma)\varphi(\lambda_0)-\varphi(\lambda_1)} < \varsigma^k. \tag{12.34}$$
Crack advance will occur for fluctuations $k+1, \cdots, m$ *iff the reverse inequality obtains.*
3. *If the peak tensile stress of the jth load fluctuation,* u_j*, is increasing for* $j = 1, \ldots, m$ *while the* ℓ_j *are decreasing and for some k we have crack advance,* $\Delta d_k > 0$*, then iff*
$$\Delta[\vartheta(\lambda_{k+1}) - \varphi(\lambda_{k+1})] > \varsigma\varphi(\lambda_k) - (1-\varsigma)\sum_{j=0}^{k}\varphi(\lambda_{k-j})\varsigma^j, \tag{12.35}$$
will the crack accelerate, i.e., $\Delta d_{k+1} > \Delta d_k$*.*
4. *If a programmed load satisfying (12.35) is run in reverse, then crack deceleration will occur.*

Lemma 3. *If* $X \sim F_X$*,* $Y \sim F_Y$ *where* $X \perp Y$*, with finite means and sample spaces* $\Re_+$ *and* $\Re$ *respectively, then*
$$\mathrm{E}[X-Y]^+ = \mathrm{E}[X-Y] + \int_0^\infty F_X(y)\bar{F}_Y(y)\,dy.$$

Proof. Using integration by parts and interchanging the order of integration as needed, we find
$$\begin{aligned}
\mathrm{E}[X-Y]^+ - \mathrm{E}Y &= + \int_0^\infty \int_{-\infty}^\infty \max(x,y)dF_Y(y)dF_X(x)\\
&= \int_0^\infty xF_Y(x)dF_X(x) - \int_0^\infty yF_X(y)d\bar{F}_Y(y),\\
&= \int_0^\infty xF_Y(x)dF_X(x) + \int_0^\infty y\bar{F}_Y(y)d\bar{F}_X(y) + \int_0^\infty \bar{F}_Y(y)F_X(y)dy.
\end{aligned}$$
where in the second equation we integrate the third term by parts and then we sum the second and third terms in the last equation. □

Exercise Set 12.H

1. An alternate expression, instead of $\mathrm{E}[X-Y]^+$, is $\mathrm{E}|X-Y|^a$ for some $a > 0$.
   (a) Show that under the assumptions of Lemma 3, with $U = X - Y$, that
$$\mathrm{E}|U|^a = \int_{t=0}^\infty [\bar{F}_U(t) - F_U(t)]\,dt^a, \qquad \text{where} \quad F_U(u) = \int_{-\infty}^\infty F_X(u+y)\,dF_Y(y).$$

(b) What changes must be made in the expression above if $X > 0$ a.s.?
(c) If $a = 2$ then $\mathrm{E}|X - Y|^a = \mathrm{Var}X + \mathrm{Var}Y + (\mathrm{E}X - \mathrm{E}Y)^2$.
Analyze the suitablity of this simple expression to reflect the "spurt and lag" phenomena exhibited in the fatigue of materials in load-order variation.

2. Suppose we have a system, structure or machine that operates in a mildly corrosive environment from which it is protected by some polymer coating. This coating is subject to random perforation over time during service. At each perforation location some chemical deterioration to the structure is initiated, such as rust, corrosion, contamination or finish imperfection the continuation of which causes cumulative damage over time. Thus, we have the protected area of the structure susceptible to perforation at some rate each point of which serves as an initiating source for some degradation process that gradually generates sufficient damage to make the structure unacceptable, thus ending the service life. Consider models for predicting service life in these possible situations:
(a) The undercoat of an automobile protecting it from the chemical action of de-icing compounds applied to roadways to improve traction during winter months.
(b) The coating on the inside of aluminum beer cans that when perforated, causes the beer to acquire a "metallic taste" over time and makes the product undesirable for consumption.
(c) The shelf life of certain household cleaners which if unsold for sufficently long "eat" through the container, start leaking, are unsaleable and foul the storage location.
(d) Describe what data you would need in each situation above and estimate the relative time and cost to obtain it.

I never looked upon ease and happiness as ends in themselves — this ethical basis for life I call the ideal of the pigsty.

One can organize (a technical staff) to apply a discovery already made, but not to make one.

It is a mistake often made in this country (America) to measure things only by how much they cost.

Few people are capable of expressing with equinimity opinions that differ from the prejudices of their social environment. Most people are even incapable of forming such opinions.

Philosophy is a mother who gave birth to and endowed all the sciences. Today, therefore one should not scorn her in her nakedness and poverty, but hope, rather, that part of her Don-Quixote ideal will live on in her children so that they all do not sink into philistinism.

Albert Einstein

CHAPTER 13

Service Life of Structures

13.1. Wear under Spectral Loading

Assume that we have an engineered structure, such as an airframe, a bridge, a deep-sea platform, a high-rise building, or a nuclear containment facility et al., which is subjected to repetitive loads of varying stress amplitude or frequency. Each structure sustains repeated duty-cycles throughout its life during which the distribution of the loading spectrum remains unchanged or changes in a known way. Moreover, it is assumed that all load distributions imposed by usage can be classified into a finite number of categories and the frequency of occurrence of each category can be measured or is known. These empirical distributions are frequently incorporated into the the specifications for the design. The incremental damage to the structure, whether it is of metallic alloy or composite material, stems from strain imparted during the imposition of load. This may range from the rearrangement of dislocations within the crystaline lattice, the fracture of polymeric bonds in the material, the microscopic extension of a fatigue crack in some metallic sub-component, the abrasion or galling of some surface or any other form of deterioration for which there is no self-recovery or healing. The component strain, for a given geometry of structure and each type of load, is a complicated but determinable function which often is realized by performing computer-aided, finite-element analysis. Archetypical examples include an airframe component sustaining gust and maneuver loads during each ground-air-ground cycle or a bridge being subjected to the repetitive loads from its daily pattern of vehicular traffic, with seasonal damage due to winter weather and de-icing compounds.

Whenever the spectrum of loads on a structure is stationary over time and can be modeled as the superposition of a finite number of renewal processes, each occurring at a rate governed by an independent stationary counting process, then, presuming linear accumulation of damage, the waiting time until the total damage to the structure exceeds a critical level in some component will be shown to have the fatigue-life distribution. Moreover, the median fatigue-life of this distribution can be calculated by the Miner–Palmgren rule and the coefficient of variation found by a corresponding formula. If we have a multicomponent structure, for which the allocation of the total load to each of its components has been determined, then the random time until a critical level of cumulative damage is first reached for each component can be jointly expressed using a multivariate fatigue-life distribution. The statistical dependency of the component loads is expressed through covariances derived from the structural load-allocations. The multivariate fatigue-life distribution, can be used to provide a confidence bound on the safe-life of the structure when the correlations between the component fatigue-lives can be derived from the structural interdependence of the design.

This procedure might have provided a measure of safety for the special vehicle described by Oliver Wendell Holmes in his *The Deacon's Masterpiece* or "*The One-Hoss Shay*" which, reportedly, had no component failure whatsoever during service and "It was built in such a logical way that it ran for a hundred years to the day and then in a moment" suffered an ultimate and total collapse. This catastrophic in-service failure would have been avoided by calculating a safe service life and retiring the shay during its 99th year.

13.2. Multivariate Fatigue Life

Let Φ be the standard normal distribution, $\Phi' = \varphi$ be the standard normal density and ξ the function defined, as in 10.8, by

$$\xi(z) = \sqrt{z} - \frac{1}{\sqrt{z}} \quad \text{for } z > 0. \tag{13.1}$$

Then X, the *fatigue life*, has distribution and density, respectively, given in 10.2 by

$$\Phi[\tfrac{1}{\alpha}\xi(t/\beta)] \quad \text{and} \quad \tfrac{1}{\alpha\beta}\xi'(t/\beta)\varphi[\tfrac{1}{\alpha}\xi(t/\beta)] \qquad \text{for} \quad t > 0; \alpha, \beta > 0. \tag{13.2}$$

NB that the scale parameter β is also the median life and the shape parameter α is virtually the coefficient of variation, for ranges of engineering application when $\alpha < 0.2$, see eqn (10.18), with

$$\text{E}X = \beta[1 + \tfrac{\alpha^2}{2}] \qquad \text{and} \qquad \text{Var}[X] = (\alpha\beta)^2[1 + \tfrac{5\alpha^2}{4}]. \tag{13.3}$$

There are cases of what is called *multisite fatigue*. This phenomenon can occur when multiplicate structural components are designed to divide equally each imposed stress during the duty cycle. When such components are nominally identical (same design, specification, and manufacturer) and they also share the same operational environment (sometimes in the presence of a notch or stress raiser) they may also undergo virtually identical fatigue loading and damage increments, thereby causing almost simultaneous fatigue failure.

A notable example of multisite fatigue fatigue failure occurred in a plane used by the Hawaiian Airlines as a shuttle between the islands. Although the plane had only accrued a moderate number of total flight hours, it had sustained many more GAG (ground-air-ground) cycles than usual. These repeated cabin pressurizations in its oceanic operating environment allowed a large section of the forward exterior cabin-wall to be blown off due to the simultaneous fatigue failure of several hundred skin fasteners. Although the plane itself was not lost, unfortunately, fatalities occurred.

13.2.1. Two-Component Load Sharing

Let us consider how this fatigue dependence might be modeled for dual structural elements designed to share equally, high stresses. Let $X_i > 0$ for $i = 1, 2, \cdots$ be the iid incremental damage shared by the pair of components during the ith duty cycle, where $\mathrm{E}X_i = 2\mu$, $\mathrm{Var}(X_i) = \sigma^2$. This shared stress induces nearly identical incremental damages to the two structural elements; denote them by

$$Y_{1,i} = (\tfrac{1}{2} - V_i)X_i, \qquad Y_{2,i} = (\tfrac{1}{2} + V_i)X_i \qquad \text{where } 0 < V_i < \varepsilon < \tfrac{1}{2} \text{ a.s.}$$

are the iid perturbations of incremental damage, with $\mathrm{E}V_i = 0$, $\mathrm{Var}(V_i) = \kappa^2$.

Since the external allocation of load is independent of the state of damage to the component it is logical to assume $V_i \perp X_i$ so we see $\mathrm{E}Y_{k,i} = \mu$ and $\mathrm{Var}(Y_{k,i}) = \sigma^2(\frac{1}{2} + \kappa^2) + 4(\mu\kappa)^2 := \delta^2$ and to define

$$S_{k,n} = \sum_{i=1}^{n} Y_{k,i}, \quad \text{with} \quad Z_{k,n} = \frac{n\mu - S_{k,n}}{\delta\sqrt{n}}, \quad \text{for } k = 1, 2. \text{ and } n = 1, 2, \cdots,$$

using previous notation in the standardized formulas to be compared. Following the arguments of Section 10.2 we obtain identical parameters for the fatigue-life lengths T_1 and T_2, namely,

$$\alpha_1 = \alpha_2 = \frac{\delta}{\sqrt{\mu\omega}} := \alpha, \quad \text{and} \quad \beta_1 = \beta_2 = \frac{\omega}{\mu} := \beta.$$

Hence the marginal distributions of the service lives of the two load-sharing components are both $\mathcal{FL}(\alpha, \beta)$. Set $U_k = \frac{T_k}{\beta_k}$ and write these transformed variates as $W_k = \frac{1}{\alpha}\xi(U_k)$, for $k = 1, 2$ we see they both will a have (marginal) distributions which is $\mathcal{N}(0, 1)$. Their correlation, say ρ, using ψ as defined in Problem 1 in EXERCISE SET 10.B and the reciprocal property, is given by,

$$\rho := \mathrm{E}[W_1 W_2] = \frac{1}{\alpha^2}\mathrm{E}[\xi(U_1)\xi(U_2)] = \frac{1}{\alpha^2}\mathrm{E}\left[\psi(U_1 U_2) - \psi(U_1/U_2)\right].$$

We now argue heuristically, since in the limit for $\varepsilon \downarrow 0$ that $T_1 = T_2$ a.s., that if we set $U = \lim_{\varepsilon\downarrow 0} T_k/\beta$, then we see by eqn (13.3), again using the reciprocity, that $\rho = \lim_{\varepsilon\downarrow 0} \mathrm{E}[W_1 W_2] = 2\alpha^{-2}\,[\mathrm{E}(U - 1)] = 1$, but for any $\varepsilon > 0$, ρ will be very near unity. NB, a positive correlation between the two variates marginally normal, is not sufficient to guarantee that the joint distribution is bivariate normal. However, it becomes a very reasonable assumption useful to calculate the waiting time until the likelihood of the extreme consequences of virtually simultaneous failure becomes intolerably large.

Exercise Set 13.A

1. If $U \sim \Phi[\frac{1}{\alpha}\xi(u)]$ for $u > 0$ show that $\mathrm{E}\sqrt{U} = 2\int_0^\infty \sqrt{(\frac{\alpha u}{2})^2 + 1}\, d\Phi(u)$ and then argue for $\alpha < .5$ it follows $\mathrm{E}\sqrt{U} \doteq 1 + \frac{\alpha^2}{8}$.

2. Let U_k for $k = 1, 2$ be bivariate fatigue lives where $U_k \sim \mathcal{FL}(\alpha_k, 1)$, where $Z_k = \xi(U_k)$ and $\mathrm{E}(Z_1 Z_2) = \rho$. Derive the joint density of U_1, U_2 and show that it fellows $\mathrm{Cov}[U_1, U_2] = (\alpha_1 \alpha_2 \rho)^2/2$.

3. Calculate the distribution of the minimum of bivariate fatigue lives when $\beta_1 = \beta_2 = 300$, and $\alpha_1 = \alpha_2 = .05$ with $\rho = .9$. What would you recommend as a safe life for this dual unit?

13.2.2. The Multivariate Fatigue-Life Distribution

Similar arguments can be made for multiple related fatigue lives. We now extend the definition to a *multivariate fatigue-life* ($\mathcal{MVFL}$) distribution.

Definition 1. If the $\boldsymbol{P} = (\rho_{i,j})$ is a correlation matrix with $\boldsymbol{P}^{-1} = (r_{i,j})$ and there exists vectors $\boldsymbol{\alpha} = (\alpha_1, \ldots, \alpha_n)$ and $\boldsymbol{\beta} = (\beta_1, \ldots, \beta_n)$ such that for the function ξ, as defined in eqn (13.1), the random vector of transformations has the following distribution:

$$\left(\frac{1}{\alpha_1}\xi(X_1/\beta_1), \cdots, \frac{1}{\alpha_n}\xi(X_n/\beta_n)\right) \sim \mathcal{N}(\boldsymbol{O}, \boldsymbol{P}); \tag{13.4}$$

then we say $\boldsymbol{X} = (X_1, \ldots, X_n)$ has a multivariate fatigue-life distribution and we write

$$\boldsymbol{X} \sim \mathcal{FL}(\boldsymbol{\alpha}, \boldsymbol{\beta}, \boldsymbol{P}) \tag{13.5}$$

with the density of $\boldsymbol{X}$ given, for all $x \xi \Re^n$, by

$$f_{\mathbf{X}}(\boldsymbol{x}; \boldsymbol{\alpha}, \boldsymbol{\beta}, \boldsymbol{P}) = |\boldsymbol{P}|^{-\frac{1}{2}} \left[\prod_{i=1}^{n} \frac{\xi'(x_i/\beta_i)}{\sqrt{2\pi}\alpha_i \beta_i}\right] \times \exp\left\{\frac{-1}{2}\sum_{i=1}^{n}\sum_{j=1}^{n} \frac{r_{i,j}}{\alpha_i \alpha_j} \xi(x_i/\beta_i)\xi(x_j/\beta_j)\right\}. \tag{13.6}$$

It is assumed the load order is random, and the sequence of loads of each given type form an iid sequence with the frequency of load occurrence of each category governed by a stationary stochastic process. (It is presumed that virtually all structures in service encounter stresses which should be considered as stochastic rather than deterministic.)

Let $N_i(t)$ count the (total) number of random load fluctuations of type i, for $i = 1, \ldots, \ell$, to which the structure has been subjected during the time interval $(0, t]$. The total damage due to loads of category i at time $t > 0$ is the sum of (random) incremental damages, say,

$$S_i(t) = \sum_{j=1}^{N_i(t)} Y_{i,j} \quad \text{for } i = 1, 2, \ldots, \ell, \tag{13.7}$$

where $Y_{i,1}, Y_{i,2}, Y_{i,3}, \ldots$ are the incremental damages to the structure due to successive repetitions of loads of type i. The random variables $Y_{1,j}, \ldots, Y_{\ell,j}$ represent the fatigue damage induced by the jth occurrence of stresses, whatever time they occur, from each of the possible various load configurations. If we consider a bridge of prestressed concrete girders, then the incremental damages (stresses) are comprised of the dead load of the structure plus the dynamic loads resulting from diurnal vehicular traffic. In the case of a deep-sea platform, the loads would stem from waves and gusts of each given type considered, as they occur during repeated applications of its duty cycle.

Assuming that all permutations of the arrival-order of load fluctuations during each duty cycle are equally likely (or what is the same, the *load-order is not important* in determining the cumulative damage to the structure), the total cumulative damage sustained by the structure up to any fixed time is then merely the sum of damage increments from each category of load fluctuations over every duty-cycle. This assumption is often referred to as the *linear cumulative damage hypothesis* (LCDH). Thus we have

Theorem 1. *Let $Y_{i,j} > 0$ be the incremental damage to a structure on the ith occurrence of a strain (load) of category $j = 1, \ldots, \ell$ where $Y_{1,j}, Y_{2,j}, \cdots$ is an iid sequence with mean and variance, respectively,*

$$EY_{i,j} = \mu_j, \quad \text{and} \quad Var[Y_{i,j}] = \sigma_j^2 \quad \text{for all} \quad i = 1, 2 \cdots. \tag{13.8}$$

Let the frequency of loads of category j during the interval $(0, t]$ be governed by a stationary counting process that is statistically independent of damage, say $N_j(t)$ for $j = 1, \ldots, \ell$, with

$$EN_j(t) = \lambda_j t \quad \text{and} \quad Var[N_j(t)] = \eta_j t. \tag{13.9}$$

Then under the LCDH the (random) waiting time, $X(\omega)$, until a fixed level of damage, say ω, is first exceeded has, asymptotically, a FL-distribution, viz,

$$\Pr[X(\omega) \le t] \cong \Phi\left[\tfrac{1}{\bar{\alpha}}\xi(t/\bar{\beta})\right] \quad \text{for} \quad t > 0, \tag{13.10}$$

where the median life, $\bar{\beta}$, and the coefficient of variation, $\bar{\alpha}$, are given by

$$\bar{\beta} = \frac{1}{\sum_{j=1}^{\ell} \lambda_j/\beta_j} \quad \text{and} \quad \bar{\alpha} = \sqrt{\bar{\beta} \sum_{j=1}^{\ell} \left(\frac{\eta_j}{\beta_j^2} + \frac{\lambda_j \alpha_j^2}{\beta_j}\right)}. \tag{13.11}$$

Here, in terms of incremental damage parameters,

$$\beta_j = \frac{\omega}{\mu_j} \quad \text{and} \quad \alpha_j = \frac{\sigma_j}{\sqrt{\omega \mu_j}} \quad \text{for } j = 1, \ldots, \ell. \tag{13.12}$$

Proof. The total cumulative damage that has occurred by time $t > 0$ is the random sum

$$Y(t) = \sum_{j=1}^{\ell} S_j(t) = \sum_{j=1}^{\ell} \sum_{i=1}^{N_j(t)} Y_{i,j}. \tag{13.13}$$

We must show that $X(\omega) = \inf\{t > 0 : Y(t) \geq \omega\}$ has asymptotically a $\mathcal{FL}$-distribution. We use Wald's lemma, which implies that for any stopping rule N which is statistically independent of an iid sequence $Y_1, Y_2, \ldots$ then the random sum $S = \sum_1^N Y_j$ has mean and variance given by

$$\text{E}S = \text{E}N \cdot \text{E}Y \quad \text{and} \quad \text{Var}(S) = (\text{E}Y)^2 \cdot \text{Var}(N) + \text{E}N \cdot \text{Var}(Y). \tag{13.14}$$

Applying Wald's lemma to each category of load we find, from eqns (13.13) and (13.14), that

$$\text{E}[S_j(t)] = t\lambda_j\mu_j \quad \text{and} \quad \text{Var}[S_j(t)] = t\left(\eta_j\mu_j^2 + \lambda_j\sigma_j^2\right), \tag{13.15}$$

and so the mean and variance of cumulative damage are both linear in time with a steady-state rate of expected accumulation of damage to the structure of ν, and a variance κ^2 (per unit of time from all categories of load). These constant quantities are given by

$$\text{E}[Y(t)] = t\nu \quad \text{and} \quad \text{Var}[Y(t)] = t\kappa^2, \tag{13.16}$$

where

$$\nu = \sum_{j=1}^{\ell} \lambda_j\mu_j \quad \text{and} \quad \kappa^2 = \sum_{j=1}^{\ell} \left(\lambda_j\sigma_j^2 + \eta_j\mu_j^2\right). \tag{13.17}$$

Failure occurs when ω, the critical accumulated damage to the structure, is first exceeded. By definition $X(\omega)$, the service life of the structure when expressed in the units of time of operation, has a distribution which may be found by the duality relation between time-of-failure and cumulative damage, namely,

$$\Pr[X(\omega) \geq t] = \Pr[Y(t) \leq \omega] = \Pr\left[\frac{Y(t) - t\nu}{\sqrt{t}\kappa} \leq \frac{\omega - t\nu}{\sqrt{t}\kappa}\right]. \tag{13.18}$$

If we standardize the cumulative damage by setting

$$Z(t) = \frac{Y(t) - t\nu}{\sqrt{t}\kappa}, \qquad \text{with} \qquad \bar{\alpha} = \frac{\kappa}{\sqrt{\omega\nu}}, \qquad \bar{\beta} = \frac{\omega}{\nu}, \tag{13.19}$$

and use the definition of the function ξ given in eqn (13.1) we have, from eqns (13.18) and (13.19),

$$\Pr[X(\omega) \geq t] = \Pr[Z(t) \leq \tfrac{-1}{\bar{\alpha}}\xi(t/\bar{\beta})]. \tag{13.20}$$

To ascertain the asymptotic behavior we modify the incremental damage summands so they become "uniformly asymptotically negligible" (Loevé's terminology, see [62]) yet keep the same coefficient of variation. Multiply all parameters by the scale parameter ϵ:

$$\omega' = \omega\epsilon, \quad \lambda_j' = \epsilon\lambda_j, \quad \eta_j' = \epsilon\eta_j, \quad \mu_j' = \epsilon\mu, \quad \sigma_j' = \epsilon\sigma \quad \text{for } j = 1, \cdots, \ell.$$

An easy calculation from (13.17) shows $\nu' = \epsilon^2\nu$ and $\kappa' = \epsilon^{3/2}\kappa$ which, by (13.19), gives $\bar{\alpha}' = \bar{\alpha}$ and $\bar{\beta}' = \bar{\beta}/\epsilon$. In this new circumstance eqn (13.18) becomes

$$\Pr[X(\omega\epsilon) \geq t] = \Pr[Y_\epsilon'(t) \leq \epsilon\omega] = \Pr\left[Z_\epsilon'(t) \leq \frac{\omega - t\epsilon\nu}{\kappa\sqrt{t\epsilon}}\right],$$

after cancellation in numerator and denomination. By setting $x = \epsilon t$ we obtain

$$\Pr[\epsilon X(\omega\epsilon) \geq x] = \Pr[Z_\epsilon'(x/\epsilon) \leq \tfrac{-1}{\bar{\alpha}}\xi(x/\bar{\beta})].$$

By the central limit theorem we know that, in distribution, as $\epsilon \to 0$; $Z_\epsilon'(x/\epsilon) \rightsquigarrow cal N(0, 1)$ hence essentially under any conditions for which the number of loads to effect failure is large, it follows that the time at which failure occurs will have, asymptotically, the FL-distribution as in eqn (13.10) with parameters given in eqn (13.11). □

NB that the median life, $\bar{\beta}$, in eqn (13.11) is nothing more than the hoary Miner-Palmgren-Langer rule [79, 1945]. Miner's rule has been controversial in application principally because of the lack of precise information about the hypotheses under which it is true. Nevertheless Miner's rule is extensively used to predict service life, under a variety of conditions when a spectrum of loads is to be applied. Unfortunately it is often interpreted as if it were a deterministic rule, i.e., ignoring all scatter in the data, rather than just the expectation of a stochastic varable. Thus its application was successful if the coefficient of variation was sufficiently small and unsuccessful otherwise. In the former case it has also been used along with several ill-founded suggestions, purportedly to give the probability of failure when the cumulative damage is estimated to be some fraction of Miner's rule.

Miner's rule must be correctly understood as measuring only central tendency in a stochastic context. It has been shown, using the renewal theorem, to hold closely in expectation under very weak assumptions concerning the nature of independent incremental damage when load order is not important, see Chapter 12 and [96, 1970]. It has also been empirically shown to hold "on the average better than any other fatigue-life rule for airframe structures" in several proprietary studies when fatigue life is defined to be the time until the initiation of a dominant (detectable) fatigue crack.

There have been many empirical studies, in wood as well as metal, showing there was too much variation in the predicted life, if Miner's rule alone was utilized. And despite several theoretical attempts to force Miner's rule as the parameter of central tendency in other distributions, principally the Weibull distribution, none have been widely successful, see [97, 1972].

Notice also that $\bar{\alpha}$ may have several valid approximations depending upon various relationships between the η_j and β_j for $j = 1, \ldots, \ell$.

Thus, the FL-distribution has been found for which the median life can be calculated under a spectrum of traffic loads by using Miner's rule and the coefficient of variation of the resulting scatter can also be determined. Hence it becomes theoretically possible to calculate "the fraction of life used," i.e, the probability of failure after any specified fraction of the characteristic life has been expended. But, as is known empirically, this probability depends upon other factors than the median life. Here we have a formula that recognizes the influence of the loading spectrum and the material properties through the corresponding cumulative-damage distribution and its associated parameters $\bar{\alpha}$ and $\bar{\beta}$.

While we talk about incremental damage per cycle it may not always be determinable just from the incremental stress range during that cycle. The incremental crack growth per cycle, represented symbolically by $\frac{da}{dn}$, is given (under a differential equation model) by the Paris–Erdogan law in metallurgy for cyclic loads [89, 1961], as

$$\frac{da}{dn} \propto (\Delta K)^m \qquad \text{for some } m > 0, \tag{13.21}$$

where $\Delta K = K_{max} - K_{min}$ is the stress intensity range per cycle. We emphasize that the fatigue life, as here defined under the LCDH, means until the initiation of a detectable fatigue crack. A crack of critical magnitude (often larger than detectable size) propagates with a mechanism different than that during the initiation phase and its rapidity vitiates both the assumptions of iid incremental damage and independence of load order.

Exercise Set 13.B

1. If the counting processes of loads, $N_j(t)$, for $j = 1, \cdots, \ell$ are Poisson variates, then $\lambda_j = \nu_j$ in eqn (13.9) and the equation for $\bar{\alpha}$ is changed. Show that $\bar{\alpha}$ satisfies

$$\min_{j=1}^{n} \left(\beta_j^{-1} + \alpha_j^2\right) < (\bar{\alpha})^2 < \max_{j=1}^{n} \left(\beta_j^{-1} + \alpha_j^2\right).$$

2. Many structures, such as bridges, are subjected to regularly varying work loads during their life. The density of traffic varies, on average, during the time of day and night, the day of the week and the season of the year. Moreover, there is usually an annual increase in the traffic. These changes also must be accounted for in the prediction of service life.

   Represent the mean annual increase in traffic load by the transformation $c(t) = e^{\epsilon t}$ for $t > 0$ in years where ϵ is the annual percentage increase in traffic load. At time t the augmented exposure results in total damage from loads of category i, of

$$S_i[c(t)] = \sum_{j=1}^{N_i[c(t)]} Y_{i,j} \qquad \text{for } i = 1, 2, \cdots, \ell.$$

   Thus the time-to-failure, call it $T^\star$, under the augmented loading, would be given by

$$\Pr[T^\star \le t] \doteq \Phi\{\frac{1}{\alpha}\xi[c(t)/\beta]\} \quad \text{for } t > 0 \text{ and asymptotically, } c(T^\star) \sim \mathcal{FL}(\alpha, \beta).$$

   (a) Does this work if t is in days and $c(t)$ contains diurnal, weekly, and seasonal variation?
   (b) Construct formulas for c representing these combinations of complexity.
   (c) The median life of $T^\star$, call it $\beta^\star$, is $\beta^\star = c^{-1}(\beta)$. Examine the shift in $\beta^\star$ for various choices of c.
   (d) Do the same study for a *safe-life* at a probability of failure of .001.
   (e) In addition to the fatigue of the girders in a bridge there is the abrasion of the road surface through traffic wear and, in certain localities, the complication from de-icing compounds. How should this be treated in reliability assessment?
   (f) There may be the occassional supra-legal overload (from military or nuclear-power transport) passing over a "worn" bridge structure. How do you deal with this?
   (g) Who should think about the vulnerability (the probability of failure) of a bridge to small earth tremors?

3. Relate all the concerns mentioned for bridges which are applicable to deep-sea platforms, steel cables in elevators, suspension bridges, or construction hoists. Are there any other concerns which you can identify that may be particular?

13.3. Correlations between Component Damage

Let us now consider a (fail-safe) structure that sustains random loads while in service but each load is shared among all of its components. The strain for each component, for a given geometry of the structure and a fixed type of load, is a complicated, but deterministic, function of the design of the structure. For prescribed loads the component allocation may be found by using a computer to perform a finite-element analysis. Unfortunately, during service the magnitudes, intensity and allocation of load to each component are stochastic in nature and at best only their distribution can be known. Moreover, here as previously the LCDH is assumed, namely, that over the time range considered there is no significant effect due to load order, which often is random, or to an acceleration of the damage process by an unanticipated change in duty cycle or service environment. Thus it is assumed the sequence of loads of each category form an iid sequence and the number of loads of each category, as a function of time, is a stationary stochastic process with independent increments. In many applications the imposed stresses occurring over time to each component within a structure are stochastically equivalent to this model.

An archetypical example would be a bridge constructed with pre-cast concrete girders which is sustaining vehicular traffic of various categories. Each category of live-load is determined by the number of axles, and their configuration, for each given type of vehicle and, as well, combinations of multiple vehicles on the bridge. Each category will, of course, have a different frequency of occurrence. Each girder in the structure will sustain a different strain for each load of each type since the strain on each girder contains not only its share of the constant dead-load but a different allocation of the live-load because of the geometry (skewness) of bridge, the gross weight of each truck, its position in-lane, and its impact (vehicular velocity).

Following the definition in eqn (13.13), we denote the cumulative damage to the kth component during the time interval $(0, t]$ by $Y_k(t)$, and the (random) cumulative damage to the structure of all n components is denoted by the vector

$$\boldsymbol{Y}(t) = (Y_1(t), Y_2(t), \ldots, Y_n(t)).$$

Each component reacts only to its allocation of the spectral load determined by the design of the structure. We assume stationarity of the damage process and utilize the notation of eqn (13.17), to define the *cumulative standardized damage* to the kth component for $k = 1, \ldots, n$ and all $t > 0$ as

$$Z_k(t) = \frac{Y_k(t) - t\nu_k}{\kappa_k\sqrt{t}}, \quad \text{where } \mathrm{E}[Y_k(t)] = t\nu_k, \text{ and } \mathrm{Var}[Y_k(t)] = t\kappa_k^2$$

Consider again the case of a multicomponent bridge structure subjected to a stationary loading spectrum. The rate of occurrence of a load of type j is governed by the counting process $N_j(t)$, where

$$\mathrm{E}N_j(t) = \lambda_j t \quad \text{and} \quad \mathrm{Var}N_j(t) = \eta_j t \quad \text{for} \quad j = 1, \dots, \ell. \tag{13.22}$$

A portion of the ith repeated load (strain) of type i is allocated to, and borne by, each of the n components of the structure. Accordingly, only a proportionate share of the incremental damage is sustained by each component. Let $Y_{i,j,k}$ denote the incremental damage sustained under the ith repetition of the jth category of load by the kth component for $k = 1, \dots, n$. Thus the iid sequence $Y_{1,j,k}, Y_{2,j,k}, Y_{3,j,k}, \dots, Y_{i,j,k}, \dots$ represents the incremental damages caused by the repetitive strains imposed by successive loads of the jth type to the kth component. The relationship between the structural loading and that of the components may be quite complex since it is specific for each structural design.

13.3.1. Covariance and Association

Thus the total accumulated damage to the kth component at any time $t > 0$ is the random sum

$$Y_k(t) = \sum_{j=1}^{\ell} S_{j,k}(t), \qquad \text{where} \qquad S_{j,k}(t) = \sum_{i=1}^{N_j(t)} Y_{i,j,k}. \tag{13.23}$$

The incremental damage to structural components has the following behavior, for all $i = 1, 2, \cdots; k = 1, \cdots, n; j = 1, \cdots, \ell$;

$$\mathrm{E}Y_{i,j,k} = \mu_{j,k} \quad \text{and} \quad \mathrm{Cov}[Y_{i,j,k}, Y_{i',j',k'}] = \mathrm{I}(i = i', j = j')\sigma_{k,k'}(j) \tag{13.24}$$

where as before $\mathrm{I}(\pi)$ is the indicator function of any relation π. This follows from the assumed independence between loads of different categories and between successive loads of the same type.

To determine the asymptotic distribution, following the development of eqn (13.19), we must evaluate the joint probability of the standardized damages, viz.,

$$\Pr\left[Z_1(t_1) \le \frac{\omega_1 - t_1\nu_1}{\kappa_1\sqrt{t_1}}, \cdots, Z_n(t_n) \le \frac{\omega_n - t_n\nu_n}{\kappa_n\sqrt{t_n}}\right]$$

and to do so we need to obtain the correlation between the standardized cumulative damages for different components, i.e., for $i, j = 1, \dots, n$

$$\rho_{i,j}(t,s) = \mathrm{E}[Z_i(t)Z_j(s)] = \frac{\mathrm{Cov}[Y_i(t), Y_j(s)]}{\kappa_i\kappa_j\sqrt{ts}},$$

to discover how it depends on t and s. Consider the covariance for $k \neq k'$,

$$\mathrm{Cov}[Y_k(t), Y_{k'}(s)] = \sum_{j=1}^{\ell}\sum_{j'=1}^{\ell} \mathrm{Cov}[S_{j,k}(t), S_{j',k'}(s)].$$

It is necessary to examine only the covariance between damages to different components due to the same category of load. We do this using conditional expectation:

$$\begin{aligned}\mathrm{Cov}[S_{j,k}(t), S_{j,k'}(s)] &= \mathrm{E}\left[\sum_{i=1}^{N_j(t)}\sum_{i'=1}^{N_j(s)} \mathrm{Cov}(Y_{i,j,k}, Y_{i',j,k'} \,|N_j(t), N_j(s))\right] \\ &= \mathrm{E}\left[\sum_{i=1}^{N_j(t)}\sum_{i'=1}^{N_j(s)} I(i = i')\sigma_{k,k'}(j)\,|N_j(t), N_j(s)\right] \\ &= \min(t, s)\lambda_j\sigma_{k,k'}(j).\end{aligned}$$

Hence we find

Lemma 1. *The correlation between the standardized cumulative damage to any two components, say k and k', at the respective times, $t, s > 0$ is*

$$\rho_{k,k'}(t, s) = E[Z_k(t)Z_{k'}(s)] = \frac{\min(t, s)}{\kappa_k\kappa_{k'}\sqrt{ts}}\sum_{j=1}^{\ell}\sigma_{k,k'}(j). \qquad (13.25)$$

Thus, the immediate

Corollary 1. *At any given time the correlation between the standardized cumulative damage to components defines an $n \times n$ covariance matrix, independent of time, we label $\mathbf{P} = \left(\rho_{i,j}\right)$, where*

$$Cov[Z_i(t), Z_j(t)] = \rho_{i,j} \quad with \quad \rho_{ii} = 1 \quad for\ i, j = 1, \ldots, n. \qquad (13.26)$$

Thus follows the main result:

Theorem 2. *Assume that a structure of n components is fatigued by a spectrum consisting of ℓ categories of loads in which $Y_{i,j,k} > 0$ denotes the incremental damage due to the ith repetition of a load of the jth category as sustained by the kth component and for which eqn (13.23) holds. Let the stationary counting process for loads of the jth category be $N_j(t)$ for $j = 1, \ldots, \ell$; they are mutually independent as well as independent of incremental damages from the loads. Their means and variances are given in eqn (13.24). Let the cumulative damage to the kth component at time $t > 0$ be the sum*

$$Y_k(t) = \sum_{j=1}^{\ell}\sum_{i=1}^{N_j(t)} Y_{i,j,k},$$

when the fatigue life of the component is given in terms of its cumulative damage $Y_k(t)$ *by*

$$X_k = \inf\{t > 0 : Y_k(t) > \omega_k\} \quad \textit{for some critical damage level} \;\; \omega_k > 0,$$

then $\boldsymbol{X} = (X_1, \ldots, X_n)$, *the vector of fatigue lives of the components, has a multivariate fatigue-life distribution as given in eqn (13.6), namely,* $\boldsymbol{X} \sim FL(\bar{\alpha}, \bar{\beta}, \boldsymbol{P})$ *and the kth component has its median life* $\bar{\beta}_k$, *and coefficient of variation,* $\bar{\alpha}_k$, *given respectively by*

$$\bar{\beta}_k = \frac{1}{\sum_{j=1}^{\ell} \lambda_j / \beta_{j,k}}, \quad \textit{and} \quad \bar{\alpha}_k = \sqrt{\bar{\beta}_k \sum_{j=1}^{\ell} \frac{\eta_j}{\beta_{j,k}} \left[\alpha_{j,k}^2 + \frac{1}{\beta_{j,k}} \right]}. \tag{13.27}$$

Here, also as before, fatigue-life parameters are in terms of the parameters of incremental damage,

$$\beta_{j,k} = \frac{\omega_k}{\mu_{j,k}} \quad \textit{and} \quad \alpha_{j,k} = \sqrt{\frac{\sigma_{k,k}(j)}{\omega_k \mu_{j,k}}} \quad \textit{for } j = 1, \ldots, \ell. \tag{13.28}$$

PROOF. By the argument given before, letting X_k denote the fatigue life of the kth component for $k = 1, \ldots, n$, we have by the duality theorem

$$\Pr[X_1 \geq t_1, \cdots, X_n \geq t_n] = \Pr[Z_1(t_1) \leq \frac{1}{\bar{\alpha}_1} \xi(\frac{\bar{\beta}_1}{t_1}), \ldots, Z_n(t_n) \leq \frac{1}{\bar{\alpha}_n} \xi(\frac{\bar{\beta}_n}{t_n})]. \tag{13.29}$$

By analogy with preceding results, in eqn (12.9) of theorem 1, we consider the kth component. Here as in Theorem 1, if each $\bar{\beta}_i$ increases and $\bar{\alpha}_i$ remains fixed for $i = 1, \ldots, n$, while the covariance between the standardized incremental damages to each component remains constant (which can be found by considering $t_1 = \cdots = t_n = t$) we obtain joint normality, asymptotically, as expressed in eqn (13.6). □

Let us assume this $\mathcal{MVFL}$ distribution can be used to represent the probability of failure at a given time to an ensemble of components of a structure due to cumulative fatigue damage. It can be used to obtain an improved value for the design life of such a multi-component structure subjected to a stationary loading spectrum or the probability of failure at a specified life. Usually the life of any structure is conservatively bounded by assuming it to be series system of independent components since the $100\gamma\%$ safe-life is the value ζ_γ such that no component fails before that time with probability γ, viz.,

$$\Pr[\min(X_1, \ldots, X_n) > \zeta_\gamma] = \gamma \quad \text{for some } \gamma \text{ near unity.}$$

For the $\mathcal{MVFL}$ distribution we have from eqn (13.29) that

$$\Pr[\min(X_1, \ldots, X_n) > t] = \Pr[Z_1 > \tfrac{1}{\alpha_1} \xi(t/\beta_1), \cdots, Z_n > \tfrac{1}{\alpha_n} \xi(t/\beta_n)],$$

where $\boldsymbol{Z} \sim \mathcal{N}(\boldsymbol{O}, \boldsymbol{P})$ with $\boldsymbol{P}^{-1} = (r_{i,j})$. This probability can be evaluated by numerical means using the integral

$$\Pr[\min_{i=1}^{n} X_i > t] = \int_{\xi(t/\beta_1)/\alpha_1}^{\infty} \cdots \int_{\xi(t/\beta_n)/\alpha_n}^{\infty} \frac{\exp\{-\frac{1}{2}\sum_{i,j} z_i r_{i,j} z_j\}}{\sqrt{(2\pi)^n |\boldsymbol{P}|}} dz_1 \cdots dz_n. \tag{13.30}$$

An efficient method of calculating such integrals is known; see [37, 1991].

One may also obtain an easily computed bound on the safe-life if we set

$$\varpi(t) = \max_{i=1}^{n}\{x_i^*(t)\}, \quad \text{where} \quad \boldsymbol{x}^*(t) = \left(\tfrac{1}{\alpha_1}\xi(\tfrac{t}{\beta_1}), \cdots, \tfrac{1}{\alpha_n}\xi(\tfrac{t}{\beta_n})\right) \boldsymbol{P}^{1/2}, \tag{13.31}$$

since we have the well-known result, $\Pr[\min_{i=1}^{n} X_i > t] \geq \Pr[\chi_n^2 \leq \varpi(t)]$.

Corollary 2. *If $\psi(x_1, \ldots, x_n)$ is a coherent system of n components having minimal path sets $\mathcal{P}_1, \ldots, \mathcal{P}_p$ where $(X_1, \ldots, X_n)$ is the vector of fatigue lives of the components, then T, the time until fatigue failure of the system, is given by $T = \max_{j=1}^{p} \min_{i \in \mathcal{P}_j} X_i$, and its distribution is given by*

$$\Pr[T > t] = \Pr\{\bigcup_{j=1}^{p} \bigcap_{i \in \mathcal{P}_j} [Z_i > \xi(t/\beta_i)/\alpha_i]\}. \tag{13.32}$$

Corollary 3. *The fatigue-life vector, $\boldsymbol{X}$, of the components of a system as defined in eqn (13.6), are associated with $r_{i,j} \geq 0$ for $i \neq j$.*

PROOF. This follows from the joint normality of a monotone transformation of each component's life and the nonnegative correlation between each of them; see [7, 1985]. □

Exercise Set 13.C

1. Three common life distributions, fatigue-life, Galton, and Weibull, when similarly parameterized for $x > 0; \alpha, \beta > 0$, are

$$F_1(x) = \Phi[\frac{1}{\alpha}\xi(x/\beta)], \qquad F_2(x) = \Phi[\frac{1}{\alpha}\ln(x/\beta)], \qquad F_3(x) = 1 - \exp[-(x/\beta)^{1/\alpha}].$$

   Let $\kappa_i(\alpha)$ be the corresponding coefficient of variation and show that it does not depend upon the scale parameter β since for $i = 1, 2, 3$ we find:
   (a) $\kappa_1(\alpha) = \frac{\alpha\sqrt{1+5\alpha^2/4}}{1+\alpha^2/2}$, (b) $\kappa_2(\alpha) = \sqrt{e^{\alpha^2} - 1}$, (c) $\kappa_3(\alpha) = \sqrt{\frac{\Gamma(1+2\alpha)}{\Gamma^2(1+\alpha)} - 1}$.

2. Despite appearances, in each case $\kappa_j(\cdot)$ for $j = 1, 2, 3$ is approximately α, when α is small. Show that the maximum discrepancy, as defined in each case, is
   (a) $\sup\{\kappa_1(\alpha)/\alpha : \text{for } 0 < \alpha < 2/\sqrt{3}\} = 5/2\sqrt{6} = 1.0206$, with $\kappa_1(1) = 1$.
   (b) $\sup\{\kappa_2(\alpha)/\alpha : \text{for } 0 < \alpha < 1\} = \sqrt{e} = 1.6487$.
   (c) $\sup\{\kappa(\alpha)/\alpha : \text{for } 0 < \alpha < 1\} = \frac{\pi}{\sqrt{6}} = 1.2825$.
   (d) Comment on why $\mathcal{W}(\alpha, \beta)$ was not included.

3. Consider an iid sequence of stresses each of which is equally allocated by configurational geometry to two identical structural components. Consequently, the engendered damage increments to each component, say $\{X_{i,j}\}_{i=1}^{\infty}$ for $j = 1, 2$ and all $i = 1, 2, \cdots$, have

$$\mathrm{E}X_{i,1} = \mathrm{E}X_{i,2} = \nu, \quad \mathrm{Var}X_{i,1} = \mathrm{Var}X_{i,2} = \kappa^2, \quad \text{and} \quad \mathrm{Corr}[X_{i,1}, X_{i,2}] = \rho > 0.$$

Define, in analogy with Section 10.2, the two variates for all $n = 1, 2 \cdots$

$$S_{n,j} = \sum_{i=1}^{n} X_{i,j}, \qquad Z_{n,j} = \frac{n\nu - S_{n,j}}{\sqrt{n}\kappa}, \qquad \text{for } j = 1, 2.$$

   (a) Calculate covariance and correlation between $S_{n,1}$ and $S_{n,2}$ and between $Z_{n,1}$ and $Z_{n,2}$.
   (b) Derive the joint distribution of the waiting times until failure of the two components by considering the asymptotic joint distribution of

$$\mathrm{P}[Z_{n,1} \geq \tfrac{\sqrt{\nu\omega}}{\kappa}\xi(\tfrac{n\nu}{\omega}), Z_{n,2} \geq \tfrac{\sqrt{\nu\omega}}{\kappa}\xi(\tfrac{n\nu}{\omega})].$$

   (c) Compare, in this case, the easily computed safe-life in eqn (13.31) with an actual bound found from eqn (13.30).

4. Let $X_i \sim \mathcal{FL}(\alpha_i, \beta_i)$ for $i = 1, 2$ be dependent rvs that are positively correlated as $\mathrm{E}[Z_1 Z_2] = \rho > 0$ where $Z_i = \frac{1}{\alpha_i}\xi(\frac{X_i}{\beta_i})$ for $i = 1, 2$. Show that this correlation is a nasc that $\mathrm{Cov}[\frac{X_1}{\beta_1}, \frac{X_2}{\beta_2}] = \frac{(\alpha_1\alpha_2\rho)^2}{2}$.

13.4. Implementation

A calculation of the useful life of every structure, the failure of which endangers public weal, should include the fatigue-life capabilities of its components. We mention two. The service lives of bridges, those which have been constructed using prestressed and partially prestressed concrete girders, are principally limited by the fatigue tolerance of its girders under the frequency and magnitudes of the loadings induced by vehicular traffic. For such bridges the prestressing steel, the precast girders, the cast-in-place slab and the tensile reinforcing bars, which determine its load capacity, are the components of principle concern.

As in other applications, the fatigue-life tests of the components must often be aborted before all the specimens have failed when the stress ranges encountered in practice are employed. This can happen for a number of reasons such as: the exhaustion of time, patience or money allocated for the test; results are required, for political reasons, before testing is completed; failure of the fatigue-testing machine; failure of the specimen due to accidental overload or *any cause* other than the one being measured.

Another interesting application arises in the assessment of the life of dental prostheses. What is observed in the clinical behavior of *pfm systems* (porcelain-fused-to-metallic-alloy) that are used in dental restorations is cumulative damage over time that occurs in a number of related modes. The eventual failure, or replacement of the dental restoration, may be due to (a) excessive gap at the margin, (b) excessive porcelain crazing, (c) the loss of veneer-to-metal integrity, (d) fatigue failure of the alloy. But restorative removal may also occur due to: (I) unacceptable porcelain aesthetics (staining), (II) unrelated periodontal disease, (III) dissatisfaction for a combination of personal reasons. Clearly, the service life of a prosthesis is a multivariate cumulative damage process (some components of which are dependent) with a complex censoring of service life. The application of the multivariate fatigue-life distribution would seem to be preferable, for diagnostic purposes, to assuming independent extreme minimum-value (Weibull) distributions. The specific statistical problems arising in these cases will not be addressed here.

13.4.1. Estimation for Small Censored Samples

Can unbiased, objective methods of estimation be utilized for *CD*-distributions when few or none of the test specimens have failed? Does the engineer always have to resort to legerdemain such as safety factors or Bayesian conjugate-priors in such instances?

There have been some controversial methods suggested to address this, including obtaining a subjective prior by omphaloskepsis (using one's gut feeling). We discuss these methods separately.

1. Seek "prior engineering information" which can be used to alter the likelihood so as to obtain acceptable estimates. Such prior information may be supplied by the engineer in several ways.
   (a) Place bounds on the parameter space because of known physical restrictions.
   (b) Introduce a Bayesian prior density for the characteristic life β and then, by assuming the calculus of probability holds for opinion, derive a posterior distribution after incorporating the likelihood of unfailed data. [NB this requires that either the characteristic life (a constant of Nature) or the modeler's opinion of its value be treated as an rv; neither assumption can be empirically verified.]
   (c) Alter the likelihood by including a fictitious observation larger than the run-outs, presuming failure would have occured had that additional test time been observed. Including this figmental failure datum allows the mle procedure to be used as before. This is a conservative procedure.
2. Unify the estimation of both characteristic life and coefficient of variation by utilizing their functional dependence upon the imposed stress regime.

The difficulty in Procedure 1 is that all of the methods mentioned are subjective and as a consequence the results will vary with the knowledge, experience, and economic interest (read bias) of the investigator and his(her) employer. [It is difficult for anyone to be objective when (s)he thinks his(her) livelihood depends on not being objective.] Another objection is that often only the time of failure is observed with no account taken of the nature of failure or of those aspects of service life, such as the material, the duty cycle, or the stress regime (which engineers know to be important). The difficulty with having only run-outs in the data, when the mle is to be used, is that the posterior probability of observing no failures is maximized if the characteristic life is infinitely large, which assumption is contrary to all but religious conjecture. This unacceptable answer occurs not from the statistical likelihood but from a failure to include in the model, as an appropriate constraint, additional information making infinite life itself unlikely.

13.4.2. Relating Cumulative-Damage Parameters to the Exposure

The shape (coefficient-of-variation) and scale (median-life) parameters, respectively α and β, governing the cumulative-damage distribution of the service-life for each component, are determined from the damage imposed by its stress regime or environmental exposure. Let $\boldsymbol{y}_k = \left(y_1, \ldots, y_{n_k}\right)$ be a data-set containing $n_k$observations under a stress regime, say ς_k, and environment, say ϑ_k, from which estimates $\hat{\alpha}_k, \hat{\beta}_k$ of the parameters are to be calculated using maximum likelihood techniques. In practice this may be impossible under $k = 1, \ldots, m$ different sets of conditions because life testing is very expensive, and so each the sample sizes $n_k$will be small, and also there may be run-outs in the set. To estimate two parameters with three observations, one of which is a run-out, will give an estimate of low precision.

What is desired are the functional relationships, label them ϱ and ϖ, which determine the distributional parameters of life under various realistic conditions, viz., $\alpha = \varrho\,(\varsigma, \vartheta)$ and $\beta = \varpi\,(\varsigma, \vartheta)$. Unfortunately a safe (or *warranteed*) life must often be certified for a product under anticipated service conditions (involving stress and environment) which may not be those under which any tests have been run! Consequently determining the realistic functional relationships between the distributional parameters α, β and the stress-environment regime (ς, ϑ) lie at the core of such a procedure. For certain distributions in special applications there have been parameter-stress relationships proposed, such as the Arrhenius, Eyring and inverse-power laws, all of which make the logarithm of the scale parameter linearly related to stress. Inference made under these assumptions for the Weibull distribution has been presented in [69]. Under certain stochastic conditions the environment can be disregarded and so both functions, ϱ and ϖ, depend only upon the stress ς for some models. These quantitative relationships can sometimes be postulated using classical deterministic theories of stress analysis to give parameter values.

Consider another simple model to do this for the $F \simeq FL\,(\alpha, \beta)$ distribution with both ϖ, ϱ functions of the stress ς only. Let us set, for $x > 0$

$$F(x) = \Phi\left[\frac{1}{\alpha_0}\left(\frac{\sqrt{x}}{\beta_0} - \frac{\varpi}{\sqrt{x}}\right)\right] = \Phi\left[\frac{1}{\alpha}\xi\left(\frac{x}{\beta}\right)\right], \tag{13.33}$$

$$\text{where} \quad \alpha = \alpha_0\sqrt{\frac{\beta_0}{\varpi}}, \beta = \beta_0\varpi.$$

In this case the scale parameter is proportional to ϖ, an arbitrary function of the stress, but there is consonance for each data-set since $\alpha^2\beta = (\alpha_0\beta_0)^2$, which makes joint estimation possible. An advantage of this formulation in terms of stress levels is that when tests of low stress regimes, approximating those stress levels encountered in service, contain no failures. This circumstance allows that information to still be utilized.

In a more general case we could have different coefficients to determine; these are the 'influence' parameters, label them ***a, b***, where $\beta = \varpi(\varsigma, \boldsymbol{a})$ and $\alpha = \varrho(\vartheta, \boldsymbol{b})$ with ϖ and ϱ known except the parameters $\boldsymbol{a}$ and $\boldsymbol{b}$ are not. All the parameters (***a, b***) must be estimated using the entire matrix $\mathcal{M}$ of information collected from all m sets of censored service-life data, each of which was obtained under a different stress regime. Let us represent this information as the matrix,

$$\mathcal{M} = \begin{bmatrix} \boldsymbol{y}_1, & \varsigma_1, & \vartheta_1 \\ \cdots & , & \cdots \\ \boldsymbol{y}_m & \varsigma_m, & \vartheta_m \end{bmatrix}.$$

Here during the jth life test the data vector,y_j, some elements of which may be run-outs, was obtained under the stress-environmental condition ς_j, ϑ_j for $j = 1, \ldots, m$.

An advantage of this formulation, in terms of statistical estimation and stress analysis, is that if some of the tests with stress regimes near those stress levels encountered in service contain no failures, such information can still be utilized. We obtain the joint log-likelihood of the unknown influence coefficients, given the test information matrix $\mathcal{M}$, a very complicated expressions, and we represent it as

$$\pounds\,(a, b|\mathcal{M}) = \sum_{j=1}^{m} \ell\left[\alpha(\boldsymbol{a}), \beta(\boldsymbol{b})\,|\boldsymbol{y}_j, \varsigma_j, \vartheta_j\right].$$

This then must be maximized computationally, using software such as *Mathematica*, with respect to the unknown influence constants $(\boldsymbol{a}, \boldsymbol{b})$.

Notice that by estimating shape and scale parameters within each of the m data-sets there would be $2m$ parameters to be estimated but the number of influence parameters to be estimated is only the sum of the dimension of $\boldsymbol{a}$ and $\boldsymbol{b}$, which will often be far smaller than $2m$.

Nature may be too complex (or perverse) in some engineering applications for ϱ and ϖ to be linear functions of their arguments and so linear regression techniques, such as that of Bhattacharyya and Fries [15], might be inapplicable. If so one must then depend upon a fuller knowledge of the chemistry or physics of failure, correspondingly abetted by some knowledge of reactions in physical deterioration or engineering knowledge of stress response, to formulate a usable quantified model. *Statistical Theory by itself may not always be sufficient.*

Exercise Set 13.D

1. If a sample $y_1+ \leq \cdots \leq y_n+$ contains only run-outs, what physical reasons can be given for the parameter space of the $\mathcal{FL}(\alpha, \beta)$ distribution to be restricted so as to produce reasonable estimates of $\hat{\alpha}$ and $\hat{\beta}$.

2. In the model of eqn (13.33) show that if ϖ is a decreasing function of stress ς thcn both the expectation of life and its variance decrease as stress increases, in accord with one's intuition.

3. Can one formulate a model similar to eqn (13.33) for the Wald distribution? Since in that case the parameter α is a coefficient of variation, are there cases where as a function of stress the physics of failure dictates that it must decrease with increasing stress?

4. Select a life distribution, e.g., Weibull, Galton, and examine the statistical estimation problems which arise if one adopts the model of the scale parameter being a log-linear function of the stress. Then compare it with a corresponding $\mathcal{CD}$-distribution.

5. Formulate a general procedure for cyclic fatigue testing of partially-precessed concrete bridge girders, which conceptually consists of three components in series, viz., pre-stressing steel, concrete and reinforcing steel (testing bridge girders to failure is very expensive).

6. * Construct a life-length prediction model for a particular situation of interest to you. Specify some class of stress regimes in servive, determine the stress regimes used in life-testing and calculate their anticipated cost of testing as well as that of warranteeing the service life which you have subsequently estimated. Does minimizing cost always produce the same answer as maximizing profits?

To any question, scientific or philosophical, one should not believe in what one hopes the answer will be, before one finds out what the answer is. This applies to the Shakespeare authorship question and to global warming as well as to Intelligent Design.

Richard Dawkins

CHAPTER 14

Strength and Durability

14.1. Range of Applicability

14.1.1. Introduction

In his notebooks Leonardo de Vinci recorded experiments demonstrating that short lengths of iron wire were stronger, in tension, than longer ones. Galileo asserted that if one makes geometrically similar structures of different dimensions the larger will be the weaker, see, e.g., Timoshenko [111, pp. 5–7]. This phenomenon is called the *size effect*. It is well known today that scale models of canal/river locks or dams, not to mention bridges, airplanes, etc., may fail to show defects that appear in the real structure. This may be because neither gravity nor the environment (often too the composition and density of air or water) can be scaled. Moreover, when failure stems from the exceedence of a critical load the conditions producing it may not be invariant with scale.

Certain aspects of this size-effect were first explained in the 1920s by Griffith's flaw theory. He deduced that the weakening of isotropic solids with their increasing size was due to the increased number of internal flaws or "discontinuities" and concluded that effective strength could be improved by their reduction; see his theory of rupture in [40]. The initial application by Weibull of Gumbel's "third asymptotic distribution of minimum values" (the distribution later named for him) was in connection with his study of rupture in solids; see [119, 1939].

The fatigue of metals is a surface phenomenon. The first noted occurrence of fatigue failure, in the life of railroad axles during the 1860s, was forestalled by milling off a few thousandths of an inch of the wheel surface, at about half the expected service life. The fatigue life of aluminum is extended if the surface is electropolished, but electropolishing large components is difficult. The fatigue life of certain materials after being coated with cosmoline for protection is different whether they are cleaned with acetone or with alcohol. The service life of polymers is often governed by surface condition, i.e, the density and severity of micropitting.

Consequently, to measure either ultimate or fatigue strength from sample coupons, which data are then used to predict reliability of components of such material in a system even when knowing the stress regime, without taking such appropriate external factors into account is to do more than risk one's own scientific reputation.

14.1.2. Reliability Analysis of Strength

The probabilistic analysis of strength becomes an important problem for components of any structure in which high reliability standards are imposed, usually because failure endangers either public safety or weal, such as would occur in nuclear power plants, high-rise buildings, long-span bridges, et al. Strength problems for static loads are often assumed to be time-invariant, and frequently cumulative damage is not considered.

The probability of failure, say p_f, is defined by an integral

$$p_f = \int_{D_f} f_{\boldsymbol{X}}(\boldsymbol{x})\, d\boldsymbol{x} = \Pr[\boldsymbol{X} \in D_f],$$

where $\boldsymbol{X}$ is the multidimensional vector of strength variates and D_f is the domain of failure for the component. It is determined from the failure criterion for the component by a limit-state function, say $g(\boldsymbol{X})$ with $D_f = [g(\boldsymbol{X}) \le 0]$. In general the evaluation of this multidimensional integral is very difficult, and much theoretical effort has gone into efficient computational methods, e.g., see the text on finite element analysis [21].

Static Strength for a Column

For illustration we reconsider a typical example from [87], the evaluation of $p_f = \Pr[.85 A_c f_c' + A_s f_y \le P]$, for a short, reinforced, concrete column where it was used *loc. cit* for a comparison of Monte Carlo computational methods. The definition of these variates is given in the table below, where "Gumbel" presumably refers to his maximal law on $\Re_+$ [see eqn (5.6)], and COV stands for the coefficient of variation.

Variable	Description	Distribution	Mean	COV
A_c	net cross-sectional area of concrete	Normal	800 [cm^2]	.05
f_c'	compress. strength	Lognormal	2.747 [kN/cm^2]	.20
A_s	area of steel	Normal	40 [cm^2]	.025
f_y	yield strength	Lognormal	27.47 [kN/cm^2]	.10
P	axial force	Gumbel	784.8 [kN]	.30

Let us rescale all variates using their means. Thus, the failure event can be rewritten, setting $a = 1.7$, $b = .714234$, as

$$[.85 A_c f_c' + A_s f_y \le P] = [a U_1 U_2 + V_1 V_2 \le b W],$$

and all U_i, V_j, W variates now have unit means with their coefficients of variation unchanged!

If all U_i, V_j are independent and nonnegative a.s., then we have

$$F_{U_1U_2}(u) = \int_0^\infty F_{U_1}(u/x)\,dF_{U_2}(x) \quad \text{for } u > 0, \tag{14.1}$$

with a similar integral for $F_{V_1V_2}$. Thus, $Z := aU_1U_2 + V_1V_2$ has a distribution given by

$$F_Z(z) = \int_0^{z/a} F_{V_1V_2}(z - au)\,dF_{U_1U_2}(u), \quad \text{for all } z > 0, \text{ with } a = 1.7. \tag{14.2}$$

Hence, we can express the the failure probability as

$$p_f = \int_0^\infty F_Z(w)\,dF_W(bw) = \int_0^\infty \bar{F}_W(bw)\,dF_Z(w), \quad \text{where } b = .714234. \tag{14.3}$$

A calculated unreliablity as low as 10^{-5} depends only upon the tail-behavior of the distributions of Z and W and not at all upon regions of central tendency! Thus the answer depends upon what distributional assumptions are made as well as its accurate computation.

The distributions of the area and strength variates chosen in the table above should be carefully noted. For example, if one took measurements of the critical diameter (or circumference) along such concrete columns, say the statistic is $\hat{D}$, then should not the net critical area variate be near $\pi\hat{D}^2/4$? How does this area become normal even when $\hat{D}$ is? While the square of a Galton variate is a Galton variate, the square of a Gaussian variate is not Gaussian! Here it is not the random variation in the measurements of a fixed object but the exact measurement of varying objects from a population that is being considered.

Exercise Set 14.A

1. You must have a "ball park" answer to the column-strength problem for an impending meeting. So assume the products U_1U_2 and V_1V_2 are both Gaussian variates:
   (a) Calculate their corresponding parameters from the table above.
   (b) Find the parameters of Z to determine F_Z, without using eqn (14.2).
   (c) Assume $bW = Y \sim \text{Wei}(\delta, b)$ with $\delta = 10/3$ then use Mathematica, or a similar program, to calculate p_f. (Identify your possible sources of error.)

2. Assume the two area-variates are truncated-Gaussian, eqn (2.29), and redo the reliability calculation assuming the Vs are $\mathcal{FL}$. What now are the sources of error?

3. Make the original distributional assumptions and perform the same computations. What do you conclude? Is having one parameter, which is the COV, with the relation to normality of the $\mathcal{CD}$-distributions as shown in eqn (10.45), anymore than a computational convenience?

4. Assume $U_i \sim \mathcal{FL}(\alpha_i, 1)$ and $V_j \sim \mathcal{FL}(\kappa_j, 1)$ where $\alpha_1 = .20, \alpha_2 = .05$, with $\kappa_1 = .10, \kappa_2 = .025$:

(a) Show the product U_1U_2 has distribution given by the integral

$$F_{U_1U_2}(u) = \int_{-\infty}^{\infty} \Phi\left\{ \tfrac{\xi(u)}{\alpha_1} \sqrt{1 + (\tfrac{\alpha_2 x}{2})^2} + (\tfrac{\alpha_2 x}{2}) \tfrac{\psi(u)}{\alpha_1} \right\} d\Phi(x).$$

(b) Perform the computations required to calculate eqn (14.1) and the corresponding density of V_1V_2; then program the computation of F_Z in eqn (14.2).

(c) Let the Gumbel-maximum distribution be $G(y) = \exp\{-(b/y)^\delta\}$ for $y > 0$ where $b = .714234$ and $\delta = 5.1843$. and then verify its cov is .30.
Is there any computational value in letting $T = (b/W)^\delta$?

(d) Calculate (14.3), for this case, when making $\mathcal{FL}$ assumptions. What happened?

(e) Would there be any value in the above by substituting Wald variates.

NB The value $p_f = 2.252 \times 10^{-5}$ was given, *loc. cit.*, as "exact".

14.1.3. Strength of an Airframe Subsystem

Consider the reliability problem for the failure of force-summed actuators in a flight-control system for a supersonic aircraft. The interaction of actuators and flight control can be depicted as follows:

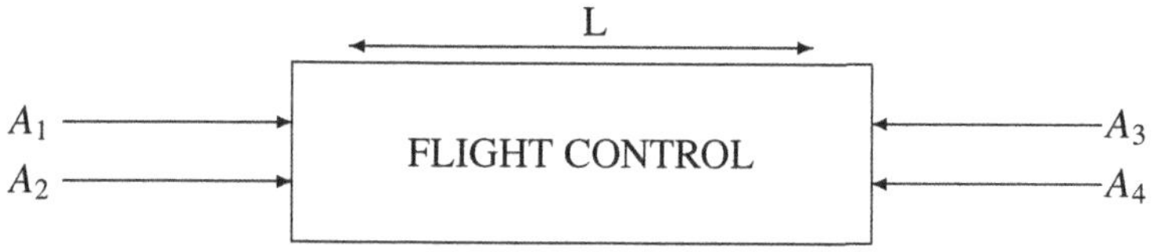

The system consists of duplicate pairs of actuators that oppose each other and an internal resistance to either motion. It is assumed that each time during flight when the jth actuator transmits a spurious signal it will have a Gaussian distribution with a nominal value $\mu_j > 0$, which is also the mean, and a standard deviation which is proportional to the nominal value. (Hence we assume the actuator magnitude follows Brownian motion.) Thus, for some $r > 0$,

$$A_{i,j} \sim \mathcal{N}(\mu_j, r^2\mu_j^2) \quad \text{for } j = 1, \cdots, 4 \text{ and } i = 1, 2, \cdots.$$

Exercise Set 14.B

1. (a) Assume the internal resistance $L = \ell + Z_k$ where Z_k is the random increased drag due to k failed actuators for $k = 0, 1, 2, 3$ with $\mathrm{E}Z_k = k\nu$ and a coefficient of variation of r.

   (i) Assume $k = 0$ so L has a fixed value ℓ, and so $Y_i = \sum_{j=1}^{2} A_{i,j} - \sum_{j=3}^{4} A_{i,j}$ are iid for $i = 1, 2, 3, \cdots$. The system fails when for the first time either Y_i exceeds ℓ or $-Y_i$ is less than $-\ell$. If $\mu_1 = \cdots = \mu_4 = 450$ and $r = 1/30$ $\nu = 150$ what value of ℓ will keep the probability of false movement per spurious signal less than .002?

(ii) Redo this calculation with one actuator failed, with two opposing actuators failed, and with three actuators failed.

(b) Now find the distribution of N, the number of spurious signals until the flight control system moves (becomes unstable). Then assuming an Epstein interarrival time between spurious signals determine the distribution of flight-time until flight-control instability occurs, assuming all the actuators are in operation.

2. If one knew the distribution of the service life of the actuators could one calculate the reliability of this flight-control unit?

3. In a fleet of 200 planes, what would be the time until a first fleet failure? Make a list of other questions you believe will likely be asked during your presentation. How many can you answer now?

14.2. Accelerated Tests for Strength

The application of reliability distributions to assess the strength of materials was discussed in an article by A. S. Watson and R. L. Smith (see [118, 1985]), and later Smith introduced a regression model for the Weibull distribution (see [108]). A study of the applicability of the Galton distribution to carbon-fiber strength was made by Own et al. in [86] and a log-linear model for the $\mathcal{FL}$-distribution was introduced by Reick and Nedelman, in [93].

Of chief interest was the development of theoretical methods enabling the observation/measurement of life data under more stringent conditions than anticipated in the duty cycle and so to shorten the testing period yet be able to estimate the service life well. One such study is W. J. Owen's and W. J. Padgett's article in [85, 1999], concerning a unification of several previous cumulative damage models for the strengths of systems, all of which were based on the $\mathcal{FL}$-distribution. These papers, by Padgett et al., given in the references in *loc. cit.*, contained, in the authors' words,"four new models that are three-parameter extensions of the Birnbaum-Saunders distribution," all of which were obtained using the same type of argument that was used to derive the cumulative damage distributions. This unified cdf for the system strength S, with a known "acceleration variable" ℓ is

$$G(s;\ell) = \Phi\left\{\frac{1}{\alpha}\left[\frac{\sqrt{s}}{\beta} - \frac{\lambda(\theta;\ell)}{\sqrt{s}}\right]\right\}, \quad \text{for } s > 0; \alpha > 0, \beta > 0,$$
$$\text{with } \lambda(\theta;\ell) > 0 \text{ for all } \theta, \ell. \qquad (14.4)$$

Here $\lambda(\theta;\ell)$ is a function with θ unknown, but ℓ is identified as a physical quantity, "gauge length," in tensile applications. These models were applied to the fatigue life of gun barrels and to the tensile-strength distribution for composite materials (carbon fibers embedded in a polymer). Here the incorporation of the gauge length with its concomitant size effect allows a prediction to be made of the distribution of service life at gauge lengths which were not tested. The four models for $\lambda(\theta;\text{ł})$, introduced and validated by the datasets presented under the different circumstances, are

$$\lambda_1(\theta,\text{ł}) = \text{ł}^{1/\theta}, \quad \lambda_2(\theta,\text{ł}) = \theta - \sqrt{2\text{ł}/\pi}, \quad \lambda_3(\theta,\text{ł}) = \theta - \ln\text{ł},$$
$$\lambda_4(\theta,\text{ł}) = \ln\theta - \frac{\sqrt{\text{ł}}}{\theta}\sqrt{\frac{2}{\pi}} - \frac{\text{ł}}{2\theta^2}. \tag{14.5}$$

Unfortunately, the mles $\hat{\alpha}$, $\hat{\beta}$, $\hat{\theta}$ can be determined only implicitly through functions of the likelihood. These equations must be solved using numerical techniques, e.g., by iteration with a nonlinear root finding method, as are in *Mathematica* and elsewhere, until convergence is obtained.

What is important here is to note that one can write for $j = 1, \cdots, k$ and $s > 0$ that

$$G(s;\text{ł}_j) = \Phi\left[\frac{1}{\delta_j}\xi(s/\nu_j)\right] \qquad \text{where } \delta_j = \frac{\alpha\sqrt{\beta}}{\sqrt{\lambda(\theta;\text{ł}_j)}}, \quad \text{and } \nu_j = \beta\lambda(\theta;\text{ł}_j). \tag{14.6}$$

So (14.4) is only a reparameterization and thus the mean and variance of S can be obtained by inspection from (10.17) and (10.18), the equations for mean and variance in the original formulation.

NB that for a number of sample observations from gauge-length ł_j, viz.,

$$X_{i,j} \sim G(\cdot;\text{ł}_j) \quad \text{for } i = 1, \cdots, n_j \text{ and } j = 1, \cdots, k,$$

that $\delta_j^2\nu_j \equiv \alpha^2\beta$ independent of $\lambda(\theta;\text{ł})$; so these samples are all *consonant*. Hence any of the applicable tests in dispersion analysis, previously discussed, would apply.

Lemma 1. *If $X_i \sim \mathcal{FL}(\alpha_i.\beta_i)$ are independent and consonant statistic variables for $i = 1, \cdots, n+1$ with*

$$X_+ = \textstyle\sum_1^{n+1} X_i, \quad \beta_+ = \sum_1^{n+1} \beta_i, \quad \alpha_{(-2)} = \left(\sum_1^{n+1} \alpha_i^{-2}\right)^{-1/2}, \quad \theta = 2\alpha_{(-2)}^2\beta_+,$$

then F_{X_+}, the cdf of X_+, can be readily computed for any $n \geq 1$ by the double integral

$$F_{X_+}(x) = \int_{y=0}^{x}\int_{u=0}^{(x-y)/\theta} \Psi_n(u)\,du\; d\Phi[\tfrac{1}{\alpha_{(-2)}}\xi(\tfrac{y}{\beta_+})], \tag{14.7}$$

where for $u > 0$ we have the weighted sum of $n \geq 1$ gamma densities given by

$$\Psi_n(u) = \frac{1}{2^n}\sum_{j=1}^{n}\binom{n}{j}\frac{u^{j/2-1}e^{-u}}{\Gamma(j/2)}.$$

PROOF. We see from eqn (10.13) that $F^{\dagger}_{X_+}$, the transform of the cdf of X_+, is under consonance,

$$F^{\dagger}_{X_+}(t) = 2^{-n}\exp\left\{\frac{1}{\alpha^2_{(\text{-}2)}}\left(1-\sqrt{1-2t\alpha^2_{(\text{-}2)}\beta_+}\right)\right\}\times\left(\frac{1}{2}+\frac{1}{2\sqrt{1-2t\alpha^2_{(\text{-}2)}\beta_+}}\right)$$
$$\times\sum_{j=0}^{n}\binom{n}{j}\left(1-2t\alpha^2_{(\text{-}2)}\beta_+\right)^{-j/2}$$
$$= 2^{-n}\textstyle\sum_{j=1}^{n}\binom{n}{j}G^{\dagger}(t;\theta,j/2)\cdot F^{\dagger}(t;\alpha_{(\text{-}2)},\beta_+,1/2),$$

where the transforms are of $F(\cdot;\alpha_{(\text{-}2)},\beta_+,p)$, the $\mathcal{CD}$-cdf in eqn (10.12), and the gamma distribution

$$G(x;\theta,\kappa) = \int_0^{x/\theta}\frac{u^{\kappa-1}e^{-u}}{\Gamma(\kappa)}\,du \quad \text{for } \theta,\kappa > 0.$$

Hence

$$F_{X_+}(x) = 2^{-n}\sum_{j=1}^{n}\binom{n}{j}\int_{y=0}^{x}G(x-y;\theta,j/2)\,d\Phi[\tfrac{1}{\alpha_{(\text{-}2)}}\xi(\tfrac{y}{\beta_+})],$$

which gives the result in eqn (14.7). □

We also state without proof the

Lemma 2. *If $X_i \sim \mathcal{CD}(\alpha_i,\beta_i,p)$ for $i = 1,\cdots,n$ are independent and consonant, then from eqns* (10.17) *and* (10.18)

$$\mathrm{E}X_i = \beta_i(1+q\alpha_i^2),\quad \mathrm{Var}X_i = (\alpha_i\beta_i)^2[1+q(2+p)\alpha_i^2] \quad \text{for } i = 1,\cdots,n,$$

and by writing $\bar{X} = \frac{1}{n}\sum_1^n X_i$, $\bar{\beta} = \frac{1}{n}\sum_1^n\beta_i$, *and* $\bar{\alpha}_{(\text{-}2)} = (\sum_1^n\alpha_i^{-2}/n)^{-1/2}$, *we have*

$$\mathrm{E}\bar{X} = \bar{\beta}[1+q\bar{\alpha}^2_{(\text{-}2)}], \qquad \mathrm{Var}\bar{X} = \frac{[\bar{\alpha}_{(\text{-}2)}\bar{\beta}]^2}{n}[1+q(2+p)\bar{\alpha}^2_{(\text{-}2)}].$$

NB the cdf of $\bar{X}$ is readily determinable numerically from problem #4 in EXERCISE SET 11.B or the method used to obtain eqn (14.7).

14.2.1. Determination of the Part of Least Accord

To determine the range of parameters in which no discrepancy is detectable, by observing ordinary samples from either a $\mathcal{W}(\alpha,\beta)$ or a $\mathcal{FL}(\delta,\nu)$ distribution, we examine the difference of the two corresponding $\mathcal{CD}$-distributions when the means and variances are equal, viz.,

$$D(x) = \Phi[\tfrac{1}{\delta}\xi(\tfrac{x}{\nu})] - \Phi[\tfrac{1}{\alpha}\xi(\tfrac{x}{\beta})] - e^{2/\alpha^2}\Phi[-\tfrac{1}{\alpha}\psi(\tfrac{x}{\beta})], \tag{14.8}$$

when

$$\alpha = \tfrac{\delta\sqrt{4+5\delta^2}}{2+\delta^2}, \quad \beta = \nu(1+\tfrac{\delta^2}{2}) \ \text{ with } \ \beta > \nu, \ \alpha > \delta \ \text{ for } \ 0 < \delta < 1. \tag{14.9}$$

Since we wish to determine $\sup_{x>0} |D(x)| = \sup_{y>0} |D(\nu y)|$ we can wlog set $\nu = 1$. We can examine the derivative of D to determine its unique maximum. Utilizing Problem 3 in 10.B

$$D^{'}(x) = \tfrac{\psi(x)}{2\delta x}\varphi[\tfrac{\xi(x)}{\delta}] - \tfrac{\psi(x/\beta)-\xi(x/\beta)}{2\alpha x}\varphi[\tfrac{\xi(x/\beta)}{\alpha}]. \tag{14.10}$$

Let $\check{x}$ be the maximizer of D such that $D^{'}(\check{x}) = 0$. In eqn (14.10) we set $y = \xi(x)$ and $D^{'}(\xi^{-1}(y) = 0$ to obtain the equation

$$\tfrac{\psi\xi^{-1}(y)}{2\delta}\varphi(y/\delta) = \tfrac{1}{\alpha}\sqrt{\tfrac{\beta}{\xi^{-1}(y)}}\,\varphi\left\{\tfrac{1}{\alpha}\xi[\tfrac{1}{\beta}\xi^{-1}(y)]\right\} \quad \text{for some} \quad y \in \Re.$$

Using the $\psi - \xi$ identities from EXERCISE SET 10.B we have, asas,

$$\sqrt{\xi^{-1}(y)}\sqrt{1+\tfrac{y^2}{4}}\,\varphi(y/\delta) = \frac{\delta\sqrt{\beta}}{\alpha}\,\varphi\left\{\tfrac{1}{2\alpha}[\psi(\beta)y - \xi(\beta)\sqrt{4+y^2}]\right\},$$

$$\sqrt{\xi^{-1}(y)}\sqrt{1+\tfrac{y^2}{4}} = \frac{\delta\sqrt{\beta}}{\alpha}\exp\left\{-\frac{1}{8\alpha^2}\big[4y^2 - 2[\beta-(1/\beta)]\right.$$
$$\left.\times\, y\sqrt{y^2+4} + 4\xi^2(\beta)\big] + \frac{y^2}{2\delta^2}\right\},$$

which becomes, by setting $z = y/2$, an equation in z, namely,

$$(z+1+\sqrt{1+z^2})\sqrt{1+z^2} = A\exp\{az^2 + bz\sqrt{1+z^2}\}, \tag{14.11}$$

with the three constants a, b, A (which are functions of δ only)

$$a = 2\left(\frac{1}{\delta^2} - \frac{1}{\alpha^2}\right) = 2\left(\frac{1-\delta^2}{4+5\delta^2}\right),$$

$$b = \frac{\beta^2-1}{\alpha^2\beta} = \left(1+\tfrac{\delta^2}{2}\right)\left(1+\tfrac{\delta^2}{4}\right) \div \left(1+\tfrac{5\delta^2}{4}\right),$$

$$A = \frac{\delta\sqrt{\beta}}{\alpha}\exp\{-\tfrac{1}{2}[\xi(\beta)/\alpha]^2\} = \frac{(1+\frac{\delta^2}{2})^{3/2}}{\sqrt{1+\frac{5\delta^2}{4}}}\exp\left\{-\tfrac{1}{4}\left[\frac{\delta^2(2+\delta^2)}{4+5\delta^2}\right]\right\}.$$

Denote the solution of eqn (14.11) by $\check{z}_\delta$ (read zee-bird subdelta), then the maximum vertical discrepancy between the two distributions is a function of $\delta > 0$ only, label it $\varepsilon(\delta)$ where

$$\varepsilon(\delta) = \Phi[2\check{z}_\delta/\delta] - \Phi[\tfrac{\psi(\beta)\check{z}_\delta - \xi(\beta)\sqrt{1+(\check{z}_\delta)^2}}{\alpha}] - e^{2/\alpha^2}\Phi[\tfrac{\xi(\beta)\check{z}_\delta - \psi(\beta)\sqrt{1+(\check{z}_\delta)^2}}{\alpha}].$$

Calculating a few values with *Mathematica* gives $\varepsilon(0.1) = \varepsilon(0.2) = 0$ but, $\varepsilon(0.3) = 3.47 \times 10^{-9}$. This calculation is the basis for the thrasonic claim, made in Section 10.4, about the interchangeability of $CD(\alpha, \beta, p)$ distributions when

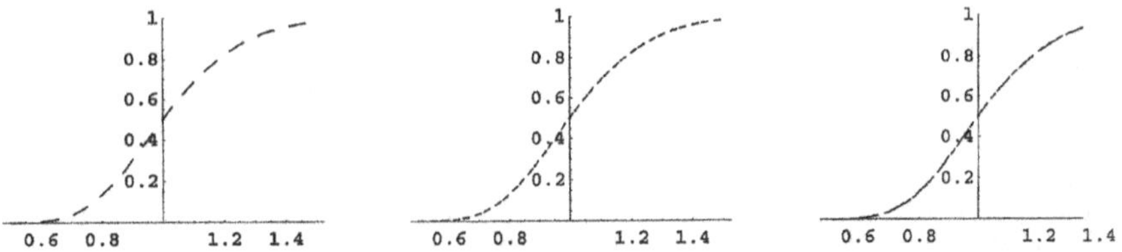

Figure 14.1. Left and center are matching $\mathcal{FL}$ and $\mathcal{W}$ distributions. Right is the combined plot.

$\alpha \leq 0.3$. See Figure 14.1 for a visual comparison when $\delta = .2$, $\nu = 1$ and α and β are as given in eqn (14.9) with equal means and variances.

Exercise Set 14.C

1. Provide the proof of Lemma 2.
2. In an optical device for measuring distances (such as the distance from the operatic stage to inaccessable points of acoustic interest such as suspended microphones) the means is by focusing on and determining the object-distance from the lens, say X, by measuring the image-distance, say Y, using the lens equation $\frac{1}{X} + \frac{1}{Y} = \frac{1}{f}$ where f is the known focal-length of the lens. (The problem is manufacturing sufficiently cost-effective and precise parts.) Presuming no measurement can be exact, let f be given and postulate a distribution of X as a function of f and Y. (Perhaps $\mathrm{E}[X\,|\,y] = \frac{1}{f} - \frac{1}{y}$). Then make assumptions assuming variability in X is due to imprecise focus by the user while Y is due tolerance allowed in machining.
   (a) Write out the distribution of X in general terms.
   (b) Consider the simplification possible if the variates has the reciprocal property.
   (c) Assume the conditional variate $X\,|\,y \sim \mathcal{W}(\alpha, \frac{1}{f} - \frac{1}{y})$ and utilize the $\mathcal{T}$-distribution.
   (d) Make some reasonable calculations if $f = 50$ mm and if $f = 100$ mm.

14.3. Danger of Extrapolation from Tests

Sherlock Holmes notoriously said, "Data, data, give me data. I can't make bricks without straw." Can anyone make a valid claim of statistically assured safety when no failure has been observed in a test at the stress regime anticipated during service and for which the component was designed? Manufacturers of airframes routinely destroy the first two units produced, one to ascertain ultimate strength and the other fatigue strength, in order to certify the unification of component testing and structural engineering theory. (An ultimate-strength test yields little or no information about fatigue service life, despite early hopes it might.) This show provides an empirical assurance of structural airworthiness to CEOs, the stockholders and the potential customers that no scientific calculations, apparently, can. (The full-scale crash testing of automobiles on TV accomplishes the same propaganda purposes.)

A reduction in full-scale service-life testing is the *sine qua non* of reliability/statistical analysis; the mirage of avoiding expensive testing arises frequently. In the admirable exposition of Markov-Chain methods and compilation of fatigue data in the book by Bogdanoff and Kozin [9, p. 198], from which we quote: (m/σ is the "mean to standard deviation ratio" i.e., the "signal to noise ratio").

> We observe that the m/σ versus stress amplitude for 6061-T6 aluminum and 1020-steel have the same general shape. This shape is maintained if we plot m/σ versus log mean life, $\cdots$ as is plotted in the book *loc. cit.* Stagg [60] noticed the same results for other types of aluminum. Moreover, Stagg [60] observed that the m/σ versus log mean life curve was independent of mean stress, that is, m/σ depends only on the mean life for a given material. This is an interesting observation. Suppose the observation was true and suppose we have found the m/σ versus log mean life for one material with mean stress = 0 over the range of mean life of interest with high accuracy. Then to determine variance under any cyclic loading condition, all we need to do is test sufficient replications to give a *good estimate of mean life;* the curve then provides a good estimate to variance without extensive testing. We use this procedure when discussing spectrum loading. *As far as we know, this idea has not been used to reduce the cost of testing* (our emphasis).
>
> We suspect that Stagg's observation is valid only for materials that exhibit stationary life behavior (no work hardening, etc.) under stationary cyclic loading and constant environment. In all events (the graphs cited show similar relationships between m/σ and log mean life) contain interesting food for thought.

Several points need comment. Firstly, while most of the comparisons given in [9] all show a closer fit of the B-models to the ecdf of the data than does a Weibull distribution, the B-models contain more parameters and often underestimate the probability of the earliest failure. If an engineer's concern is *safe-life*, it is the "toe of the distribution" that is of primary importance. In many applications, beside the Space-Shuttle or nuclear-power reactors, it is virtually the only one.

Secondly, almost any spectrum of loads can be rearranged so as to vary the mean fatigue life considerably. If there are compressive loads intermixed with tensile loads, this variation can be much increased even while the mean stress remains constant.

Lastly, let us clarify the meaning of the suppositional curve of "m/σ versus log mean life." Clearly there is a direct functional relationship between m/σ and $\log m$ so the independence meant is not functional. Can it mean that the complete-sample estimators of the mean and the signal-to-noise ratio are stochastically independent as are the sample mean and sample variance for the Gaussian distribution? This would be a theorem about the assumed life distribution and would not depend upon observation. Or could it mean simply that the data cited supported an underlying life distribution where the coefficient of variation was functionally independent of some parameter controlling the mean, such as are all the distributions in the first problem in EXERCISE SET 13.C? Perhaps the distribution may be similar to Gumbel's extreme laws where the variance is always the same so the reciprocal of the coefficient of variation is indeed proportional to the mean and therefore a simple function of the log mean. Of course under almost any spectrum of loads, the mean fatigue life, even fixing the material, is not a simple function of the mean stress, but is a complicated function of both the loads and their order.

14.3.1. Relating Parameters to the Exposure

There have not been many stochastic models introduced with distributional parameters of component life which are functions of the applied strain. This is usually done in specific deterministic models but specificity is deceitful because a distribution that will hold under changes in physical chemistry often will not apply to changes in vibrational frequency. The more specific the model, usually, the smaller the range of predictive capability. Consequently when the conditions for suitability are unknowingly exceeded and prediction fails, the model is often abandoned as if it were in error, not just misapplied.

The Pagett Models Using the Wald Distribution

Let us examine the models introduced by Padgett with strain as given in the form in eqn (14.6). Suppose we have complete samples $X_{i,j} \sim \mathcal{W}(\delta_j, \nu_j)$ for $i = 1, \cdots, n_j$ where, for $j = 1, \cdots, k$,

$$\delta_j = \frac{\alpha\sqrt{\beta}}{\sqrt{\lambda_j}}, \quad \nu_j = \beta\lambda_j, \quad \text{and we write for short } \lambda_j := \lambda(\theta, \mathfrak{t}_j).$$

Here the mean life is directly proportional to, and the dispersion is inversely proportional to what we call the the "strain" function, viz., $\lambda(\theta, \mathfrak{t}_j)$.

From results in previous chapters we know

$$\hat{\nu}_j := \bar{X}_{+,j} = \tfrac{1}{n_j}\textstyle\sum_{i=1}^{n_j} X_{i,j} \sim \mathcal{W}(\tfrac{\delta_j}{\sqrt{n_j}}, \nu_j) \quad \text{for } j = 1, \cdots, k, \quad (14.12)$$

$$\widehat{\delta_j^2} := S_j^2 = \left(\frac{n_j - 1}{n_j}\right)\left[\frac{\bar{X}_{+,j}}{\bar{X}_{-,j}} - 1\right] \quad \text{for } j = 1, \cdots, k, \quad (14.13)$$

are minimum variance undiased estimators of $\beta\lambda_j$ and $\alpha^2\beta/\lambda_j$, respectively.

We also check that the $X_{+,j} \sim \mathcal{W}(\frac{\delta_j}{\sqrt{n_j}}, n_j\nu_j)$ for $j = 1, \cdots, k$ are consonant since $\delta_j^2\nu_j = \alpha^2\beta^2$. Hence, asas,

$$X_{+,+} \sim \mathcal{W}(\kappa, \varpi), \qquad \text{where } \kappa = \frac{\alpha\sqrt{\beta}}{\sqrt{\sum_1^k n_j\lambda_j}}, \text{ and } \varpi = \beta\sum_1^k n_j\lambda_j.$$

From these observations we seek to construct easily computable estimates of α, β, θ.

Consider the statistics from the sample observations, from which we form the method-of-moments estimators; remember here $\lambda_j = \lambda(\theta, \text{ł}_j)$.

$$\bar{x}_{+,j} = \beta\lambda_j, \qquad s_j^2 = \frac{\alpha^2\beta}{\lambda_j} \qquad \text{for } j = 1, \cdots, k \text{ and } x_{+,+} = \beta\textstyle\sum_1^k n_j\lambda_j.$$

We now summarize this information in the form

$$\tfrac{1}{k}\textstyle\sum_1^k \ln[\bar{x}_{+,j}] = \ln\beta + \tfrac{1}{k}\sum_1^k \ln[\lambda_j], \tag{14.14}$$

$$\tfrac{1}{k}\textstyle\sum_1^k \ln[s_j^2] = \ln\beta + \ln\alpha^2 - \tfrac{1}{k}\sum_1^k \ln[\lambda_j], \tag{14.15}$$

$$\ln[x_{+,+}] = \ln\beta + \ln[\textstyle\sum_1^k n_j\lambda_j]. \tag{14.16}$$

To eliminate $\ln\beta$ we subtract eqn (14.14) from eqn (14.16) to obtain

$$\ln[x_{+,+}] - \tfrac{1}{k}\textstyle\sum_1^k \ln[\bar{x}_{+,j}] = \ln[\sum_1^k n_j\lambda_j - \tfrac{1}{k}\sum_1^k \ln[\lambda_j],$$

which is a function only of the parameter θ. Denote the solution as $\check{\theta}$, which can be obtained by a numerical root-finding algorithm, such as is in *Mathematica*. The root $\check{\theta}$ satisfies the equation in θ, namely,

$$\frac{\sum_1^k n_j\lambda(\theta, \text{ł}_j)}{\left[\prod_{j=1}^k \lambda(\theta, \text{ł}_j)\right]^{1/k}} = \frac{x_{+,+}}{\left[\prod_{j=1}^k \bar{x}_{+,j}\right]^{1/k}}. \tag{14.17}$$

Then from eqn (14.14) with $\check{\theta}$ known we have

$$\check{\beta} = \left[\prod_{j=1}^k \frac{\bar{x}_{+,j}}{\lambda(\check{\theta}, \text{ł}_j)}\right]^{1/k}. \tag{14.18}$$

Finally, from eqn (14.15) we find

$$\check{\alpha} = \left[\prod_{j=1}^k \frac{s_j\lambda(\check{\theta}, \text{ł}_j)}{\sqrt{\bar{x}_{+,j}}}\right]^{1/k}. \tag{14.19}$$

These results are intended for (good) initial estimates for the iteration to obtain the mle's.

Exercise Set 14.D

1. Write out and set up the iteration procedure to determine the mles for α, β, θ using the $\mathcal{FL}$-model of Padgett from eqn (14.4).
2. Comment on the possible error, if any, that was made in shifting from a $\mathcal{FL}$-distribution to a $\mathcal{W}$-distribution as in 14.3.2.
3. Under what conditions can you use the method-of-moments and Lemma 2 to provide the same procedure to extrapolate values of ł for an $\mathcal{FL}$-model as for the $\mathcal{W}$-distribution.
4. The purpose of the Padgett model is to predict a strength distribution for a gauge length ł beyond the values which are tested. List situations in which you think this approach would be satisfactory and others about which you would be reasonably cautious?

14.4. Fracture Mechanics and Stochastic Damage

Fracture mechanics is the method of analysis used in engineering to determine both the reason for, and quantification of both the micro- and substructural changes in a material which cause nucleation of permanent damage via the creation of microscopic flaws, their growth and coalescence into a dominant crack, its stable propagation until precipitate instability leads ultimately to structural failure.

The *empirical* crack growth law, due to Paris, Gomez, and Anderson (see [89]), gives the change in fatigue-crack length per cycle from a pre-existing flaw or defect in an otherwise elastic material, as

$$\frac{da}{dN} = C(K_{max} - K_{min})^m \quad \text{for some } C, m > 0,$$

where K_{max}, K_{min}, respectively, are the maximum and minimum stress intensity factors during the cycle, a is the crack-length and N is the number of cycles. Moreover, Suresh in [109, p.158] asserts that "this is the most widely used form of characterizing fatigue-crack growth rate for a vast spectrum of materials and test conditions." But it is a mnemonic; it is not a differential equation to be analyzed.

During the first phase of fatigue, until crack initiation, the physical damage from a duty cycle consists merely of the piling up of discontinuities in the crystalline lattice along the grain boundaries, most of which are reversed during stress relaxation. Eventually rupture starts occurring at some grain boundaries and submicroscopic cracks start propagating; this process remains stable until these cracks coalesce into a dominant crack of critical size. Then damage accumulation accelerates rapidly until structural collapse. This hypothesis is supported in the literature by examining fatigue-crack growth using electonmicroscopy. Of course, cumulative damage can occur in different ways in different materials under diferent environments. But this mode of failure, metallic fatigue, presently has the most extensive documentation and study.

This process may be modelled stochastically by letting the time between accumulations of a fixed increment of damage be the random variable and considering $\{X_i\}$ as the waiting time between the occurrance of the $(i-1)^{\text{st}}$ increment and the ith increment.

Assume $X_0 \sim \mathcal{FL}(\alpha, \beta)$, for some $\alpha, \beta > 0$, is the waiting time until crack initiation, i.e., the rupture of grain boundaries begins. Consider a monotone decreasing non-negative sequence $\{a_i\}_0^\infty$, where

$$1 = a_0 \gg a_1 \geq \cdots \geq a_n > a_{n+1} > \cdots > a_{n+m} \gg a_{n+m+1} \gg a_{n+m+2}.$$

Let $\{X_i\}$ for $i = 0, 1, 2, \cdots$ be independent rvs with $X_i \sim \mathcal{W}(\alpha/\sqrt{a_i}, a_i\beta)$ being the rv time increment for cumulative damage to increase by a unit amount, for all $i \geq 1$. NB that for all $i = 0, 1, 2, \cdots$ the $\{X_i\}$ are consonant.

Define $S_j = \sum_{i=0}^{j} X_i$ for any $j \geq 0$, with $A_j = \sum_{i=0}^{j} a_i$; so $S_j \sim \mathcal{FL}(\kappa_j, \nu_j)$ is the time until damage accumulates to level jth for $j \geq 1$ where $\kappa_j = \alpha/\sqrt{A_j}$ and $\nu_j = A_j\beta$. Then $\sum_{i=1}^{n} X_i$ is the time between crack initiation until an inspectable

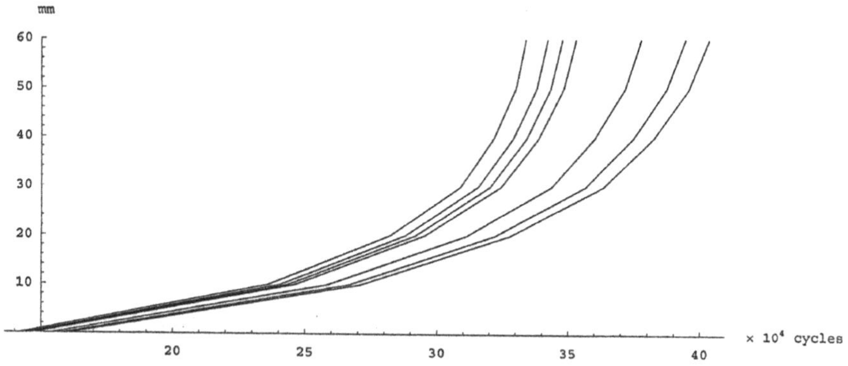

Figure 14.2. Simulated time (in cycles) showing crack extension (in mm).

level of damage, at n, is reached at time $S_n \sim \mathcal{FL}(\kappa_n, \nu_n)$. The time of stable crack growth is between S_n and S_{n+m} when the critical crack size is reached, which too has an $\mathcal{FL}$-distribution. After S_{n+m} the crack growth is precipitated until structural failure occurs.

A simulation of this model of damage increment accumulation at random times for 1cm crack extensions of $\{a_i\}$ can be seen in Figure 14.2. With random replication, it is indistinguishable from those crack-growth models depicted using Marhov Chains, see [9, p. 248].

Exercise Set 14.E

1. The numbers n and m are known to Nature. How can the analyst deal with this? How would making them random effect the model?

2. One sees $\sum_{i=1}^{m} X_{n+i} \sim \mathcal{W}(\delta, \mu)$, where $\delta = \alpha/\sqrt{A_{m+n} - A_n}$ and $\mu = (A_{m+n} - A_n)\beta$, which is the rv that must be used to calculate the inspection period during service. How should one set this period and what guarantee could be provided as to its effectiveness?

3. If a sample $y_{1+} \leq \cdots \leq y_{n+}$ contains only run-outs, when from physical reasons the parameter space of the $\mathcal{FL}(\alpha, \beta)$ distribution is within the rectangle $0 < \alpha \leq \alpha_0$ and $0 < \beta \leq \beta_0$, find the mle's $\hat{\alpha}$ and $\hat{\beta}$.

4. Formulate a general procedure for cyclic fatigue testing of partially precessed concrete bridge girders, which conceptually consists of three components in series, viz., prestressing steel, concrete and reinforcing steel. Since the the testing of full-scale beams is impractical and only data from smaller scale models will be available, along with the source of the failure determined, the size effect should be considered. You are asked to determine, for an annually increasing traffic load, what data should be accumulated so as to predict the service life of number of such bridges without waiting for in-service failure. (Remember the wear due to surface abrasion from traffic which eventually exposes the re-bar to the environment.) The quantitative relationships between the distributional parameters α, β of the service life and the stress regime are the core of this problem.

Pronunciation will be varied by levity or ignorance ··· Illiterate writers will at one time or other by publick infatuation rise into renown, who, not knowing the original import of words use them with colloquial licentiousness; they confound distinction and forget propriety. ··· But if the changes that we fear be thus irresistible ··· it remains that we retard what we cannot repel, that we palliate what we cannot cure.

Samuel Johnson

How has the experimenter decided when to stop cleaning up and massaging the data? Was the decision made before it was determined what effect those efforts had on the outcome? It is all too easy to fall into the trap of correcting until the answer looks right. To avoid this error requires an intimate acquaintance with the rules of scientific evudence and procedure. Moreover it requires not just honesty but a sense that honesty requires exertion!

Richard P. Feynman

CHAPTER 15

Maintenance of Systems

15.1. Introduction

There are many situations when a number of manufacturing systems (of either electronic or machine parts) are scheduled to be in operation performing specified duties, e.g., the manufacture of consumer goods or to provide service contracted for within a scheduled time. These situations require that a fixed number of machines be in operation and so this "fleet" must be repaired or replaced in a timely manner. There are other situations where the service or manufacturing process becomes so technically intricate that no part can be disrupted without extreme cost. This complication indicates that applicable mathematical models for repairable systems should have been studieded with results readily available for cognizant managers of operational planning.

By a repairable system, or structure, we mean a structure whose service must be interrupted, even if it is fail-safe, soon after a component failure in order to initiate maintenance, repair, or replacement. The utilization of such a component can be represented by a sequence of intervals of service (up-time) alternating with intervals of maintenance-repair or replacement (down-time). The operational cycle now consists of pairs of such nonnegative rvs, viz.,

$$(U_1, D_1), (U_2, D_2), \ldots, (U_n, D_n), \ldots.$$

Let $T_i = U_i + D_i$ for $i = 1, 2, \ldots$ denote the *duty cycle* where $U_i \sim F_{U_i}$ for $i = 1, 2, \ldots$, which may not always have the same cdf, but it always denotes the time of operation during the ith duty cycle while D_i denotes the down-time during that duty cycle.

15.2. Availability

Let $X(t)$ be the random binary indicator of system operation at time $t > 0$, as defined above and set $S_k = \sum_{i=1}^{k} T_i$, where $T_i = U_i + D_i$, for $k = 1, 2, \ldots$, with $S_0 := 0$. We then have

$$X(t) = \sum_{k=0}^{\infty} \mathrm{I}(0 < t - S_k < U_{k+1}) \quad \text{for } t > 0. \tag{15.1}$$

Such a stochastic process is depicted in Figure 15.1.

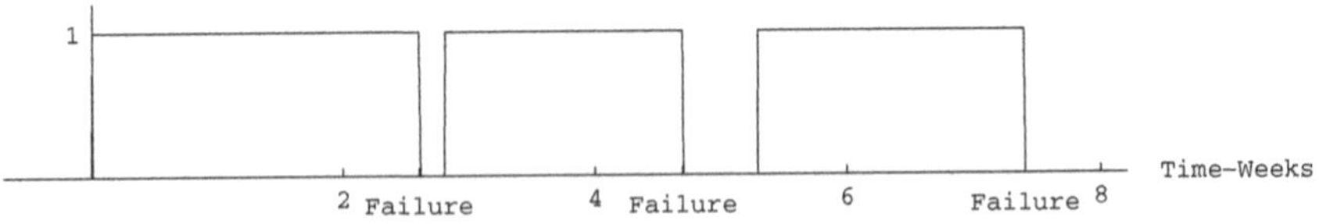

Figure 15.1. Graph of the indicator of a systems operation.

Thus, we define the *availability* of any system undergoing repair after each period of service using the indicator of its operation, viz., $X(t)$, to be the real function given by its expectation:

$$A(t) := \mathrm{P}[X(t) = 1] = \mathrm{E}[X(t)] \quad \text{for } t > 0. \tag{15.2}$$

The *nonavailability* of the system is $\bar{A}(t) = 1 - A(t)$ at any time $t > 0$. The *average availability* per unit time over the interval $(0, t)$ is

$$A_{av}(t) := \frac{1}{t}\int_0^t A(s)\,ds. \tag{15.3}$$

The *ultimate availability*, when the limit exists, is given by $A(\infty) := \lim_{t\to\infty} A(t)$, and the *ultimate average availability* is $\lim_{t\to\infty} A_{av}(t)$, which will exist virtually always.

We now make two important comments:

Remark 1. *Let $W(t)$ be the total time in service for the system during the interval $(0, t)$ then $W(t) = \int_0^t X(s)\,ds$. Thus, the average expected total service-time during $(0, t)$ is given by the average availability, in virtually all practical circumstances.*

Proof. Note that

$$\mathrm{E}\left[\frac{W(t)}{t}\right] = \frac{1}{t}\mathrm{E}\int_0^t X(s)\,ds = \frac{1}{t}\int_0^t \mathrm{E}[X(s)]\,ds = \frac{1}{t}\int_0^t A(x)\,dx = A_{av}(t),$$

in all those circumstances when Fubini's theorem allows us to change the order of integration. □

Remark 2. *If $A(\infty)$ exists then $A_{av}(\infty) = A(\infty)$.*

Proof. Let $A(\infty) = a$ then, for each $\epsilon > 0$ there is a $y > 0$ so large that $|A(t) - a| \leq \epsilon$ for all $t > y$. Hence for any $t > y$ we have $A(t) \leq a + \epsilon$ and

$$A_{av}(t) = \frac{1}{t}\left[\int_0^y A(s)\,ds + \int_y^t A(s)\,ds\right] \leq \frac{U(y)}{t} + \frac{(a+\epsilon)(t-y)}{t}.$$

Since y is fixed, we have $\lim_{t\to\infty} A_{av}(t) \leq a + \epsilon$. A comparable argument can be made to show that $\lim_{t\to\infty} A_{av}(t) \geq a - \epsilon$. This completes the proof since $\epsilon > 0$ was arbitrary. □

We now follow the model of varying duty cycles given in 12.7.2 to obtain the basic iid availability

Theorem 1. *If the duty cycles, $T_i = U_i + D_i \sim G$, and the up-times, $U_i \sim F$, are both iid for $i = 1, 2, \ldots$ and the entailed renewal function for G is given by $G_\sharp(t) := \sum_{n=1}^{\infty} G^{n*}(t)$ for $t > 0$, then the availability is*

$$A(t) = \bar{F}(t) + \int_{u=0}^{t} \bar{F}(t-u)\,dG_\sharp(u). \tag{15.4}$$

When the cdf G is nonlattice, the key renewal theorem of Theorem 6 in 12.3, implies eqn (15.4), which is used under nearly all conditions in availability calculations.

Corollary 1. *Under the conditions when eqn (15.4) holds, the ultimate availability is the ratio*

$$A(\infty) = \frac{E[U]}{E[T]} = \frac{\nu}{\nu+\delta}, \quad \text{when } E[U] = \nu \quad \text{and} \quad E[D] = \delta. \tag{15.5}$$

Exercise Set 15.A

1. Show the down-time indicator of a repairable system of the type in eqn (15.1) is

$$\bar{X}(t) = \sum_{k=0}^{\infty} \mathrm{I}\{0 < t - S_k - U_{k+1} < D_{k+1}\} = 1 - X(t) \quad \text{for } t > 0,$$

   so the total time in repair of the system during $(0, t)$ is given by $\int_0^t \bar{X}(s)\,ds$.
2. Use eqn (15.1) to obtain eqn (15.4).
3. Consider a bailing machine for compressing and packaging hops that often must be repaired during harvest because of accidents which affect one particular component. The processing must be stopped during repair but the wages continue for the operating crew standing by. Thus, net revenue in dollars during harvest consists, for all $t > 0$, of income of $c_0X(t)$ and upon each breakdown a part replacement cost of c_1 plus an opportunity cost of $c_2\bar{X}(t)$ during downtime. Express the expected revenue as a function of the availability. What further information do you need to predict expected revenue from hops harvest?

15.2.1. Application of Tauberian Properties

An example will now be given of the usefulness in availability theory of the Tauberian properties existing between a function and its transform. We first provide a didactic proof of the Corollary above, using results discussed in 12.3.3, or as given in the Appendix.

The Laplace transform of A and the $\mathcal{LS}$-transform, label them, respectively, $A^\flat$ (sometimes $\mathcal{L}A$) and $A^\dagger$, are connected by $sA^\flat(s) = A^\dagger(s)$, from which, importantly, we can obtain $A^\dagger(0) = A(\infty)$, which then gives us the ultimate availability.

Taking the Laplace transform of eqn (15.4) shows, omitting the arguments of the functions,

$$A^\flat = \bar{F}^\flat + \bar{F}^\flat \cdot \mathcal{L}G_\sharp = \frac{\bar{F}^\flat}{1 - G^\flat} = \frac{\bar{F}^\flat}{1 - F_{\mathrm{U}}^\flat \cdot F_{\mathrm{D}}^\flat}. \tag{15.6}$$

Here we presume $U \perp D$ which implies $G^\flat = F_{\mathrm{U}}^\flat \cdot F_{\mathrm{D}}^\flat$. Hence the $\mathcal{LS}$-transform becomes, by a simple expansion,

$$A^\dagger(s) = \bar{F}^\flat(s) \left\{ \frac{s}{1 - [1 - s\delta - o(s)][1 - s\nu - o(s)]} \right\} \quad \text{for } s > 0.$$

Since $\bar{F}^\flat(s) \to \nu$ as $s \downarrow 0$, we obtain

$$\lim_{s \downarrow 0} A^\dagger(s) = \frac{\nu}{\delta + \nu} = A(\infty).$$

15.2.2. System Availability

Let us now consider a coherent system φ of n components. Let $X_i(t)$ for $t > 0$ be the service indicator of the ith component in duty under stationary service and repair, i.e., iid service times $U_{i,j}$ and iid maintenance/repair times $D_{i,j}$ for $j = 1, 2, \ldots$. The availability of the i component then is given by $A_i(t) = \mathrm{E}X_i(t)$ for $t > 0$ and $i = 1, \ldots, n$. Denoting the joint service of the components by the vector $\boldsymbol{X}(t) = (X_1(t), X_2(t), \ldots, X_n(t))$ it follows that the availability of the system, φ, is

$$A_\varphi(t) := \mathrm{E}\varphi[\boldsymbol{X}(t)] = \hbar[A_1(t), \ldots, A_n(t)] \quad \text{for all } t > 0.$$

Here $\hbar[\cdot]$ is the reliability function of the system φ, as discussed previously in 4.3 and 4.5.

Remark 3. *For any coherent system of n components with reliability function $\hbar$, under stationary service and repair with $\mathrm{E}U_{i,j} = \nu_i$ and $\mathrm{E}D_{i,j} = \delta_i$ when the ultimate availability for the ith component is $A_i(\infty)$ for $i = 1, \ldots, n$, then the ultimate availability for system φ is*

$$A_\varphi(\infty) = \hbar[A_1(\infty), \ldots, A_n(\infty)] = \hbar\left[\frac{\nu_1}{\nu_1 + \delta_1}, \ldots, \frac{\nu_n}{\nu_n + \delta_n}\right].$$

NB availability by itself is often insufficient. Little useful management information can be obtained from the availability without introducing the relative costs of downtime, repair, and replacement. Under certain circumstances, e.g., when every service life is Epstein, there is no advantage in replacing any unfailed component in service. In many more instances when manufacturing very complex equipment, nearly all working components wear and in-service failures are very costly. In electronic chip manufacuring downtimes are scheduled and every system component is replaced which will attain a total service time exceeding the manufacturers recommended life before the next scheduled downtime. But when all working component lives are terminated by extrinsic "accidents", i.e., lives are Epstein, how much can the system life be extended by having spares on hand?

Systems with Spares

Suppose we have a series system of n components with one standby replacement unit, all of which have independent Epstein service lives. The life of the ith unit is $T_i \simeq \text{Ep}(\lambda_i)$ for $i = 1, \ldots, n+1$. What is the reliability of this redundant system, assuming replacement time is negligible?

The time until the first component failure is $S_0 = \bigwedge_{i=1}^{n} T_i \simeq \text{Ep}(\Lambda_0)$, among the original n units where $\Lambda_0 = \sum_1^n \lambda_i$. Then $S_0 \sim F_0$ with sdf $\bar{F}_0(t) = e^{-\Lambda_0 t}$ for $t > 0$. Let the time between the first and second failures be $S_1 \sim F_1$ which, as a consequence of no wear with Epstein lived components, has an sdf mixture,

$$\bar{F}_1(t) = e^{-\Lambda_0 t} \sum_{j=1}^{n} p_j e^{(\lambda_j - \lambda_{n+1})} \quad \text{for } t > 0, \quad \text{where } p_j = \frac{\lambda_j}{\Lambda_0}.$$

Hence $S_0 + S_1$ is the time until the second system failure, and its sdf by eqn (3.31) is given as

$$\begin{aligned}\overline{F_0 * F_1}(t) &= \sum_{j=1}^{n} p_j \left[\frac{\Lambda_j e^{-\Lambda_0 t}}{\Lambda_j - \Lambda_0} + \frac{\Lambda_0 e^{-\Lambda_j t}}{\Lambda_0 - \Lambda_j} \right] \\ &= e^{-\Lambda_0 t} \left\{ 1 - \sum_{j=1}^{n} \frac{\lambda_j}{\lambda_j - \lambda_{n+1}} \left[1 - e^{(\lambda_j - \lambda_{n+1})t} \right] \right\}. \end{aligned} \qquad (15.7)$$

The sdf for two standby units can be derived similarly by considering the mixture generated by the termination of each $T_{1,j}$ followed by the second replacement. However, for three or more standby units it is better to formulate the problem as one in Markov theory and use matrices, realizing that matrices simplify notation but not computation.

15.3. Age Replacement with Renewal

Consider a particular subsystem connected in series within a production system. The system undergoes regularly scheduled maintenance (repair or replacement) after each service period of duration τ, if it has not failed previously. After each maintenance the subsystem is considered renewed. What influence does this single subsystem, which has service life $V_i \sim F$ and a repair-replacement down-time of D_i, independent of τ, have on the system? Thus, the up-time is $U_\tau = \min(V, \tau)$

so for fixed $\tau > 0$ the expected up-time is

$$\mathrm{E}U_\tau = \int_0^\tau \bar{F}(x)\,dx + \tau\bar{F}(\tau), \tag{15.8}$$

which is an increasing function of τ only as long as $\tau q(\tau) \le 2$, where q is the hazard rate of V. The ultimate availability using this age repair-replacement policy is

$$A_\tau(\infty) = \frac{\mathrm{E}U_\tau}{\mathrm{E}U_\tau + \mathrm{E}D}.$$

Suppose the cost, in dollars, of each in-service failure-repair is c_1 while the cost of a scheduled maintenance-repair is $c_2 \ll c_1$. The expected cost of maintenance-repair for each duty cycle, which period of time includes both service and repair times, V_i and D_i, is

$$c_1 \int_0^\tau \bar{F} + c_2\bar{F}(\tau) + c_3\mathrm{E}D,$$

where $c_3 \ge 0$ is the lost opportunity cost per unit down-time. The expected revenue per duty cycle is $c_4\mathrm{E}U_\tau$, which depends only upon the production level per unit of service time.

If we want to maximize, as a function of the period τ of the age replacement policy, the expected revenue per unit of expected cost of each duty cycle we consider the ratio, label it $R(\tau)$, of these two quantities

$$R(\tau) := \frac{c_4\mathrm{E}U_\tau}{c_1 \int_0^\tau \bar{F} + c_2\bar{F}(\tau) + c_3\mathrm{E}D}. \tag{15.9}$$

NB this ratio becomes proportional to the ultimate availability only when $c_1 = c_2 = c_3 \propto c_4$.

The optimal age-replacement period, say $\hat{\tau}$, is seen to be the maximizer of $R(\cdot)$, and in certain cases when F is IHR can be found as the solution of the equation in t, with hazard rate q,

$$R(t) = \frac{c_4[2 - tq(t)]}{c_1 - c_2q(t)} \quad \text{for some} \quad t > 0. \tag{15.10}$$

Suppose we seek to maximize the net expected revenue per duty cycle, namely, the maximum of the function of $t > 0$

$$c_4\mathrm{E}U_t - \left\{c_1 \int_0^t \bar{F}(t) + c_2\bar{F} + c_3\mathrm{E}D\right\} \quad \text{for} \quad t > 0.$$

One finds the maximizer here, say $\check{t}$, to be the solution of the equation in the variable t

$$2c_4 - c_1 = (c_4t - c_2)q(t) \quad \text{for some} \quad t > 0. \tag{15.11}$$

Availability theory is often a theory in search of an application since it ignores the difference between the costs of replacement-repair due to failure in-service and the cost of replacement-repair at scheduled maintenance periods. Availability for many coherent structures under several distributions for service and repair times is presented in the *Russian Handbook*; see [56]. The theory is complete only when Epstein distributions are assumed. Such no-wear assumptions were made for nuclear reactors in the USA in the infamous Nuclear Regulatory Commission *Reactor Safety Study* in 1975. It garnered much criticism and was soon thereafter replaced.

Exercise Set 15.B

1. Is the assumption of independence used in eqn (15.6) limiting in scope or can a proof in this way be given without making such an assumption? If so, do it.
2. Let a system consist of n identical components be in series, with k standby units available, all of which have the same Ep(λ) or $\mathcal{W}$ei(α, β) distribution. For these two cases, derive usable formulae for the system life-length, presuming k is small.
3. Assume a component has a life which is the minimum time until an 'ancident', which has an interarrival time which is Ep(λ), and its wear-out which has a $\mathcal{FL}(\alpha, \beta)$ distribution. If the cost of maintenance/storage for a standby unit is \$c per unit time, while the cost of having an accidental failure with no unit on hand is \$C, derive a rule of thumb, involving distributional parameters and costs, which can be used to decide when a standby unit should be kept on hand.
4. Find conditions on the hazard rate q in eqn (15.10) that will ensure the optimal age-replacement period $\hat{\tau}$ exists uniqely. Do the same with eqn (15.11).
5. Assume a certain subsystem has costs of repair replacement of $10c_2 = 5c_4 = 2c_3 = c_1$ when $V_i \sim \mathcal{W}\text{ei}(\alpha, \beta)$. Select a value of α of practical usefulness and find the optimal replacement period $\hat{\tau}$ in eqn (15.10) and $\check{\tau}$ in eqn (15.11). Which would you recommend? Why? Under what circumstances would you recommend neither?
6. What change would have to be made to the the sdf of the series system in eqn (15.7) if $T_i \sim \mathcal{W}\text{ei}(\alpha, \beta_i)$ for $i = \ldots, n+1$? Under what practical conditions might one argue that this Weibull approximation to the life is sufficiently accurate? How could this be verified?

15.3.1. A Single Machine with Repair

To fix ideas consider a problem in Industrial Engineering. A machine cutting-tool has a blade that dulls with usage; when should it be replaced? The service life is determined by wear forestalling output with acceptable tolerances or finish quality. The initial cost of the component is much higher than the cost of early renewals, but the cost of successive renewals increases. Let $C[k]$ be the cost of the kth renewal for $k \geq 0$ associated with iid renewed service life $V_{k+1} \sim F$. We define

$$C[n] = \begin{cases} c_0 & \text{for } n = 0, \\ c_1 + n\varepsilon_1 + n^{2\downarrow}\varepsilon_2 & \text{for } n \geq 1. \end{cases} \tag{15.12}$$

This usage-renewal cycle is repeated until it is economically disadvantageous. One criterion used to decide when the blade is replaced is the expected cost per unit of run-time. Assuming down-time is negligible, the expected cost is, setting $\mathrm{C}(t) := \mathrm{E}C[N(t)]$,

$$\mathrm{C}(t) = \sum_{k=0}^{\infty} C[k] \int_{[0,t)} \bar{F}(t-x)\, dF^{k*}(x), \quad \text{with } F^{0*} \text{ the unit step-function at zero.} \tag{15.13}$$

Two simple cases are: (1) $c_0 \gg c_1$ with $\varepsilon_1 = \varepsilon_2 = 0$, and (2) $c_0 \gg c_1$ with $\varepsilon_1 > 0$ and $\varepsilon_2 = 0$.

In Case (1) the optimal replacment time τ is the minimizer of

$$\frac{c_0 \bar{F}(t) + c_1 F_\sharp(t)}{t} \quad \text{for } t > 0, \quad \text{where } F_\sharp := \textstyle\sum_{n=1}^{\infty} F^{n*}. \tag{15.14}$$

In Case (2) the optimum replacment time is the minimizer for $t > 0$, of

$$\frac{c_0 \bar{F}(t) + c_1 \bar{F} * F_\sharp(t) + \varepsilon \bar{F} * F_{2\sharp}(t)}{t} \quad \text{where } F_{2\sharp} := \textstyle\sum_{k=1}^{\infty} k F^{k*}. \tag{15.15}$$

In either case the finite optimal replacement time τ, when it exists, must satisfy the equation $\tau C^{'}(\tau) = C(\tau)$ where $C(\cdot)$ was defined in eqn (15.13). Clearly, if the V_i are NBUE with $\mathrm{E}V_i = \mu$, we know by eqns (12.23) and (12.24) that for Case (1),

$$\frac{c_0 \bar{F}(t) - c_1}{t} + \frac{c_1}{\mu} \leq \frac{c_0 \bar{F}(t) + c_1 F_\sharp(t)}{t} \leq \frac{c_0 \bar{F}(t)}{t} + \frac{c_1}{\mu} \quad \text{for all } t > 0,$$

so no finite minimum exists.

Increasing renewal cost is a complication, since $F_\sharp$ and $F_{2\sharp}$ can be explicitly determined only in a few cases. So sometimes an approximation is employed, namely,

$$\mathrm{C} \approx \sum_{k=0}^{n} C[k] \bar{F} * F^{k*}, \quad \text{for a small positive integer } n. \tag{15.16}$$

For easier analysis only those distributions closed under convolution are used, namely, the Gaussian and Gamma families, but now the Wald distributions, more applicable to such conditions of wear, might be applied, before resorting to brute force computation.

Exercise Set 15.C

1. Does the approximation in eqn (15.16) give a conservative or nonconservative answer? When it is known from physical reasons that the component can be renewed at most n, times, why is eqn (15.16) an approximation?
2. Since for $\mathcal{LS}$-transforms $F_\sharp^\dagger = F^\dagger \div (1 - F^\dagger)$ and $F_{2\sharp}^\dagger = F^\dagger \div (1 - F^\dagger)^2$, by assuming $F \simeq \mathrm{Ep}(\lambda)$, show $F_\sharp(t) = \lambda t$ and $F_{2\sharp}^\dagger(t) = \lambda t + (\lambda t)^2/2$. Use these results to verify, for Case (2), that τ is the solution of the equation in t, $e^{\lambda t}[1 + \lambda t] = c/(c_0 + c)$.
3. Let $F \simeq \mathcal{W}(\alpha, \beta)$ and use eqn (12.16), omitting $\varepsilon(y)$, to determine $F^\dagger$. What else must be derived to utilize eqn (15.15)? Do it.

4. Suppose there are 3 independent subsystems in series in the system that require renewal-replacement, and all have linearly increasing renewal costs, but with different service lives $T_i \sim G_i$ for $i = 1, 2, 3$. What should be done in this circumstance?

15.4. The Inversion of Transforms

By Theorem 4.18, closure of the IHR distributions under convolution, it follows that many IHR distributions can be well approximated by a suitable sum of Epstein variates. For example, many failures are caused by a sequence of untoward events that each occur following an Epstein rwt. Suppose such an IHR distribution is needed to study its life behavior having a specified mean β and coefficient of variation α. One might proceed as follows:

Let $X_i \simeq \mathrm{Ep}[a\lambda_i]$ be independent for $i = 1, \ldots, n$ for some $\lambda_i, a > 0$ and some integer n. Then we know by eqn (3.31) that $Y = \sum_{i=1}^{n} X_i \sim G$ with sdf given by

$$\bar{G}(t) = \sum_{i=1}^{n} \left\langle \frac{\boldsymbol{\lambda}}{j} \right\rangle \exp\{-a\lambda_i t\} \quad \text{for } t > 0 \text{ where } \boldsymbol{\lambda} = (\lambda_1, \ldots, \lambda_n). \tag{15.17}$$

Thus, we have

$$\mathrm{E}[Y] = \sum_{i=1}^{n} \frac{1}{a\lambda_i} = \beta, \quad \mathrm{Var}[Y] = \sum_{i=1}^{n} \frac{1}{(a\lambda_i)^2}, \quad \alpha^2 = \frac{\mathrm{Var}[Y]}{(\mathrm{E}[Y])^2}. \tag{15.18}$$

Clearly, we can match the mean and coefficient of variation in various ways, with many choices remaining to particularize special behavior.

We now examine the behavior of the transform of the distribution of a sum of Epstein rwts.

Lemma 1. *If $X_i \simeq \mathrm{Ep}(\lambda_i)$ for $i = 1, \ldots, n$ are independent with $\sum_{i=1}^{n} X_i \sim F$, then letting σ_i for $i = 0, \ldots, n$ be the symmetric polynomials in $\{\lambda_1, \ldots, \lambda_n\}$, i.e,*

$$\sigma_0 = 1, \quad \sigma_1 = \sum_{i=1}^{n} \lambda_i, \quad \sigma_2 = \sum_{i>j}^{n} \lambda_i\lambda_j, \quad \sigma_3 = \sum_{i>j>k}^{n} \lambda_i\lambda_j\lambda_k, \ldots, \quad \sigma_n = \prod_{i=1}^{n} \lambda_i,$$

then the $\mathcal{LS}$-transform of the renewal function, $F_\sharp^\dagger$, is given

$$F_\sharp^\dagger(s) = \frac{\sigma_n}{\sum_{i=0}^{n-1} s^{n-i}\sigma_i} = \frac{\varsigma_n}{s} - \frac{\sum_{i=0}^{n-2} s^{n-2-i}\varsigma_i}{\sum_{i=0}^{n-1} s^{n-1-i}\varsigma_i}, \tag{15.19}$$

where, for $i = 0, \ldots, n$,

$$\varsigma_i = \sigma_i/\sigma_{n-1} \quad \text{with } \varsigma_{n-1} = 1 \text{ and } \quad \varsigma_n = \left(\sum_1^n \lambda_i^{-1}\right)^{-1}. \tag{15.20}$$

Using Mathematica *or a similar program, the exact partial-fraction decomposition of eqn (15.19) can always be found if $n \leq 5$.*

PROOF. By definition $F^{\dagger}(s) = \prod_{i=1}^{n} \frac{\lambda_i}{\lambda_i + s}$, so from eqn (12.13) the result for $F_{\sharp}^{\dagger}$ follows asas. □

We illustrate the extent of this procedure for $n = 3$. (Section 12.3 contains the result for $n = 2$.) We have

$$F_{\sharp}^{\dagger}(s) = \frac{\sigma_3}{s(s^2 + \sigma_1 s + \sigma_2)} = \frac{\sigma_3}{\sigma_2}\left[\frac{1}{s} - \frac{s + \sigma_1}{s^2 + \sigma_1 s + \sigma_2}\right].$$

We now consider the three possible cases and for notational convenience write $2a = \sigma_1$.

Case I: $\sigma_2 > a^2$ and write $b = \sqrt{\sigma_2 - a^2}$, so

$$F_{\sharp}^{\dagger}(s) == \frac{\sigma_3}{\sigma_2}\left[\frac{1}{s} - \frac{s + a}{(s + a)^2 + b^2} - \frac{a}{(s + a)^2 + b^2}\right],$$

where each term is a standard Laplace transform as given in the *CRC Handbook* or in Exer.16.C, q.v. So the inverse Laplace–Stieltjes transform is

$$F_{\sharp}(t) = \frac{\sigma_3}{\sigma_2}\left\{t - \int_0^t e^{-ax}[\cos bx + \frac{a}{b}\sin bx]\,dx\right\},$$

which after integration and simplification becomes

$$F_{\sharp}(t) = \frac{\sigma_3}{\sigma_2}\left\{t - \frac{1}{b\sigma_2}\left\{2ab - e^{-at}[2ab\cos bt + (a^2 - b^2)\sin bt]\right\}\right\}.$$

Case II: $\sigma_2 < a^2$ and write $c = \sqrt{a^2 - \sigma_2}$, with $u_1 = a + c$ and $u_2 = a - c$, to give

$$F_{\sharp}^{\dagger}(s) = \frac{\sigma_3}{\sigma_2}\left\{\frac{1}{s} - \frac{1}{s + u_2} - \frac{\sigma_1 - u_1}{u_2 - u_1}\left[\frac{1}{s + u_1} - \frac{1}{s + u_2}\right]\right\}.$$

Thus from tables of standard Laplace transforms we have

$$F_{\sharp}(t) = \frac{\sigma_3}{\sigma_2}\left\{t - \int_{x=0}^t e^{-u_1 x} + \tfrac{\sigma_1 - u_1}{u_2 - u_1}\left[e^{-u_1 x} - e^{-u_2 x}\right]dx\right\},$$

which again after integration becomes (a form that may be simplified further),

$$F_{\sharp}(t) = \frac{\sigma_3}{\sigma_2}\left\{t - \frac{1}{u_1}(1 - e^{-u_1 t}) + \tfrac{\sigma_1 - u_1}{2cu_1}[1 - e^{-u_1 t}] - \tfrac{\sigma_1 - u_1}{2cu_2}[1 - e^{-u_2 t}]\right\}. \tag{15.21}$$

Case III: $\sigma_2 = a^2$ so we again write it as a sum of identifiable terms, viz.,

$$F_{\sharp}^{\dagger}(s) = \frac{\sigma_3}{\sigma_2}\left\{\frac{1}{s} - \frac{1}{s + a} - \frac{a}{(s + a)^2}\right\}.$$

We recognize the last term as the transform of the Gamma distributions to obtain

$$F_{\sharp}(t) = \frac{\sigma_3}{\sigma_2}\left\{t - \frac{1 - e^{-at}}{a} - \frac{1}{a}\int_0^{at} xe^{-x}\,dx\right\}. \tag{15.22}$$

This procedure is commonly used for inversion, but we can do better by making use of eqn (3.31).

Theorem 2. *Under the assumptions and notation of Lemma 1, if F is the distribution of the sum of independent Epstein variates then the renewal function is given by*

$$F_\sharp(t) = \varsigma_n \int_0^t \sum_{j=1}^{n} \left\langle {z \atop j} \right\rangle (1 - e^{-z_j x})\, dx, \tag{15.23}$$

where $\mathbf{z} = (z_1, \ldots, z_{n-1})$*, with the* z_i *determined by the roots in the factorization of*

$$1 - F^\dagger(s) = \sum_{i=0}^{n-1} s^{n-1-i}\sigma_i = \prod_{i=1}^{n-1}(s + z_i).$$

The result $F_\sharp(t)$ *is real since the* z_i *are real or occur in complex conjugate pairs.*

Proof. In the complex factorization of the denominator in the transform of the renewal function $F_\sharp^\dagger = F^\dagger \div (1 - F^\dagger)$, note that $\varsigma_n := \sigma_n/\sigma_{n-1}$ in eqn (15.20) is the reciprocal of the integral $\int_0^\infty \bar{F}(t)\,dt$, since $\prod_1^{n-1} z_i = \sigma_{n-1}$. Thus, $F_\sharp^\dagger$ can be made into the product of Laplace transforms of Epstein variates, which by identifying with the Lagrange coefficients of eqn (3.31), gives the renewal function immediately. Clearly the fundamental theorem of algebra allows us to factor it into products of linear and quadratic terms, each quadratic term consisting of a complex root and it conjugate. These factors can always be individually inverted and then be convolved, at least by numerical integration, into a survival distribution which is then integrated. □

For example, consider the two cases of extreme factorization

$$F_\sharp^\dagger(s) = \frac{\sigma_n}{\sum_{i=0}^{n-1} s^{n-i}\sigma_i} = \begin{cases} \dfrac{\varsigma_n}{s} \displaystyle\prod_{i=1}^{n-1} \dfrac{\nu_i}{s + \nu_i} & \text{for some } \nu_i > 0, \text{ or} \\ \dfrac{\varsigma_n}{s} \left(\dfrac{\nu_0}{s + \nu_0}\right)^{n-1} & \text{for some } \nu_0 > 0. \end{cases} \tag{15.24}$$

In eqn (15.24) the two inversions are immediate:

$$F_\sharp(t) = \begin{cases} \varsigma_n \displaystyle\int_{u=0}^t \sum_{j=1}^{n-1} \left\langle {\nu \atop j} \right\rangle (1 - e^{-\nu_j u}\, du & \text{for all } t > 0, \text{ or} \\ \varsigma_n \displaystyle\int_{y=0}^t \int_{u=0}^{\nu_0 y} \dfrac{u^{n-2}e^{-u}}{\Gamma(n-1)}\, du\, dy & \text{for all } t > 0, \end{cases} \tag{15.25}$$

However, if the numerical values of λ_i are known, *Mathematica* can obtain the σ_i and factor $\sum_{i=0}^{n-1} s^{n-1-i}\sigma_i$ into products of linear and quadratic terms as specified in the fundamental theorem of algebra. In general, we might have the factors, for some $\nu_i, a_k, b_k > 0$,

$$F_\sharp^\dagger(s) = \frac{\sigma_n}{\sum_{i=0}^{n-1} s^{n-i}\sigma_i} = \frac{\varsigma_n}{s}\left(\frac{\nu_0}{s+\nu_0}\right)^{n_1} \cdot \prod_{i=1}^{n_2} \frac{\nu_i}{s+\nu_i} \cdot \prod_{k=1}^{n_3} \frac{1}{(s+a_k)^2 + b_k^2}, \tag{15.26}$$

which can be inverted individually and then convolved numerically, at least.

Exercise Set 15.D

1. Show that eqns (15.21) and (15.22) are special cases of the formulas in eqn (15.25). Check, in both cases, that $F_\sharp(t)/t \to \varsigma_n$ as $t \to \infty$.

2. Is eqn (15.18) better than letting $X \sim \mathcal{G}am(\kappa, \beta)$, where the dispersion is $\alpha^2 = 1/\kappa$, which then can be set by choosing κ appropriately?

3. A classical problem in statistics is to find the exact distribution of a linear combination of Chi-square variates. Let X have sdf $\sum_{j=1}^{n} \binom{\boldsymbol{\lambda}}{j} e^{-\lambda_j x}$ and Y have sdf $\sum_{i=1}^{m} \binom{\boldsymbol{\nu}}{i} e^{-\nu_j y}$ and so write down the sdf for $aX + bY$ for $a, b > 0$. What remains to be done to answer the problem for Chi-square variates?

4. Verify, in Case I for $n = 3$ and $\sigma_2 > a^2$, where $a = 2\sigma_1$, $b = \sqrt{\sigma_2 - a^2}$, with $z = a + \imath b$, that

$$\int_0^t \frac{z(1-e^{-\bar{z}x})}{z-\bar{z}} + \frac{\bar{z}(1-e^{-zx})}{\bar{z}-z}\,dx = \int_0^t \{1 - e^{-ax}[\cos bx + \tfrac{a}{b}\sin bx]\}\,dx,$$

as was obtained previously.

5. Prove that if $Y \sim \sum_1^n X_i \sim F$, with independent $X_i \simeq \text{Ep}(\lambda_i)$ for $i = 1, \ldots, n$ then the $\mathcal{LS}$-transform of the weighted renewal function, $F_{2\sharp} := \sum_{n=1}^{\infty} nF^{n*}$, satisfies $F_{2\sharp}^\dagger = F_\sharp^\dagger + (F_\sharp^\dagger)^2$.

6. Consider shifted Epstein variates; let $X_i \simeq \text{Shep}(\nu_i, \lambda_i)$, for $i = 1, \ldots, n$ with sdfs $\bar{F}_i(x) = \exp\{-\lambda_i(v = \nu_i)^+\}$ for $x > 0$, where $x^+ = \max(x, 0)$ for $x \in \Re$.
   (a) Show that for all $x > 0$ we have

$$\overline{F_1 * F_2}(x) = \frac{\lambda_1}{\lambda_1 - \lambda_2}\exp\{-\lambda_2(x-\nu_1-\nu_2)^+\} + \frac{\lambda_2}{\lambda_2 - \lambda_1}\exp\{-\lambda_1(x-\nu_1-\nu_2)^+\}.$$

   (b) What would be the sdf of $Y = \sum_1^n X_i \sim G$?
   (c) How could one find the renewal function $G_\sharp$?

15.5. Problems in Scheduled Maintenance

Determination of the proper maintenance schedule in many industries is not just a statistical problem; it often requires cost accounting for replacement plus inventory expense. This is especially true in the nuculear industry where standby units must be maintained, serviced, and checked. Let us consider two common problems: (1) when managers must decide how often a production run on a machine is to be halted for scheduled maintenance-replacement or (2) when its run should be terminated.

In the production of integrated-circuit chips, for use in computers and communication devices, manufacturers, like Intel, are faced with the following decision: during the production run the chip-fabrication machines are scheduled for regular periodic maintenance during the manufacturing process in which certain components are to be serviced or replaced (usually before expiration of their manufacturer's warranteed life). If a such a machine component fails during operation, it causes an unsheduled interruption, and the cost is high. But excessive wear within the irrepairable internal core of the chip-making machine, intrinsic to the operation, can only be detected subsequently in the reduction of the performance parameters of the product; an actualization might cause economic consequences that are catastrophic. Here we dare not ignore the possible closure due to obsolescence or reduced demand.

In such circunstances it is both the risk, expected economic loss, and the danger, the probability of catastrophe, that should be accounted for. Planning only to maximize expected profit often fails to consider either. One ought always to attack the real problems.

15.5.1. A Problem with Unscheduled Fleet Maintenance

The derivation of cumulative damage distributions was based on finding the asymptotic distribution of the time when stationary random incremental damage had accumulated until it reached a critical level. In maintenance one might wish to consider the distribution of cumulative damage, for units in a fleet, under the regular advance of service time. The CD-distributions were obtained by considering the sum of damage increments as being asymptotically Gaussian. However, in practice the units may be exposed to such different operating environments and duty cycles that if $W_{i,j}$ is the wear from the ith repetition by the jth operator (or the jth service environment) that to average over j may neither be close to normality nor may it be reasonable to consider. (Can the same units operating in Alaskan winters have the same maintenance schedule as those in Florida summers?) For some systems the duty cycles imposed by different owner-operators may be so different that no single indicator can be utilized for regular scheduled service.

Often unscheduled repair is initiated when a symptom of incipient failure is detected. Such symptoms can be qualitative, and known only to the technician as an unusual sound of operation (such as howling of the power steering at low speeds) or quantitative, as detected by the diagnostic computer systems for maintenance now used to analyze measurements of internal operating characteristics (such as excessive heat). A general theory for such diverse circumstances has yet to be formulated.

Consider the problem of availability when selected system components receive scheduled maintenance-replacement after every service period of length τ. However the primary operating unit, in series within the system, cannot be repaired since it sustains irreversible cumulative damage over time. It has a service life due to wear off. Suppose there is an *affiliated* variate, say Y_t, which can be repeatedly observed at any time t at small cost and is a symptom related to the cumulative damage extant, not to the time of operation. To be useful there must be some known quantitative distributional relationship between these two variates.

The $\mathcal{FL}$-distributions when suitably transformed are Gaussian variates for which regression theory is already known. None of the few other multivariate distributions known seems to be applicable to the problem. If $T \sim \mathcal{FL}(\alpha, \beta)$, with α, β unknown, and $Y_t \sim \mathcal{FL}(\delta, \nu_t)$ at time t are "affiliated" by which we mean "strongly associated" so that

$$\mathrm{E}\,[\xi(Y/\nu_t)\xi(T/\beta)] = \alpha\delta\rho \quad \text{for some } 0 \ll \rho \leq 1.$$

Some functional relationship, based on physical principles, such as rate of heat generation being related to dullness of a machine cutting tool blade, must be used as a model to proceed farther. If we assume that $Z_1 = \frac{1}{\delta}\xi(Y/\nu)$ and $Z_2 = \frac{1}{\alpha}\xi(X/\beta)$ are jointly Gaussian variates with correlation coefficient ρ. Thus, the conditional density of Z_2 given Z_1 is

$$f(z_2|z_1) = \frac{1}{\sqrt{2\pi(1-\rho^2)}} \exp\{-\tfrac{1}{2(1-\rho^2)}\,[z_2 - \rho z_1]^2\},$$

from which we might find the distribution of T given Y_t, if more physically meaningful assumprtions could be made.

15.5.2. A Problem with Scheduled Fleet Maintenance

Consider the following cumulative-damage problem that arose during fleet maintenance by a US airline carrier: The MD-80 series model jet aircraft has a 'jack-screw' controlling the elevator in the horizontal stabilizer (resembling a small wing) mounted at the top of the vertical fin on the tail. This jack-screw is a component in a controlling subsystem which alters this horizontal elevator and fixes the planes attitude during flight. It is not fail-safe, i.e., it is a series component in the plane's operation since its in-flight failure will cause loss of airplane control.

In 1997 during scheduled major maintenance on an MD-83 airplane a mechanic had issued an order to replace the jack-screw, noting it was too worn to remain on the plane. However, the plane was nearing a 'deadline' to be returned to service and a replacement jack-screw was not in inventory. His supervisor reran a 'wear test' and found 'end play' to be within the maximum tolerable limits and cancelled the replacement order. The plane was returned to service and despite several complaints from pilots over control of the planes attitude during the next two years AA-flight 261, plunged into the ocean in January 2000 causing the death of all on board. The subsequent investigation revealed fatigue-failure of the jack-screw threads due to a lack of lubrication. The increased stress from reduced lubrication caused sudden thread-stripping fatigue failure of the bead, not just slackness in end-play due to wear. After the crash the FAA ordered that jack-screws be lubricated every

650 flight-hours rather than the hithertofore lubrication every 2250 flight-hours. This new inspection period lowered the likelihood the jack-screw would become severely worn before detection.

In October 2005, see [78], the same airline and the same supervisor were investigated by the FAA for incidents involving insufficient lubrication of the jack-screw in three subsequent maintenance inspections. In Jan. 10, 2005, three mechanics reported finding an ungreased jack-screw in an MD-83 plane. In Oct. 6, 2005, a jack-screw was found to have lubrication only at the ends, not in the middle where it is primarily used and friction occurs.

As in many accidents there may be several proximate causes of failure and not a single cause but *in your opinion* rank in importance the possible contributory causes for the loss of Flight 261, as well as the subsequent oversights in airframe maintenance.

1. The MD Corporation for not having designed a fail-safe horizontal stabilizer.
2. The manufacturer was at fault for having, very likely, passed a defective worm-gear, since the inspection and check of hardness of the metal was done statistically (only on every fifth unit). This check should have been made on every unit since the bead on the worm-gear must have the proper degree of hardness, if too hard, the metal is brittle and will break; if too soft the metal will wear too quickly.
3. The Reagan Administration for having deregulated the US airlines, thus providing an incentive for airlines to increase their profitability by reducing maintenance expenses.
4. The AA-mechanic for not citing a reason as being "lack of lubrication" instead of just ordering "Replace nut and perform, E.O.8-55-10-01."
5. The supervisor who cancelled the replacement order, after checking the endplay only.
6. The FAA for not having enough inspectors to ascertain all specified maintenance procedures were carried out.
7. Congress for having cut funds to the FAA thereby forcing a reduction in numbers of inspectors.

Knowledge, i.e., science, liberates mankind from superstition and from the irrational fear of retribution from this or that diety. Indeed, one of the greatest gifts of science is the continuing elimination of the supernatural.

James D. Watson

Pure mathematics is a beautiful profundity wrapped in forbidding technicality. However its legitimacy depends principally on applications in the sciences and statistics, which are footed in the mire of empiricism. The wonder is that the laws of Nature can be expressed so simply and elegantly in mathematical terms.

John Rainwater

Thus, from the war of nature, from famine and death, the most exalted object which we are capable of conceiving, namely, the production of the higher animals, directly follows. Their is grandeur in this view of life, with its several powers, having been originally breathed into a few forms or into one; and that, whilst this planet has gone cycling on according to the fixed law of gravity, from so simple a beginning endless forms most beautiful and most wonderful have been, and are being, evolved.

From *The Origin of Species*, 1859, Charles Darwin

The logical conclusion to natural selection as the mechanism that has produced mankind is that the science of ethics is but a corollary to the science of evolution.

Herbert Spencer

CHAPTER 16

Mathematical Appendix

16.1. Integration

16.1.1. Stieltjes Integrals

We are all familiar with the ordinary Riemann integral studied in calculus. In our work it becomes quite useful to apply a somewhat more general definition of the integral called the *Stieltjes integral.*

Definition. Over the interval (a,b]=$\{x \in \Re : a < x \leq b\}$, the Stieltjes integral of φ with respect to F is defined by the expression, setting $\Delta F(x_k) = F(x_k) - F(x_{k-1})$,

$$\int_{(a,b]} \varphi(x)\, dF(x) = \lim_{\|\pi\| \to 0} \sum_{k=1}^{n} \varphi(\xi_k) \Delta F(x_k),$$

where $\pi = \{x_0, x_1, \dots .x_n\}$ is a *partition* of the half-closed interval $(a, b]$, where

$$a = x_0 < x_1 < \cdots < x_n = b \text{ with } x_{k-1} \leq \xi_k \leq x_k \quad \text{for } k = 1, \dots, n,$$

and

$$\|\pi\| = \max_{1 \leq i \leq n} \{\Delta x_i\} \text{ with } \Delta x_i = x_i - x_{i-1} \quad \text{for } i = 1, \dots, n.$$

The question of when the integral exists is answered, for most practical instances, by the

Theorem. *If φ is continuous and F is nondecreasing on the interval $(a, b]$, then the Stieltjes integral of φ with respect to F always exists.*

Since we are primarily interested in integrating with respect to functions F which are themselves cdfs this definition is precisely what is needed. Moreover, this definition reduces to what we know when a density exists. This we see in the next

Remark. *If F is differentiable on $(a, b]$, with derivative F', then*

$$\int_{(a,b]} \varphi(x)\, dF(x) = \int_a^b \varphi(x) F'(x)\, dx.$$

In fact, virtually all the usual properties of the ordinary (Riemann) integral are true in this new context. Without providing arguments, we list some relevant properties of this integral.

Properties. For any real constants c_0, c_1

$$\int_{(a,b]} dF(x) = F(b) - F(a). \tag{16.1}$$

$$\int_{(a,b]} c_0\varphi(x)\, d[F(x) + c_1] = c_0 \int_{(a,b]} \varphi(x)\, dF(x). \tag{16.2}$$

Since, in the following discussion, all integrals are over the interval $(a, b]$, we omit these limits of integration as well as the arguments of the functions.

Properties. For any positive integers n, m

$$\int (\sum_{i=1}^{n} \varphi_i)\, d[\sum_{j=1}^{m} F_j] = \sum_{i=1}^{n}\sum_{j=1}^{m} \int \varphi_i\, dF_j. \tag{16.3}$$

If φ is continuous at c, we can write without ambiguity

$$\int_a^b \varphi\, dF = \int_a^c \varphi\, dF + \int_c^b \varphi\, dF, \tag{16.4}$$

presuming all integrals exist.

If F is increasing, then

$$\varphi_1 \le \varphi_2 \quad \text{implies} \quad \int \varphi_1\, dF \le \int \varphi_2\, dF, \tag{16.5}$$

and

$$\left| \int \varphi\, dF \right| \le \int |\, \varphi \,|\; dF \le [F(b) - F(a)] \sup_{(a,b]} |\, \varphi \,| \,. \tag{16.6}$$

In practice both integration by parts and change of variables are useful.

Properties. If φ and F have no points of discontinuity in common on $(a, b]$ then

$$\int_{(a,b]} \varphi\, dF = \varphi(x)F(x)\Big|_{x=a}^{x=b} - \int_{(a,b]} F\, d\varphi. \tag{16.7}$$

For any one-to-one transformation ψ of $(a, b]$ onto $(c, d]$, we have

$$\int_{(a,b]} \varphi(x)\, dF(x) = \int_{(c,d]} \varphi[\psi^{-1}(y)]\, dF[\psi^{-1}(y)]. \tag{16.8}$$

Exercise Set 16.A

1. Evaluate the following integrals:
(a)

$$\int_0^\infty e^{tx} dF(x) \quad \text{for any } t < 0, \text{ where } F(x) = \sum_{j=1}^{\infty} \frac{p^{j-1}}{1-p} \mathrm{I}(x \geq j) \quad for \quad x > 0.$$

(b)

$$\int_0^1 x \, dU(x), \quad \text{where} \quad U(x) = \int_0^x \frac{t^3}{3!} \, dt.$$

(c)

$$\int_0^\infty x^2 \, dG(x), \quad \text{where } G(x) = \exp\{-(x/\beta)^2\} \quad \text{for } x > 0; \beta > 0.$$

2. If $[x]$ denotes the largest integer not exceeding x, evaluate:

$$(a) \int_0^\pi \cos x \; d\,[\sin x] \quad (b) \int_{\pi/6}^{\pi/4} x \; d\,[\tan x] \quad (c) \int_0^5 e^x \; d\,(x + [x])$$

$$(d) \int_{-1}^{1} [|x|] \; d\,[1 + e^{-1/x}]^{-1}.$$

3. Let $x \vee y = \max(x, y)$ and $x \wedge y = \min(x, y)$, so $F(x) = x \wedge 1$ for $x > 0$ is the cdf of a Unif(0, 1) rv; evaluate:
(a) $F^{2*}(x) = \int_0^x [(x - y) \wedge 1] \, d(y \wedge 1) \quad \text{for } x > 0.$
(b) $F^{3*}(x) = \int_0^{2 \wedge x} [(x - y) \wedge 1] \, d\, F^{2*}(x) \quad \text{for } x > 0.$
(c) Graph F^{2*} and F^{3*} and conjecture the graph of F^{4*}.

4. Compare the difficulty above with determining for all $x \in \Re$

$$f_2(x) = \int_{-\infty}^{\infty} f_1(y) f_1(x - y) \, dy \quad \text{and} \quad f_3(x) = \int_{-\infty}^{\infty} f_1(y) f_2(x - y) \, dy,$$

where $f_1(x) = \mathrm{I}(0 < x < 1)$ for $x \in \Re$. to show that

$$f_3(x) = \begin{cases} \frac{x^2}{2} & \text{for } 0 \leq x \leq 1, \\ -x^2 + 3x - \frac{3}{2} & \text{for } 1 \leq x \leq 2, \\ \frac{(3-x)^2}{2} & \text{for } 2 \leq x \leq 3. \end{cases}$$

16.2. Probability and Measure

For many purposes the Riemann–Stieltjes integrals previously discussed are adequate. Students in applied mathematics are often averse to the lengthy prerequisite study of measure theory to understand the more advanced theories of integration; but the convergence theorems entailed by such study are needed. There are different opinions as to the best pedogogy that should then lead either to the *Generalized Stieltjes* (also called the Kurzweil and Henstock Integral) or the *Lesbegue–Stieltjes*

Integral. Some recent authors have addressed this problem; see [122], [8], and [23].

We now present a précis of the latter viewpoint. Consider any quantitative experiment, the outcome of which will be some measurement that is a real number. The basic event, called *the sample space*, label it $\mathcal{X}$, is the space into which the measurement must fall and it is always some subset of the real line. The *field of events* (which encompasses all possible event outcomes corresponding to the experiment) must be closed with respect to countable set operations, i.e., any countable number of unions, intersections or complementations on a set of events results in another event. This entire set of events is called a σ-field; label it $\mathcal{B}$.

Let μ denote any measure on $\mathcal{B}$ that assigns finite values to finite intervals. Let $\mathcal{B}$ be the complete σ-field of events measurable by μ. $\mathcal{B}$ is called a *Lesbesgue–Stieltjes field*, and μ is a σ-finite measure on $\mathcal{B}$, which is called a *Lebesgue–Stieltjes measure*. The relation on an interval $\{x \in \Re : a < x \leq b\}\}$ for the function $\mu(a, b] = F(b) - F(a)$ determines, to within an additive constant, a function F that is finite, nondecreasing, and continuous from the right. It is a *cumulative distribution* corresponding to μ. Of course, any such function F suitably normalized also determines a probability measure, and vice versa. Thus, a probability space is a triple $(\mathcal{X}, \mathcal{B}, \mathrm{P})$ where $\mathcal{X}$ is the *sample space*, $\mathcal{B}$ is the σ-field of events, and P is the probablity measure of any event in $\mathcal{B}$, usually defined in terms of a cdf.

Let g be any $\mathcal{B}$-measurable function, i.e., the antecedents of measurable sets are themselves measurable, which is integrable, i.e., the integral $\int g \, d\mu$ exists. If F is the distribution corresponding to μ, then this integral is also denoted by $\int g \, dF$ and the integral

$$\int_{(a,b]} g \, d\mu \quad \text{is also denoted by} \quad \int_a^b g \, dF.$$

In the case F is absolutely continuous on the interval $[a, b]$ it can be shown that the Lebesgue–Stieltjes integral becomes the Riemann integral since

$$\int_a^b g \, dF = \int_a^b g(x)F'(x) \, dx.$$

A *random variable*, say X, on a sample space, is just a measurable function and its identification is often the point of beginning, with the problem being to determine its distribution.

Thus we state for a fixed measure μ, or its associated cdf F, the
MONOTONE CONVERGENCE THEOREM:

If $0 \leq X_n \uparrow X$ *a.e-*μ *then* $\int X_n d\mu \uparrow \int X d\mu$.

LEBESGUE'S DOMINATED CONVERGENCE THEOREM:

If $|X_n| \leq Y$ *a.e. with* Y *integrable and if* $X_n \overset{a.e.}{\to} X$, *or* $X_n \overset{\mu}{\to} X$, *then* $\int X_n d\mu \to \int X d\mu$.

FUBINI'S ITERATED INTEGRALS THEOREM:
On $\mathcal{X} \times \mathcal{Y}$, with the product measure $\pi = \mu \times \nu$, if f on $A \times B$ is nonnegative or π-integrable then

$$\int_{\mathbf{x} \in A \times B} f(\mathbf{x}) d\pi(\mathbf{x}) = \int_A \left[\int_B f(x, y) d\nu(y) \right] \mu(x)$$
$$= \int_B \left[\int_A f(x, y) d\mu(x) \right] d\nu(y).$$

Exercise Set 16.B

1. Prove Euler's summation formula, namely, that if f is a integrable function then
$$\sum_{j=1}^{n} f(j) = \int_0^n f(x)\,dx - \frac{1}{2}[f(n) - f(0)] + \int_0^n B(x) df(x),$$
where, after defining $B(x)$ its Fourier expansion is given,
$$B(x) := x - [x] - \frac{1}{2} = -\sum_{k=1}^{\infty} \frac{\sin(2k\pi x)}{k\pi}.$$
The function $B(x)$, when defined only for $0 < x < 1$, is called the *Bernoulli polynomial.*
2. Prove iff $g(t) = \int_{\alpha(t)}^{\beta(t)} f(t, y)\, dy$ is differentiable wrt t then then its derivative is
$$g^{'}(t) = \int_{\alpha(t)}^{\beta(t)} \frac{\partial f(t, y)}{\partial t}\, dy + f[t, \beta(t)]\beta^{'}(t) - f[t, \alpha(t)]\alpha^{'}(t).$$
NB this is *Leibnitz's Rule* and if $\alpha^{'}$ and $\beta^{'}$ are continuous then so is $g^{'}$.
3. Show, for any real functions F, G of bounded variation where the *convolution* is
$$G{*}F(x) := \int_{-\infty}^{\infty} G(x - y)\, dF(y) \quad \text{for } x \in \Re.$$
   (a) $G{*}F = F{*}G$ iff $G(-\infty) = F(-\infty) = 0$.
   (b) $(G{*}F)' = G'{*}F$ if G' exists a.e.
   (c) $F{*}(G + a) = F{*}G$ for any constant a.
   (d) The function $1(x) := \mathrm{I}(x \geq 0)$ for $x \in \Re$ serves as an identity, i.e., $G{*}1 = 1{*}G = G$.
   (e) $(G{*}F){*}H = G{*}(F{*}H)$.
4. Thus, conclude if $\mathcal{F}$ is the set of cdfs with support on the set $\mathcal{S} \subset \Re$, then $(\mathcal{F}, *)$ forms a commutative semigroup with identity, when $0 \in \mathcal{S}$.

16.3. Distribution Transforms

If $X \sim F$, then the *moment generating function* (mgf) of the rv X, say m_X, is defined by

$$m_X(t) := \mathrm{E}e^{tX} = \int_{-\infty}^{\infty} e^{tx}\, dF(x), \tag{16.9}$$

when this integral exist for all t in some neighborhood of 0.

Suppose $X > 0$ has a density f_X. Then its *Laplace transform* is defined by

$$f_X^\flat(s) := \mathrm{E}e^{-sX} = \int_0^\infty e^{-sx} f_X(x)dx. \qquad (16.10)$$

It is often used in engineering science, where it is frequently denoted by $\mathcal{L}g(s)$ for any function g on $[0, \infty)$. Clearly, the Laplace transform is related to the mgf.

Theorem. *If the mgf of X, say m, exists and so does $\mathrm{E}X^r$ for some positive integer r then*

$$\frac{d^r m(t)}{dt^r} := m^{(r)}(t) \quad and \quad m^{(r)}(0) = \mathrm{E}X^r; \qquad (16.11)$$

and if $\mathrm{E}X^r$ *exist for all* $r = 0, 1, \ldots,$ *then,*

$$m(t) = \sum_{r=0}^{\infty} \frac{\mathrm{E}(X^r)t^r}{r!}. \qquad (16.12)$$

The Laplace–Stieltjes transform of a function F is denoted by $F^\dagger$ and is defined by

$$F^\dagger(s) = \int_{-\infty}^{\infty} e^{-st}\, dF(t), \qquad (16.13)$$

when the integral exists for s in some neighborhood of zero. NB $F^\dagger = \mathcal{L}F'$ when F' exists.

In reliability theory we have the immediate but important

Theorem. *If F is a life distribution on* $[0, \infty)$*, then $F^\dagger$ always exists.*

Another important function is the *factorial power* given for any $x \in \Re$ by

$$(x)^{r\downarrow} = \prod_{j=1}^{r}(x + 1 - j) \quad \text{for} \quad r = 1, 2, \ldots\ ; \qquad (16.14)$$

then $\mathrm{E}(X^{r\downarrow})$ is called the *rth factorial moment* of X. The *factorial moment generating function* (fmgf), say g_X, when it exists is related to the mgf by

$$g_X(t) := \mathrm{E}t^X = \mathrm{E}e^{X \ln t} = m_X(\ln t).$$

Successive derivatives of g_X, evaluated at $t = 1$, give the factorial moments of X.

Theoretically, the most useful of the transforms, because it always exists, is the *characteristic function*, denote it by $F_X^\ddagger$. It is defined, for all $t \in \Re$, by

$$F_X^\ddagger(t) = \int_{-\infty}^{\infty} e^{\imath tx} dF_X(x) = \mathrm{E}[\cos(tX) + \imath \sin(tX)]. \qquad (16.15)$$

Here $\imath = \sqrt{-1}$. In mathematics the characteristic function is called the *Fourier transform* of the density F_X', when the density exists, otherwise $F^\ddagger$ denotes the *Fourier–Stieltjes transform* of F.

Since the integral in eqn (16.15) always exists and is absolutely and uniformly convergent it follows that $F_X^{\ddagger}$ is always a continuous function on $\Re$. The characteristic function is also useful because of its inversion theorem.

Theorem. *Let F and $F^{\ddagger}$ denote, respectively, the distribution and characteristic function of some random variable X. If $x+y$ and $x-y$, for some $x \in \Re$ and $y > 0$, are continuity points of the cdf F, then*

$$F(x+y) - F(x-y) = \lim_{a\to\infty} \frac{1}{\pi} \int_{-a}^{a} \frac{\sin(y\,t)}{t} e^{-\imath tx} F^{\ddagger}(t)dt. \tag{16.16}$$

From this follows the simpler inversion lemma.

Lemma. *If the characteristic function $F^{\ddagger}$ is known and the corresponding density F' exists and is continuous, then*

$$F'(x) = \frac{1}{2\pi} \int_{-\infty}^{\infty} e^{-\imath tx} F^{\ddagger}(t)dt. \tag{16.17}$$

It is the constant 2π in eqn (16.17) that marks the difference in the definitions of Fourier transforms as used in various disciplines. The one used here is favored in mathematics, statistics, and systems engineering. But the default setting in *Mathematica* for its "Fourier transform" is the one used in modern physics for the transform-and-its-inverse pair. This pair is, respectively,

$$\hat{f}(s) = \frac{1}{\sqrt{2\pi}} \int_{-\infty}^{\infty} f(t)e^{\imath ts}\,dt \quad \text{and} \quad f(t) = \frac{1}{\sqrt{2\pi}} \int_{-\infty}^{\infty} \hat{f}(s)e^{-\imath st}\,ds.$$

Of course, as in our case with life distributions, different limits of integration may apply as from 0 to ∞ rather than from $-\infty$ to ∞.

Other transforms prove to be useful in special circumstances; one is the *Mellin transform* defined by $\mathrm{E}X^t$, for all t for which it exists. It is used mostly in engineering. Another is the *cumulant generating function* (cgf), say $K_X(t)$, defined in terms of the characteristic function as

$$K_X(t) = \ln c_X(t). \tag{16.18}$$

Succesive derivatives of $K(t)$ evaluated at $t = 0$ are called *cumulants*.

Exercise Set 16.C

1. If $F(t) = t^a$ for $t > 0, a > -1$, show that $F^{\dagger}(s) = \Gamma(a+1)s^{-a}$ for $s > 0$.
2. If $F(t) = \sin(bt)$ for $t > 0, b \in \Re$, show that $F^{\dagger}(s) = \frac{bs}{b^2+s^2}$
3. If $F(t) = \cos(bt)$ for $t > 0, b \in \Re$, show that $F^{\dagger}(s) = \frac{b^2}{b^2+s^2}$ for $s \leq b$.
4. If $F(t) = \sinh(bt)$ for $t > 0, b \in \Re$, show that $F^{\dagger}(s) = \frac{bs}{s^2-b^2}$ for $s \leq b$.

5. If $F(t) = \cosh(bt)$ for $t > 0, b \in \Re$, show that $F^\dagger(s) = \frac{b^2}{s^2-b^2}$ for $s \le b$.

6. For Laplace transforms (a) Prove the shifting theorem and the derivative theorem: If f is given by $\mathcal{L}f(s) = \int_0^\infty e^{-st} f(t)\,dt$ and $g(t) = e^{\alpha t} f(t)$, then $\mathcal{L}g(s) = \mathcal{L}f(s-\alpha)$. If $h(t) = t^n f(t)$, then $\mathcal{L}h(s) = (-1)^n \frac{d^n \mathcal{L}f(s)}{(ds)^n}$.

7. For Laplace transforms show:

$$\mathcal{L}f(s) = \frac{b}{(s-a)^2+b^2} \text{ iff } f(t) = e^{at}\sin(bt) :$$

$$\mathcal{L}f(s) = \frac{s-a}{(s-a)^2+b^2} \text{ iff } f(t) = e^{at}\cos(bt).$$

$$\mathcal{L}f(s) = \left(\frac{a}{s+a}\right)^k \quad \text{for } k > 0 \quad \text{iff} \quad f(t) = \frac{a^k t^{k-1} e^{-at}}{\Gamma(k)}.$$

8. Show the following operations hold:
   (a) $(F+G)^\dagger = F^\dagger + G^\dagger$.
   (b) $(aF)^\dagger = aF^\dagger$ for any constant a.
   (c) $(F'')^\dagger(s) = s^2 F^\dagger(s) - sF'(0+) - 1$.
   (d) If $F' = f$ exists, then $f^\dagger(s) = -f(0+) + sF^\dagger(s)$.
   (e) If $F(t) = \int_0^t f(x)\,dx$, then $F^\dagger(s) = s^{-1} f^\dagger(s)$.

9. Prove the result of Meng [75]: If $Y \sim F$ and $g^\dagger$ is the Laplace–Stieltjes transform of g and the Fubini theorem holds then,

$$\mathrm{E}g^\dagger(Y) = \int_0^\infty F_Y^\dagger(x)\,dg^\dagger(x).$$

   (a) Exercise 16C.1 implies $y^{-\kappa} = [1/\Gamma(\kappa+1)] \int_0^\infty e^{-xy}\,dx^\kappa$ for $\kappa > 0$ so show

$$\mathrm{E}[Y^{-\kappa}] = \frac{1}{\Gamma(\kappa+1)} \int_0^\infty F_Y^\dagger(x)\,dx^\kappa.$$

   (b) Let $X \sim \mathcal{FL}(\alpha, \beta)$ and use this result to verify $\mathrm{E}(1/X)$.

The initial papers detailing the discoveries of both Fourier of *Fourier Analysis* and Fegenbaum of *Chaos Theory* were rejected by referees for several years, until eventually other persons became interested then they ultimately received the recognition their genius deserved. But how many others were not so lucky? Do you know of J.J. Waterston? In 1843 and 1846 he had abstracts of papers published but his papers, giving the first modern kinetic-theory definition of temperature with its important consequences, were rejected! "The system of secret refereeing, still in almost exclusive use today, has protected the anonymity of eminent referees, while burying the genius of Waterston in permanent oblivion."[1] Fortunately, in an incident similar to Louiville's post mortem recognition of Galois, Lord Rayleigh discovered a Waterston manuscript and caused it to be published, in 1893, in the very journal to which it had been submitted and rejected, 48 years earlier.

[1] From p.303 of C. Truesdell's *An Idiot's Fugative Essays on Science*, Springer-Verlag, 1984.

16.4. A Compendium of Discrete Distributions

$\aleph_a^b = \{$ integers : $a \leq j \leq b\}$ and $0 < p < 1, q = 1 - p$ with or without affixes.

Name	**pdf**	**Mean**	**Variance**	**mgf**
Bernoulli	$p^x q^{1-x}$ for $x = 0, 1$	p	pq	$pe^t + q$
	$X \sim \mathcal{B}\text{er}(1, p)$ indicates success in a single trial			
Binomial	$\binom{n}{x} p^x q^{n-x}$ for $x \in \aleph_0^n$	np	npq	$(pe^t + q)^n$
	$X \sim \mathcal{B}\text{in}(n, p)$ counts successes in n Bernoulli trials			
Geometric	qp^{x-1} for $x \in \aleph_1^\infty$	$\frac{1}{q}$	$\frac{p}{q^2}$	$\frac{qe^t}{1-pe^t}$
	$X \sim \mathcal{G}(p)$ counts Bernoulli trials until first failure			
De Moivre	$\sum_{j=1}^k \left\langle {q_k \atop j} \right\rangle p_j^x q_j$ for $x \in \aleph_k^\infty$	$\sum_1^k \frac{1}{q_j}$	$\sum_1^k \frac{p_j}{q_j^2}$	$\prod_1^k \frac{q_j e^t}{1-p_j e^t}$
	$W_k = \sum_1^k X_j \sim \mathcal{D}\text{eM}(\boldsymbol{q_k})$ when $X_j \sim \mathcal{G}(p_j)$ are independent			

If we draw with replacement from an urn containing m different objects, then

S_n, the number of draws to obtain n different objects, has pdf

$\frac{\binom{m}{n}\{x-1\}_{n-1}}{m^x}$ for $x \geq n$	$\sum_1^k \frac{m}{m+1-j}$	$\sum_1^k \frac{m(j-1)}{(m+1-j)^2}$	

If we toss at random n indistinquishable balls into m enumerated cells, then

X_0, the number of unoccupied cells, has pdf

$\frac{\binom{m}{m-x}\{n\}_{n-x}}{m^n}$ for $x \in \aleph_0^m$	$m(1-\frac{1}{m})^n$	$\sum_1^2 m^{j\downarrow}(1-\frac{j}{m})^n - m^2(1-\frac{1}{m})^{2n}$	

Name	**pdf**	**Mean**	**Variance**	**mgf**
Pascal	$\binom{x-1}{r-1} p^r q^{x-r}$ for $x \in \aleph_r^\infty$	$\frac{r}{p}$	$\frac{rq}{p^2}$	$\left(\frac{pe^t}{1-qe^t}\right)^r$
	$X \sim \mathcal{P}\text{as}(r, p)$ counts Bernoulli trials until the rth success			
Hypergeometric	$\frac{\binom{n}{x}\binom{m-n}{k-x}}{\binom{m}{k}}$ for $x \in \aleph_0^k$	$\frac{kn}{m}$	$\frac{kn}{m}(1-\frac{n}{m})(\frac{m-k}{m-1})$	
	$X \sim \mathcal{H}\text{yp}(k, n, m)$ counts number of type-1 in a sample of size k from a population with n elements of type-1 and m otherwise			
Poisson	$\frac{e^{-\lambda}\lambda^x}{x!}$ for $x \in \aleph_0^\infty$	λ	λ	$\exp\{\lambda(e^t - 1)\}$
	$X \sim \mathcal{P}\text{oi}(\lambda)$ counts number of rare events when their density is λ			

16.5. A Compendium of Continuous Distributions

$\Re$ denotes the real line, $\Re_+$ denotes $[0, \infty)$, and $\aleph$ is the set of positive integers.

Name	pdf or sdf	Mean	Variance	mgf
Uniform-pdf	$\frac{1}{b-a}$ for $a \le x \le b$	$\frac{a+b}{2}$	$\frac{(b-a)^2}{12}$	$\frac{e^{bt}-e^{at}}{(b-a)t}$

$X \sim Uni[a, b]$ indicates a point equally likely in interval $[a, b]$

Epstein-pdf	$\lambda e^{-\lambda x}$; for $x \in \Re_+$	$\frac{1}{\lambda}$	$\frac{1}{\lambda^2}$	$(1-\lambda t)^{-1}$

$X \sim Ep(\lambda)$ is the wait until success when there is no bonus for waiting.

Erlang-sdf	$\sum_{j=1}^{k}\binom{\lambda_k}{j}e^{-\lambda_j x}$ for $x \in \Re_+$	$\sum_1^k \frac{1}{\lambda_j}$	$\sum_1^k \frac{1}{\lambda_j^2}$	$\prod_1^k \frac{1}{1-\lambda_j t}$

If $X_j \sim Ep(\lambda_j)$ are $\perp$ then $Y_k = \sum_1^k X_j \sim Erl(\lambda_k)$.

Beta-pdf	$\frac{\Gamma(\alpha+\beta)x^{\alpha-1}(1-x)^{\beta-1}}{\Gamma(\alpha)\Gamma(\beta)}$ for $x \in [0, 1]$	$\frac{\alpha}{\alpha+\beta}$	$\frac{\alpha\beta}{(\alpha+\beta+1)(\alpha+\beta)^2}$	
Gamma-pdf	$\frac{x^{\kappa-1}e^{-x/\beta}}{\beta^\kappa\Gamma(\kappa)}$ for $x \in \Re_+$	$\kappa\beta$	$\kappa\beta^2$	$(1-\beta t)^{-\kappa}$

$X \sim Gam(\kappa, \beta)$ is called $\chi^2(k)$ when $\beta = 2$ and $\kappa = k/2$ with $k \in \aleph$

Weibull-sdf	$e^{-(x/\beta)^\alpha}$ for $x \in \Re_+$	$\beta\Gamma(1+\frac{1}{\alpha})$	$\beta^2[\Gamma(1+\frac{2}{\alpha})-\Gamma^2(1+\frac{1}{\alpha})]$	

$X \sim Wei(\alpha, \beta)$ is also Gumbel's asymptotic minimum distribution.

Gumbel-cdf	$\exp\{-\exp(\frac{x-\eta}{\alpha})\}$ for $x \in \Re$	$\eta - \gamma\alpha$	$\frac{\pi^2}{6}\alpha^2$	

$X \sim Gum(\eta, \alpha)$ is Gumbel's asymptotic maximum distribution.

Define φ, Φ, ξ, ψ as in Chapter 10; Let $\eta_{(r)} = (\Sigma_1^k \eta_i^r)^{1/r}$ if $r \ne 0$ for any set $\{\eta_i\}_{i=1}^k$

Gaussian pdf	$\frac{1}{\sigma}\varphi(\frac{x-\mu}{\sigma})$ for $x \in \Re$	μ	σ^2	$e^{t\mu+\frac{t^2}{2}\sigma}$

If $X_i \sim \mathcal{N}(\mu_i, \sigma_i^2)$ are $\perp$, then $X_{(1)} \sim \mathcal{N}[\mu_{(1)}, \sigma_{(2)}^2]$ with $X_{(1)} \perp X_{(2)}^2$.

$$X \sim \mathcal{N}(\mu, \sigma^2) \text{ iff } \left(\frac{x-\mu}{\sigma}\right)^2 \sim \chi^2(1)$$

Cumulative damage distributions defined on $\Re_+$ with $\alpha, \beta > 0, 0 \le p \le 1, q = 1 - p$.

$CD(\alpha, \beta, p)$-pdf	$\frac{\psi(x/\beta)-(p-q)\xi(x/\beta)}{2\alpha x}\varphi[\frac{1}{\alpha}\xi(x/\beta)]$	$\beta(1+q\alpha^2)$	$(\alpha\beta)^2[1+q(2+p)\alpha^2]$	
$CD(\alpha, \beta, p)$-cdf	$\Phi[\frac{1}{\alpha}\xi(x/\beta)]+(p-q)e^{2/\alpha^2}\Phi[\frac{-1}{\alpha}\psi(x/\beta)]$	$\exp\left\{\frac{1}{\alpha^2}(1-\sqrt{1-2t\alpha^2\beta}\right\}\times\left[p+\frac{q}{\sqrt{1-2t\alpha^2\beta}}\right]$		

Wald: $W(\alpha, \beta) := CD(\alpha, \beta, 1)$, Tweedie: $T(\alpha, \beta) := CD(\alpha, \beta, 0)$,

Fatigue life: $FL(\alpha, \beta) := CD(\alpha, \beta, 1/2)$.

$$X \sim CD(\alpha, \beta, p) \text{ iff } \frac{1}{\alpha^2}\xi^2(X/\beta) \sim \chi^2(1);$$

$$X \sim FL(\alpha, \beta) \text{ iff } \frac{1}{\alpha}\xi(X/\beta) \sim \mathcal{N}(0, 1).$$

If $X_i \sim W(\alpha_i, \beta_i)$ are $\perp$ and $\alpha_1^2\beta_1 = \cdots = \alpha_n^2\beta_n$, then

$$X_{(1)} \sim W(\alpha_{(-2)}, \beta_{(1)}).$$

If $X_i \sim T(\alpha_i, \beta_i)$ are $\perp$ and $\frac{\alpha_1^2}{\beta_1} = \cdots = \frac{\alpha_n^2}{\beta_n}$, then $X_{(-1)} \sim T(\alpha_{(-2)}, \beta_{(-1)})$.

NB if $\alpha < .3$ there is no significant difference between the $CD(\alpha, \beta, p)$ distributions, for any $p = 0, .5, 1$, when the means and variances are equal!

Bibliography

[1] Aalen, O. (1976), Nonparametric Inference in Connection with Multiple Decrement Models, Scand. J. Statist., **3**, 15–27.

[2] Abramowitz, M., and Stegun, I.A. (1965), *Handbook of Mathematical Functions*. New York: Dover Publications

[3] Altshuler, B. (1970), Theory for Measurement of Competing Risks in Animal Experiments, *Math. Biosci.*, **6**, 1–11.

[4] Aroian, L.A., and Robison, D.E. (1966), Sequential life tests for the exponential distribution with changing parameter. *Technometrics* **8,** 217–227

[5] Atkins, P.W. (1986), *Physical Chemistry*, Third Edition, Chapter 28, pp. 687–712, W.H. Freeman and Company, New York, NY.

[6] Bain, L.J., and Engelhardt, M. (1987), *Introduction to Probability and Mathematical Statistics*. Boston: Duxbury Press.

[7] Barlow, R.E., and Proschan, F. (1981), *Statistical Theory of Reliability and Life Testing*. Silver Spring, MD.: To Begin With

[8] Bartle, R.G. (1996), Return to the Riemann integral, *Amer. Math. Monthly* **103**, 625–632.

[9] Bogdanoff, J.L., and Kozin, F. (1985), *Probabilistic Models of Cumulative Damage*, John Wiley & Sons, New York, NY.

[10] Bauer, D. R. (1994), Chemical Criteria for Durable Automotive Topcoats, *Progress in Organic Coatings*, 23:105 (QD380.p75).

[11] Bauer, D.R., Mielewski, D.F., and Gerlock, J.L. (1992), Photooxidation kinetics in crosslinked polymer coatings. *Polymer Degradation and Stability* **38,** 57.

[12] Bauer, D.R. (1994), Chemical criteria for durable automotive topcoats. *Progress in Organic Coatings* **23,** 105.

[13] Bergman, B. (1985), On reliability theory and its applications. *Scandinavian Journal of Statistics* **12,** 1.

[14] Bernau, S.J. (1988), The evaluation of a Putnam integral. *The Amer. Math. Monthly* **95,** 935.

[15] Bhattacharyya, G.K. and Fries, Arthur (1982), Inverse Gaussian Regression and Accelerated Life Tests,*Survival Analysis*, Eds. J. Crowley and R. Johnson,IMS LECTURE NOTES-MONOGRAPH SERIES, Vol.2, 101–117.

[16] Bhattacharyya, G.K. and Fries, Arthur (1982), Fatigue failure models – Birnbaum Saunders vs. inverse Gaussian. *IEEE Transactions on Reliability* **R-31**, (5), 439–440.

[17] Billingsley, P. (1979), *Probability and Measure*, John Wiley & Sons, Chicago, IL.

[18] Birnbaum, Z.W. and Saunders, S.C. (1969), A new family of life distributions. *Journal of Applied Probability* **6**, 319–327.

[19] Birnbaum, Z.W. and Saunders, S.C. (1969), Estimation for family of life distributions with applications to fatigue. *Journal of Applied Probability* **6**, 328–347.

[20] Brémaud, P. (1988), *An Introduction to Probabilistic Modeling*. New York: Springer-Verlag.

[21] Brenner, S and Scott, L.R. (1991), *The Mathematical Theory of Finite Element Methods*, New York: Springer-Verlag.

[22] Capellos, C., Bielski, B.H.J. (1972) *Kinetic Systems*, Wiley-Interscience, New York, NY.

[23] Carter, Michael and van Brunt, Bruce (2000), *The Lesbegue-Stieltjes Integral: A Practical Introduction,* Springer, New York.

[24] Castellan, G.W. (1995), *Physical Chemistry*, Second Edition, Chapter 31, Chemical Kinetics, Addison-Wesley Publishing Co., Reading, MA.

[25] Chen, Y.Y., Hollander, M. and Langberg, N.A. (1982), Small-sample results for the Kaplan–Meier estimator, *Theory and Methods Section of J. Amer. Stat. Assoc.*, **77**, 141–144.

[26] Chhikara, R.S., and Folks, J.L. (1989), *The Inverse Gaussian Distribution: Theory, Methodology, and Applications*. New York: Marcel Dekker.

[27] Cone, Marla (2005), *Silent Snow: The Slow Poisoning of the Artic*, Grove Press.

[28] Cramer, Harald (1951) *Mathematical Methods of Statistics,* Princeton University Press, Princeton, NJ.

[29] de Rijk, W. G., Tesk, J. A., Penn, R. W., and Marsh, J. (1990), Applications of the Weibull method to statistical analysis of strength parameters of dental materials. *Progress in Biomedical Polymers,* eds. C.G. Gebelien and R.L. Dunn, New York: Plenum Press.

[30] De Temple, D.W., Wang, S.H. (1991), Half-Integer Approximations for the Partial Sums of the Harmonic Series, *Journal of Mathematical Analysis and Applications*, **160**, pp. 149–156.

[31] Doksum, K.A., and Hóyland, A. (1992), Models for variable stress accelerated life testing experiments based on Wiener processes and the inverse-Gaussian distribution. *Technometrics* **34**, (1), 74.

[32] Efron, B. (1967), The Two Sample Problem with Censored Data, *Proc. of Fifth Berkeley Symp., IV*, 831–853.

[33] Feller, William (1968), *An Introduction to Probability Theory and Its Applications*, Third Edition, John Wiley & Sons, New York, NY.

[34] Freudenthal, A.M. and Heller, R.A. (1959), On stress interaction in fatigue and a cumulative damage rule. *J. Aero. Sci.* **26**, 431–442.

[35] Fano, U. (1950), Chapter 1, Principles of Radiological Physics, in Radiation Biology.

[36] Freudenthal, A.M. and Shinozuka, M. (1961), Structural safety under conditions of ultimate load-failure and fatigue. *WADD Technical Report*, 61–77.

[37] Genz, A. (1991), Numerical computation of multivariate normal probabilities. Report EECS-91-004, Wash. State Univ., School of Elec. Eng. & Comp. Sci., Pullman, WA 99164-2752

[38] Gompertz, B. (1825), On the nature of the function expressive of the law of human mortality and a new mode of determining value of life contingencies, Philos. Trans. Roy. Soc. London, **115**, 513–585.

[39] Green, A.E.S., Sawada, T., and Shettle, E.P. (1974), The middle ultraviolet reaching the ground. *Photochemistry and Photobiology*, 251–259.

[40] Griffith, A. A. (1924), The Theory of Rupture, *Proceedings of the First International Congress for Applied Mechanics*, 55–63.

[41] Guillet, J. (1985), *Polymer Photophysics and Photochemistry*, New York: Cambridge University Press.

[42] Gumbel, E. J. (1958), *Statistics of Extremes*, New York: Columbia University Press.

[43] Halmos, Paul R. (1985), *I Want to Be a Mathematician: An Automathography*, Springer-Verlag, p. 116.

[44] Hjorth, U. (1980) A Reliability Distribution with Increasing, Decreasing, Constant and Bathtub-Shaped Failure Rates, *Technometrics*, **22**, 99–107.

[45] Jacobsen, M. (1988), Right-Censoring and the Kaplan-Meier and Nelson-Aalen Estimators: Summary of Results. Contemporary Mathematics, **80**, 61–65.

[46] Jacobsen, M. (1989), Right-censoring and martingale methods for failure time data, *Annals of Statistics*, **17**, 1133–1146.
[47] Johnson, N.L., and Kotz, S. (1970), *Continuous Univariate Distributions-1*, New York, John Wiley & Sons.
[48] Jørgensen, B. (1982), *Statistical Properties of the Generalized Inverse Gaussian Distribution*. New York: Springer-Verlag.
[49] Jørgenson, B., Seshadri, V., and Whitmore, G.A. (1991), On the mixture of the inverse-Gaussian distribution with its complementary reciprocal. *Scand. J. Stat.*, **18**, 77-89.
[50] Karlin, S. (1968). *Total Positivity*. Vol. I. Stanford, CA.: Stanford University Press.
[51] Katsev, P.G. (1968), Statistical methods in the study of cutting tools. (in Russian) *Mashinostroyeniye*
[52] Kempthorne, Oscar (1952), *Design and Analysis of Experiments*, John Wiley & Sons, New York, NY.
[53] Kaplan, E.L. and Meier, P. (1958), Nonparametric Estimation from Incomplete Observations, *J. Am. Stat. Assoc.*, **53**, 457–481.
[54] Khaleel, M.A. (1993), Probabilistic Fatigue of Partially Prestressed Concrete Bridges. Ph.D. Thesis, Washington State University.
[55] Konvalina, J. (2000), A unified interpretation of the binomial coefficients, the Stirling numbers, and the Gaussian coefficients. *The American Mathematical Monthly*, **107**, 901–910.
[56] Kozlov, B. A., Ushakov, I.A. (1970), *Reliability Handbook*, New York: Holt Rinehart and Winston, Inc.
[57] Kreyszig, Erwin (1988), Advanced Engineering Mathematics, Sixth Edition, New York: John Wiley & Sons.
[58] Lasota, A., and Mackey, M.C. (1994), *Chaos, Fractals and Noise*. 2nd ed. New York: Springer-Verlag.
[59] Lawless, J.F. (1982), *Statistical models and methods for lifetime data*. New York: Wiley
[60] Little, R.E., and Ekvall. J.C., Eds. (1979), Statistical analysis of fatigue data, ASTM, STP744.
[61] Loader, C.R. (1991), Inference for a hazard rate change point. *Biometrika* **78**, 749.
[62] Loéve, M. (1955), *Probability Theory*. New York: Van Nostrand.
[63] Lorden, G. and Eisenberger, I. (1973), Detection of failure rate increases, *Technometrics* **15**, 167–175.
[64] Lockheed Missles and Space Company Inc. (1991), Reliability engineering assessment of C3 primary electronic battery P/N 2763536. Unclassified report.
[65] Luo, D. and Saunders S.C. (1994), Bias and mean-square error for the Kaplan–Meier and Nelson–Aalen estimators, *Jour. Nonpar. Statist.*, Vol. 3, pp. 37–51.
[66] McLeod, R.M. (1980), *The Generalized Riemann Integral*, Carus Mathematical Monographs no. 20, Mathematical Association of America, Washington D.C.
[67] Makeham, W.M. (1860), On the law of mortality and the construction of Annuity Tables, *Jour. Inst. Actuar.*, **8**, 301–310.
[68] Mann, N.R. (1985), *The Keys to Excellence – The Story of the Deming Philosophy*, Prestwick Publishing.
[69] Mann, N.R., Schafer, R.E. and Singpurwalla, N.D. (1984), *Methods for Statistical Analysis of Reliability and Life Data*. New York: Wiley.
[70] Marshall, A.W. and Olkin, I. (1967), A multivariate exponential distribution, J.A.S.A, **62**, 30–44.
[71] Mathews, D.E., Farewell, V.T., and Pyke, R. (1985), Asymptotic score-statistics processes and tests for constant hazard against a change-point alternative. *Ann. Stat.* **13**, 583–591

[72] Martin, J.W. (1984), A stochastic model for predicting the service life of photolytically degraded poly(methyl methacrylate) films. *Journal of Applied Polymer Science* **29**, 777–794.

[73] Martin, J.W. (1992), Quantitative characterization of spectral ultraviolet radiation-induced photodegradation in coating systems exposed in the laboratory and the field. *NIST* Publication.

[74] McMillan, J.C. and Pelloux, R.M.N. (1967), Fatigue crack propagation under program and random loads. *ASTM STP 415, American Society for Testing Materials*, 505–532.

[75] Meng, Xiao-Li (2005), From unit root to Stein's estimator to Fisher's k statistics; If you have a moment, I can tell you more. Staistical Science, **20**, 141–162.

[76] Michael, J.R., Schucany, W.R. and Haas, R.W. (1976), Generating random variables using transformations with multiple roots. *Amer. Statist.* **30**, 88–90

[77] Mielewski, D.F., Bauer, D.R., and Gerlock, J.L. (1993), The role of hydroperoxides in the photo-oxidation of cross-linked polymer coatings containing hindered amine light stabilizers. *Polymer Degradation and Stability,* **41**, 323–331.

[78] Miletich, S. (2005), *The Seattle Times*, New Woes for Alaska Jack-Screw Servicing, Thursday, October 6, 2005.

[79] Miner, M.A. (1945), Cumulative damage in fatigue. *J. Appl. Mech.* **12**, A159–A164.

[80] Minogue, C.D. (1991), *A Unifying Generalization of the Inverse-Gaussian, Reciprocal Inverse-Gaussian and Birnbaum-Saunders Distributions*. Unpublished PhD Thesis, Washington State University.

[81] Mises, R.V. (1939), Über Aufteilungs-und Besetzungswahrscheinlichkeiten, *Revue de la Faculté des Sciences de l'Université d'Istanbul,* N.S., **4**, pp. 313–324.

[82] Mosteller, F. (1965), *Fifty Challenging Problems in Probability (with solutions)*, Reading, MA: Addison-Wesley Publishing Co.

[83] Nelson, W. (1969), Hazard plotting for incomplete failure data, J. Qual. Tech., **1**, 27–52.

[84] Nguyen, H.T., Rogers, G.S., Walker, E.A. (1984), Estimation in change-point hazard rate models. *Biometrica* **71**, 299–304.

[85] Owen, W.J., Padgett, W.J. (1999), Accelerated test models for system strength based on Birnbaum–Saunders distributions. *Lifetime Data Analysis,* **5**, 133–147.

[86] Own, S-H., Subrumanian, R.V., Saunders, S.C. (1986), A Bimodal Lognormal Model of the Distribution of Strength of Carbon Fibres: Effects of Electrodeposition of Titanium Dioxyacetate, *J. Mater. Sci.*, **21**, 3912–3920.

[87] Ouypornprasert, W. (1989), Efficient computational methods for structural reliability analysis based on conditional importance sampling functions. *ZAMM·Z. angew. Math. Mech.* **69**, T69–T71.

[88] Palmgren, A. (1924), Die Lebensdauer von Kugellagern. *Z. Ver. Dtsch. Ing.* **68**, 339–341

[89] Paris, P.C., Erdogan, F. (1963), A critical analysis of crack propagation laws, *Journal of Basic Engineering* **85**, 528–34.

[90] Parzen, E. (1967), On models for the probability of fatigue failure of a structure. *Time Series Analysis Paper*, 532–548, Holden-Day, San Francisco.

[91] Peterson, A. V. (1977) Expressing the Kaplan-Meier Estimator as a Function of Emperical Subsurvival Functions, J. Amer. Statist. Assoc., **72**, 854–858.

[92] Ravichandran, N. (1990), *Stochastic Methods in Reliability Theory*. New Dehli: Wiley Eastern Ltd.

[93] Reick, J. R. and Nedelman, J. R. (1991), A log-linear model for the Birnbaum–Saunders distribution, *Technometrics*, **33**, 55–61.

[94] Saunders, S.C. (1965), *Proceedings of the Fifteenth Conference on the Design of Experiments in Army Research Development and Testing*, On Confidence Limits for the Performance of a System when few failures are encountered. ARO-D report 70-2.

[95] Saunders, S.C. (1968), On the determination of a safe-life for distributions classified by failure rate. *Technometrics,* **10**, 361–377.
[96] Saunders, S.C. (1970), A probabalistic interpretation of Miner's rule. *Siam J. on Applied Math.* **19**, 251–265.
[97] Saunders, S.C. (1972), On the probabilistic determination of scatter factors using Miner's rule in fatigue. *Probabilistic Aspects of Fatigue* ed. R.A. Heller, A.S.T.M. STP- 511.
[98] Saunders, S.C., and Khaleel, M. (1992), The estimation of parameters for the fatigue-life distribution under spectrum loading, Scientific Consulting Services, Inc.
[99] Saunders, S.C., and Moody, M. E. (1986), Great Expectations or Playing the Odds in the State Lotteries, *Northwest Science*, **61**, 239–248.
[100] Saunders, S.C., and Myhre, J. (1984), On the behavior of certain maximum likelihood estimators from large, randomly censored samples. *JASA* **79**, 294–301.
[101] Schrödinger, E. (1915), Zur theorie der Fall- und Steigversuche an Teilchen mit Brownischer Bewegung. *Physikalishe Zeitschrift* **16**, 289–295.
[102] Schwarz, C.M., and Samanatha, M. (1991), An inductive proof of the sampling distribution for the MLEs of the parameters in an inverse-Gaussian distribution. *The Amer. Statist.* **45**, 223–225.
[103] Science News, (2005), Fertility and Pollution, October 8, **168**, p. 230.
[104] Shorack, G.R. and Wellner, J. (1986), *Empirical Processes with Applications to Statistics*, New York, Wiley.
[105] Shuster, J. (1968) On the inverse Gaussian distribution function, *Journal of the American Statistical Association*, 1514–1517.
[106] Standard Mathematical Tables (1978), *Chemical Rubber Company Handbook.* West Palm Beach, FL.: CRC Press.
[107] Singpurwalla, N.D. and Wong, M. (1982), Fourier Integral Estimates of the Failure Rate Function and Its Mean Square Properties, *Survival Analysis*, Eds. Crowley, J. and Johnson, R. A., IMS Lecture Notes- Monograph Series, 41–55.
[108] Smith, R. L. (1991), Weibull regession models for reliability data, *Reliability Engineering and System Safety*, 55–76.
[109] Suresh, S. (1992), *Fatigue of Materials*, Cambridge University Press.
[110] Tanner, M. A. and Wong, W.H. (1983), The estimation of the hazard function from randomly censored data using the kernel method, *The Annals of Statist.*, **11**, 989–993.
[111] Timoshenko, S. P. (1953), *History of Strength of Materials*, New York, McGraw-Hill.
[112] Turnbull, B.W. (1974), Nonparametric Estimation of a Survivorship Function with Doubly Censored Data, *J. Amer. Statist. Assoc.*, **69**, 169–174.
[113] Turnbull, B.W. (1976), The empiric distribution function with arbitrarily grouped, censored and truncated data, J. of Royal Stat. Soc., Series-B, **X**, 290–295. **69**, 169–174.
[114] Tweedie, M.C.K. (1945), Inverse statistical variates. *Nature* **155**, 453.
[115] Tweedie, M.C.K. (1957a), Statistical properties of inverse-Gaussian distributions I. *Ann. Math. Statist.* **28**, 367–377.
[116] Tweedie, M.C.K. (1957b), Statistical properties of inverse-Gaussian distributions II. *Ann. Math. Statist.* **28**, 696–705.
[117] Wald, A. (1944), On cumulative sums of random variables. *Ann. Math. Stats.* **15**, 283–296.
[118] Watson, A.S. and Smith, R.L. (1985), An examination of statistical theories for fibrous materials in the light of experimental data, *J. Mater. Sci.*, **20**, 3260–3270.
[119] Weibull, W. (1939a), A Statistical Theory of the Strength of Materials, No. 151; (1939b) The Phenomenon of Rupture in Solids, No. 153; Generalstabens Litigrafisca Anstalts Förlag, Stockholm.

[120] Widder, D.V. (1946), *The Laplace Transform*. Princeton, NJ.: Princeton University Press, pp. 180, 197.

[121] Yao, Yi-ching. (1987), A note on testing for constant hazard against a change-point alternative. *Ann. Inst. Statist. Math* **39**, Part A, 377–383.

[122] Yee, L. P. and Výborný, R. (2000) *The Integral: An Easy Approach after Kurzweil and Henstock,* Cambridge University Press.

[123] Zigangirov, K. Sh. (1962), Expression for the Wald Distribution in terms of the Normal Distribution, Radiotekhnika Electronika, **7**, 145–148.

[124] Zhou, M. (1991), Some properties of the Kaplan–Meier estimator for independent nonidentically distributed random variables, The Annals of Statist., **19**, 2266–2274.

Index

Springer Series in Statistics

(continued from p. ii)

Jolliffe: Principal Component Analysis, 2nd edition.
Knottnerus: Sample Survey Theory: Some Pythagorean Perspectives.
Kolen/Brennan: Test Equating: Methods and Practices.
Kotz/Johnson (Eds.): Breakthroughs in Statistics Volume I.
Kotz/Johnson (Eds.): Breakthroughs in Statistics Volume II.
Kotz/Johnson (Eds.): Breakthroughs in Statistics Volume III.
Küchler/Sørensen: Exponential Families of Stochastic Processes.
Kutoyants: Statistical Influence for Ergodic Diffusion Processes.
Lahiri: Resampling Methods for Dependent Data.
Le Cam: Asymptotic Methods in Statistical Decision Theory.
Le Cam/Yang: Asymptotics in Statistics: Some Basic Concepts, 2nd edition.
Liu: Monte Carlo Strategies in Scientific Computing.
Longford: Models for Uncertainty in Educational Testing.
Manski: Partial Identification of Probability Distributions.
Mielke/Berry: Permutation Methods: A Distance Function Approach.
Pan/Fang: Growth Curve Models and Statistical Diagnostics.
Parzen/Tanabe/Kitagawa: Selected Papers of Hirotugu Akaike.
Politis/Romano/Wolf: Subsampling.
Ramsay/Silverman: Applied Functional Data Analysis: Methods and Case Studies.
Ramsay/Silverman: Functional Data Analysis.
Rao/Toutenburg: Linear Models: Least Squares and Alternatives.
Reinsel: Elements of Multivariate Time Series Analysis, 2nd edition.
Rosenbaum: Observational Studies, 2nd edition.
Rosenblatt: Gaussian and Non-Gaussian Linear Time Series and Random Fields.
Särndal/Swensson/Wretman: Model Assisted Survey Sampling.
Santner/Williams/Notz: The Design and Analysis of Computer Experiments.
Saunders: Reliability, Life Testing and the Prediction of Service Lives.
Schervish: Theory of Statistics.
Shao/Tu: The Jackknife and Bootstrap.
Simonoff: Smoothing Methods in Statistics.
Singpurwalla and Wilson: Statistical Methods in Software Engineering: Reliability and Risk.
Small: The Statistical Theory of Shape.
Sprott: Statistical Inference in Science.
Stein: Interpolation of Spatial Data: Some Theory for Kriging.
Taniguchi/Kakizawa: Asymptotic Theory of Statistical Inference for Time Series.
Tanner: Tools for Statistical Inference: Methods for the Exploration of Posterior Distributions and Likelihood Functions, 3rd edition.
van der laan: Unified Methods for Censored Longitudinal Data and Causality.
van der vaart/Wellner: Weak Convergence and Empirical Processes: With Applications to Statistics.
Verbeke/Molenberghs: Linear Mixed Models for Longitudinal Data.
Weerahandi: Exact Statistical Methods for Data Analysis.
West/Harrison: Bayesian Forecasting and Dynamic Models, 2nd edition.